A great deal of Britain's heritage and local history is to be found in its taverns and inns which, over the centuries, have formed a central part of life in both town and country. Inns offered a welcome to locals and strangers alike, a staging post for coaches, and sustenance and rest for travellers and horses. As this book shows, a great many inns chose the special and easily recognizable name 'The White Horse' – a reflection, perhaps, of the high esteem in which white horses have always been held in this country.

Araminta Aldington and Sally Potter have compiled an historical vignette of the White Horse inns of Britain and the Republic of Ireland in this remarkable gazetteer. They have researched the history of over 1,500 buildings, their owners, and those who visited them. Many characters are to be found meeting or drinking in their 'local' White Horse inn, and while writing about them, the authors regale the reader with a fascinating variety of incidents and facts, many of which have passed into folk law.

The narrative is accompanied by over six hundred line drawings by Vi Sprawling. Together they form a charming record of an aspect of town and country life which is gradually disappearing. The roll-call of famous personalities associated with White Horse inns over the centuries makes this a useful reference work for both the local historian and the general reader.

GW00711475

# WHITE HORSE INNS

## An Historical Vignette

Araminta Aldington

Damon & Jo
With love & best wishes for
Christmas 2009.
Sally

# WHITE HORSE INNS

## An Historical Vignette

*Araminta Aldington*
*& Sally Potter*

Librario

Published by
**Librario Publishing Ltd.**

ISBN : 978-1906775124

Copies can be ordered from retail
or via the internet at :

**www.librario.com**

or from :

Brough House
Milton Brodie
Kinloss
Morayshire
IV36 2UA

Tel / Fax : 01343 850178

Typesetting and layout by Steven James
**www.chimeracreations.co.uk**

Printed and bound in Great Britain
Cpod, a division of The Cromwell Press Group, Trowbridge, Wiltshire

*To our children,
and their children,
in the hope they, too, may be
blessed with an inspiring quest.*

# CONTENTS

The Importance of a White Horse     9

The Importance of a Public House     11

Introduction     13

Acknowledgements     15

The Authors / The Artists     17

**THE INNS :**

Channel Islands     19

England     23

Isle Of Man     485

Northern Ireland     489

Southern Ireland     495

Scotland     503

Wales     509

Trade Directories     539

Notable Visitors Index     559

Places Index     563

'There is a Tavern in the Town'     577

# THE IMPORTANCE OF
# A WHITE HORSE

## By the late Lord Deedes M.C.

Araminta Aldington has been researching the place of the white horse in our history ever since I first came to know her forty or so years ago. If you are going to spend half a lifetime on such a study, make sure your findings are original. And the first thing to be said about her work is that it is original.

The next point to observe is that Araminta, like many who climb mountains, explore deserts or sail the seas, has been temporarily diverted from her main course. During her studies she was struck – and who could fail to be struck? – by the number of Inns which go by the name of the White Horse.

Why? It is the sort of question to which those who submit themselves to Mastermind would be well advised to say "Pass" and not lose valuable time hazarding a guess.

Yet it is a question that digs deep into our folklore and calls for an answer. I am not going to spoil the plot by revealing just how many White Horse inns have been tracked down by her; but it is several times the figure I first thought of.

This book, it must be understood, is merely a foretaste of what is to come. Having been diverted by the ubiquitous White Horse Inn and having discovered their whereabouts, the author has thought it right to share her discoveries with us and devote a small work to them; a window into the social life of our times.

It is to be seen as a *hors d'oeuvres* to the main work, *Behold a White Horse*, which will for the first time unveil the place of the white horse in our history, and explain the extraordinary hold this animal has for centuries exercised over our imaginations.

*23rd July 2001*

# THE IMPORTANCE OF
# A PUBLIC HOUSE

We have been given permission to quote from the Introduction to *The Pub is the Hub* by H.R.H. Prince Charles :

*'Rural communities, and this country's rural*
*way of life, are facing unprecedented challenges.*
*Now, perhaps more than ever in their history,*
*they must draw on their resourcefulness and*
*resilience, built up over centuries . . . '*

*'Even the country pub, which has been*
*at the heart of village life for centuries,*
*is disappearing in many areas . . . '*

*'Finding new uses for village pubs, many of which*
*are the only remaining service in the community,*
*has therefore become one part of Business in*
*the Community's Rural Action programme*
*under the campaign headline 'Make the Pub the*
*Hub'. By providing new services from the pub,*
*such as a post office or a shop, not only keeps*
*an essential service in the village or brings a*
*new one in, but increases the income of the pub*
*itself, giving it a more secure future . . . '*

The mention by Prince Charles in 2001 of 'post office or a shop' has been one of the life-lines we have used to discover the correct site of the inn.

We owe a great debt of gratitude to the many busy post masters, post mistresses and kind people who have provided us with local knowledge – and often the name of the Local Historian which has been invaluable.

Alas by 2008 both pubs and post offices were fast disappearing . . .

# INTRODUCTION

*"Where rain and wind would wash and swing
the crudely painted sign"*

*The Village Inn* from *A Few Late Chrysanthemums*
John Betjeman, 1954

The Romans left two Latin words with us when they returned to Italy early in the fifth century – 'taberna', a hut or cottage, hence tavern, and 'innitor' to lean or rest, to support oneself. To be able to 'inn', to find rest and sustenance at some wayside house has been necessary for the traveller whether on foot or horseback the world over for thousands of years.

On these islands where the lumbering wagons pulled by oxen would have made the tracks almost impassable in places, a traveller would have been looking out for a pole decked with ivy or evergreen in front of a dwelling which would tell him where he would be welcome. The pole was gradually replaced by the inn-sign post carrying the name of the publican who with his wife would be brewers of beer or ale. The sign would have some eye-catching crudely painted image, perhaps the crest of the land owner or a picture depicting a local legend or mythical beast, such as the honest, pure, lucky white horse.

The drovers with their flocks of sheep, cows and geese would have needed friendly landlords who would take payment in advance for food for their dogs should they have to find their own way home from the market; these amazing dogs would travel hundreds of miles using their homing instinct.

If the traveller was a pilgrim, a monastery would give him shelter; in later years these establishments supplied lodging for the better-off with their retinues of servants. The stabling and fodder for their many horses and the good beer and ale brewed by the monks, would have engendered much needed revenue for the monastery.

For centuries the inn has been the centre of town and country life, a public house where every stratum of life has been enacted from cudgel fights and shove ha'penny to darts and whist, to the hearing of legal court cases and official dinners; occasionally inns were used as mortuaries after some local disaster had taken place.

An innkeeper often had a second trade, usually carried on in an adjoining building, while his wife ran the inn and brewed the ale – he might be a blacksmith or a farmer with a few acres for growing fodder or bedding for the travellers' horses, he often sold other provisions as well.

Eventually the inn became of great importance in the middle of the 17ᵗʰ century when the Royal Mail coaches needed regular staging places for the change of horses. These stopping places were highly organised, running to an exact timetable. It was not until 1840 when the trains opened up the whole country with a quicker form of travel that a great many inns lost their custom, or became hotels – later for the motor-driving public.

The sign of the White Horse became popular after 1714 when the Hanoverian King George I's pennant was seen fluttering from the ship upon his arrival at Greenwich displaying a galloping white horse upon a scarlet background. But many taverns and alehouses were displaying a painted white horse on their boards, and wording such as 'Whit Hors' on their tokens long before; in London's Tooley Street in Southwark there was one in 1492. In Suffolk alone there were fourteen* inns carrying this sign dating from 1518 to 1712.

It would seem that fate has taken a hand in this book being published when pubs are closing almost weekly, so it would be impossible to give a true picture of those still carrying the sign of a white horse. Originally 1,517 had been trading at some time – there may well have been more. Several have become private houses; larger buildings becoming banks or shops, where the origin of the building is seldom known. The most upsetting are those which have had their ancient sign changed to some modern horror of no historic meaning.

The research for this book began some fifty years ago as a small part of a far larger tapestry emphasising the unconscious influence that a horse which is white plays in our lives. It grew into an all consuming interest, telling of a cosmos of daily life which had been enacted day after day over the years in the bars and saloons of public houses in countless towns and villages.

Let's hope that in some magical way this book may conjure up a realisation that the integrity and importance of the vital life of 'The Local' to all ages in the community must be preserved.

*Badwell Ash, Beyton, Bury St. Edmunds, Corton, Finningham, Framlingham, Hadleigh, Ipswich, Laxfield, Old Felixstow, Rickinghall Superior, Stoke Ash, Syleham and Tattingstone.

# ACKNOWLEDGEMENTS

Two very different contemporaries from the 18<sup>th</sup> century provided some fascinating details of these early taverns and ale houses:

**James Bell** (1769 – 1833) was born in Jedburgh, a Calvinist son of a Presbyterian minister of the Relief Church. An asthmatic all his life, he was given a small annuity by his father and was apprenticed as a weaver. He set up as a manufacturer of cotton goods in Glasgow where he taught Greek and Latin to university students, writing and publishing geographical books. His last work in four volumes was published in 1836, 'A New Comprehensive Gazetteer of England and Wales', giving the population of towns and parishes from the 1831 census. This has enabled us to pin-point small ale-houses in hamlets now engulfed by modern towns. Unable to travel he retired to Campsie, where he died three years before these, his last books, were published. *Oxford Dictionary of National Biography.*

**Walley Chamberlain Oulton** (1770 – 1820), a playwright from Dublin, whose plays were performed in both Covent Garden and Drury Lane in London. In 1805 he published 'The Traveller's Guide, or an English Itinerary'. Recording visits to a number of White Horse Inns. *Baker's Biography Dramatica, 1812; Biographical Dictionary of Living Authors 1816.*

\* \* \*

It is impossible to mention everyone who has helped us compile this book, but several people need to be thanked by name; without their help it would have been a very different size, consisting of a row of dates for each entry – hardly compulsive reading!

In 1980 three young ladies, **Jane Morris**, **Caroline Berkeley** and **Jane Williams** spent several weeks in the Guildhall Library, trawling through the Trade Directories which gave us the framework of where to search further.

**Nicholas Redman, F.S.A.**, whilst Archivist of Whitbread, was able to give detailed brewery records; an enthusiast and firm believer of the importance of an inn with the sign of a white horse hanging outside – which needed to be photographed immediately.

**Leslie H. Douch** was Curator of the Museum of the Royal Institution of Cornwall from 1950 until he retired in 1988, although he continued to help in the Library until he died, aged 80. He provided many excerpts from The Gentleman's Magazine and West Country newspapers, painstakingly weeding out any mention of the inn.

**Christopher Wooler** of Preston gave us much useful hand-written archival material concerning those pubs in Yorkshire and Lancashire.

**John Emberton**, with the help of his computer, has valiantly tracked down the many illusive White Horse Inns in the West Midlands and Warwickshire.

**Keith S. G. Hinde O.B.E.**, has provided a fund of knowledge of East Anglia, together with photographs, which has been invaluable.

**County Archivists** have been most helpful, together with **Local History Societies**, many of which did not exist some forty years ago.

\* \* \*

Both of our husbands and families have been stalwarts in their support over the years, encouraging us never to lose faith in the project. In the last few years, spasmodic help has also been provided by three of Araminta's grandchildren: **Sophie** and **Amelia**'s diligence and meticulous work contributed to the final draft, while **Philip** has sought out all the White Horse Inns in Norfolk.

**John Grundy** is the computer expert who has been on hand to deal with any crisis we may have had, every two weeks picking up the artist's original line drawings – reducing them to miniatures for inclusion in the text and keeping the back-up material in his office; his daughter **Victoria** became an expert in finding Local History Societies.

**Claire Eaglesham** has amalgamated miniatures and text, but also used modern satellite facilities to send the book to our publisher in distant northern Scotland.

**Mark Lawson** of **Librario Publishing Ltd**. was persuaded to take on this mammoth task – unaware at the time of all the meticulous details involved. His kindness and long-suffering nature, together with **Steven**'s designer skills, have allowed this book to see the light of day, in a very different century to which these houses flourished.

Without the interest, kindness and generosity of **Colin Badcock** this book could not have reached your hands.

# THE AUTHORS

Araminta Aldington and Sally Potter worked together on *A History of the Jacob Sheep* published in 1989. For the last forty years they have been collecting material for this book, *White Horse Inns : An Historical Vignette*, and for a further book, *Behold a White Horse*.

# THE ARTISTS

## Violet Sprawling

The excellent line-drawings of the pubs by Vi Sprawling, which illustrate the book, were taken from photographs, newspaper cuttings and old prints. There are nearly six hundred in total, a fantastic achievement – alas they have all had to be reduced in size.*

Vi had no formal training, but when she turned sixty, she decided to take an A-level in Art and Design. This was firstly to prove that she could, and secondly because the course was half price for pensioners!

She is a member of the Ashford Art Society in Kent.

## Caroline Gould

The book cover is designed by Caroline Gould and depicts the various activities which have taken place in pubs over the centuries.

After qualifying with a Fine Art degree Caroline taught in schools and adult centres. She takes commissions and exhibits paintings in group shows as opportunities arise.

## Amelia Roberts

The pen and ink drawings which represent the regions covered by this book were done by Amelia, Araminta's granddaughter.

She studied from 2001-2004 at the Charles H. Cecil Studios in Florence, Italy and then attended City and Guilds of London Art School. She takes commissions and exhibits in London and Cornwall.

*Copies of Violet Sprawling's original line drawings can be obtained from Librario Publishing, Brough House, Milton Brodie, Kinloss, Morayshire IV36 2UA.*

# CHANNEL ISLANDS

## JERSEY

*The Channel Islands are part of the British Islands
where the Loyal Toast is to 'The Queen, Our Duke'.*

*Jersey Cattle were first imported from Normandy in c1700.
The Herd Book dates from 1866.*

# JERSEY

*In 1640, Charles II (1630-1685) was welcomed to Jersey
during his exile. In recognition of this, he awarded George
Carteret, a governor and bailiff, a large portion of land
from the American Colonies, New Jersey which became part
of the USA in 1787.*

*The French author Victor Hugo (1802-1885) lived there
in exile during the 1850s, first in Jersey and then Guernsey.
Here, he wrote some of his most famous work including
'Les Misérables'.*

**St. Helier, St. Clements Road** 1939 listed as trading in Kelly's Directory.

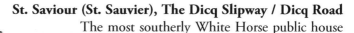

**St. Saviour (St. Sauvier), The Dicq Slipway / Dicq Road**
The most southerly White Horse public house
in the British Isles.

The 18th century building was listed under
the Dicq Inn in 1720 when it played a large
part in the political turmoil of this island
between the Charlots, supporters of Charles
Lempriere, Bailiff of Jersey (1714-1806) and
the Magots who were in opposition. Before an election it was the custom for
candidates to give dinners to their supporters at hostelries.

1790, December – *The Gazette* reported that a dinner was to be given at the
Dicq Inn by Gideon Ahier, a Charlot, in order to persuade the landlord and
others to vote for him :

> 'A plate of codfish was to be served for supper, and in order to enhance
> the feast, someone in the Charlot party brought roasted pullets wrapped
> up in pocket handkerchiefs which could afterwards be used as serviettes.'

1791, 1st January – an account of the dinner was published suggesting that the
landlord supported the opposition, Mr. John Godfrey. The codfish had been
stewed in sour milk, the pullets had been drenched in vinegar and served in filthy
handkerchiefs and stale cider had been added to fresh. This resulted in people
changing sides to support Gideon Ahier who won the election.

c. 1930, named White Horse probably after the whisky. A white horse was
painted on the chimney.

1961 the house was purchased by Ann Street Brewery of St. Helier.

Mrs. Joan Stevens, F.S.A.

# ENGLAND

*"When you have lost your inns,
drown your empty selves
for you will have lost the last of England."*

'This and That and the Other'
'On Inns'

Hilaire Belloc
1912

# BEDFORDSHIRE

*'A large proportion of the female population is employed in the
plaiting of straw, for which Dunstable in particular is famous,
and in the manufacture of thread-lace; the spinning of hemp
and cotton formerly flourished, but has of late years declined
very much. A considerable number of mats are made.'*
JB.

**Arlesey, 243 High Street** The building dates
from c.1600.

The inn was visited by both King George
II and Queen Victoria.

1877, Friday 28th September – an auction
of local property was held here.

Landlord. 1726 BCR; 1869-1940 K.

## Bedford

At one time there was a White Horse Street.

*1-3 Midland Road (formerly Well Street)* This
ancient inn stood at the corner with Harpur Street
and is recorded in a deed of 1557.

1869 listed as trading under Well Street in Kelly's
Directory.

1929, May – the house was demolished and a
Marks and Spencer store was built on the site,
now situated in a pedestrian area in the town
centre. The licence was transferred to a new White Horse
in Newnham Lane. BCR (FN343), BBPR. 1823 P; 1830-1850 S; 1914 K.

*84 Newnham Avenue (formerly Newnham Lane)*
1929 the licence was transferred from the White
Horse in Midland Road.

By 1938 the name of the road had been
changed to Newnham Avenue.

1950 the house was bought by the brewers
Charles Wells.

2003 the house was refurbished.

1940 K. BCR.

**Biggleswade, 1 High Street** Little is known of a White Horse inn which was trading in the High Street in the 17th century. By 1823 it must have ceased to trade as the sign 'White Horse' had replaced that of 'The Last[1]' – a larger established inn, so named by the original owner George Magoone who was a cordwainer[2], or shoemaker.

1671 Standing in Shortmead Street – a continuation of the High Street – The Last was recorded in the Hearth Tax Return as having three hearths.

1683, 28th March – John Croot, a tanner, surrendered out of court a messuage in Shortmead Street in the occupation of George Magoone as a mortgage in £51 10s 0d. *Presumably Mr. Croot supplied the leather for Mr. Magoone's shoes.*

1687 at a Court of 25th and 26th October, Thomas Magoone, alias 'Scotchman' took over the messuage and tenement, when he died in 1721 he left the inn and goods and chattels to his wife Ann. His will described the house as 'abutting West upon the Church Gate', hence 'Church Street – on the corner of Shortmead Street'.

An inventory of the contents of the inn was drawn up at the time of his death, which included the following:

| | |
|---|---:|
| Money in Thomas Magoon's purse. | £5  0s 0d |
| Items in the hall: a table and 4 chairs, 6 pewter dishes, 1 duz pewter plates, 2 formes, 1 Joyn stoole. | £1 10s 0d |
| The Parlor: 2 little tables, 9 chairs, 2 pairs of hand irons, a fire shovel and a paire of tongs. | £1  0s 0d |
| The Roome next the hall: 1 bed and bedding, 1 chest of drawers, 1 clock, 1 press cupboard, 1 coffer. | £7  2s 0d |
| The Kitchen: 1 litle table, 2 pottage potts, 2 kittles, 1 boiler, 3 tubbs, 1 starm, a small parsell of coals. | £2 13s 6d |
| The seller: 3 litle barrels, 1 tubb. | 9s 6d |
| The chamber over the hall: 2 old bedsteds onely | 3s 0d |
| The chamber over the Parlor: 1 table, 1 coffer | £3  1s 6d |
| The barn: a litle fire wood. | 10s 6d |
| Debts owing to Thomas Magoon | £2 12s 6d |
| Debts he owed | £6  4s 6d |
| **Total** | **£30 7s 0d** |

1738 James Bailey became the owner; when he died in 1751 he left the inn to his wife Mary.

1768 Thomas Bailey, their son, a draper, took over.

1771, 27th February – the inn was sold to a brewer, Samuel Wells of Biggleswade.

1822-1831 Thomas Ayres, became the landlord, he was also a shoemaker, which would suggest that customers once again had their shoes made or mended at the inn, but it was in his tenure the inn is listed as White Horse at 1 The High Street.

1839 John Robins called at the White Horse on Tuesdays, having carried goods from Potton, some three miles away, to Bedford and Northampton.

1890-1894 the landlord, William Wells, was also a grocer, his shop being at the inn.

1897 travellers, including cyclists using the Great North Road, today's A1 road, could stay here as there were 'letting rooms'; ham and beef were supplied at 6d. a plate.

1898 Arthur Stoten, a tailor, is recorded as landlord. This same year the inn appeared in a sale catalogue where it was described as a brick, plaster and tiled house "containing Tap Room, Kitchen, Parlour, Tailor's Shop, Bar, Bagetelle Room and good Cellars; seven Rooms with loft over, Yard to rear with Stable and Loft, and a long Stable with entrance from Church Road and a large Clubroom over with separate Staircase. There was also a garden in Chapel Fields".

1899 a travelling bazaar was advertised at the White Horse with 25,000 articles for sale at 6½d.

c. 1950 a photograph shows the inn painted white.

1982 alterations were made to the Clubroom, used by the Oddfellows Friendly Society, but the original wooden timbers were retained; by this date it had become an hotel, but with no stabling. LHS – Biggleswade, BCR (CRT 130 B/G 52), BCR (PBWP1721/9) and KP. pp.52-56. 1823 P; 1869 Shortmead Street -1890 K.

**Blunham, 1 Park Lane** Previously known as Back Lane leading to the River Ivel.

1767 John Ravens became the tenant of a cottage and land here which was held by the Manor of Blunham. He had two daughters, Elizabeth who married Benjamin Palmer – they lived in the cottage prior to 1785 – and Mary, who had moved to the cottage by 25th July with her husband Benjamin Eyre. In 1794 Elizabeth Palmer sold her share of the cottage, close and orchard to her sister Mary; by 29th January 1801 Benjamin Eyre had died.

By 22nd August 1805 Mary had built a second cottage on the orchard site – this was to become the White Horse; she sold both cottages to Isaac Pendred.

1811 Isaac sold the property to John Dale.

1832 the first beerhouse licence was granted with John Elwood as licensee.

By 1836 the house had been bought by Thomas Strickland, a brewer at Potton; he died in 1873 and, due to a Chancery suit regarding his estate's administration, it was sold the following year to Alfred Richardson, a brewer, and was recorded as a public house for the first time.

1894 the brewers Newland and Nash were recorded as the owners; in 1922 they sold it to the brewers Wells and Winch.

1934 the licensee, William Howkins, moved to The Salutation public house and Charles Albone took over.

1944, 9th August – John Harding became licensee. The website shows a photograph, dated c.1957 of him with his wife Emily and their dog standing outside the front door. John remained the licensee until December 1960 when the house was demolished, the site becoming the car park of the new public house being built by the brewers Greene King, which was situated at the corner of Park Lane and The Hill overlooking the old Brick Fields.

1961, 4th January – the newly built White Horse public house was opened; on 11th January Alec Hughes Taylor moved from The Salutation to become licensee.

1985, 25th July – Roy and Susan Walsh were presented with '200 Years of History' of the inn on their first anniversary as publicans.

1986, 11th September – the inn was closed to become a private house and Roy and Susan moved to The Salutation. Bill Exley and updated by Colin Hinson – local historians.

1832 BCR; c.1880 rebuilt; 1900 Sandy - 1940 K.

**Broom, 30 Southall Road** Situated on the village green.

c. 1775, according to the 1876 Licensing Return for Biggleswade Division.

By 2005 three generations of one family had run this pub.

A Grade II listed building. 1850 S; 1869-1940 K.

## Dunstable

1742, Monday 12th April - the first stage coach passed through at a speed of 6½ miles an hour; at the height of the coaching season eighty coaches a day used this route. North of Dunstable, where the Roman Road passes over Chalk Hill, seven or eight horses were required to pull the coaches to the top, often passengers having to walk.

*Church Street* A coaching inn originally named The Kings Head possibly in honour of King Edward VI (1547-1553).

1645 it is said that King Charles I stayed here en route for the battle which took place near the village of Naseby, some fifty miles further north. He was supposed to have tethered his horse to the mushroom shaped stone of igneous rock which then stood in front of the inn. This stone was later removed and placed next to the entrance steps.

1963 due to road improvement the building was

demolished, but the old mounting block which had become known as 'The White Horse Stone' was saved by John Lunn, M.B.E., one time Headmaster of Beecroft School, who placed it in the school's Quadrangle.

c.1819 the house took the name 'White Horse' from the one in High Street North when it was renamed 'The Anchor'. VCH. 1823/4 P; 1830 S; 1839 P; 1898 K.

**High Street North** 1537 the local Prior wrote to the King Henry VIII's secretary, Thomas Cromwell, Earl of Essex (c.1485-1540) complaining that the King would not stay 'in my poor house which I have made ready to receive him'. His Majesty stayed at the White Horse Inn.

By 1819 this 16th century building had been demolished except for the gateway which became the entrance to The Anchor Inn. VCH Vol. 3 p.355 and NCB.

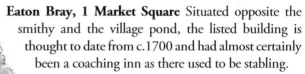

**Eaton Bray, 1 Market Square** Situated opposite the smithy and the village pond, the listed building is thought to date from c.1700 and had almost certainly been a coaching inn as there used to be stabling.

An undated Victorian photograph shows bicyclists with ladies in long dresses posing outside, with one gentleman leaning against his penny farthing cycle. A notice, 'Vote for Duke' is pasted over the front door.

c.1913 a photograph shows a family group standing outside the inn, it also shows that it was a Free House with Good Stabling and advertised 'Fine Ales, Wines and Spirits' on a board along the ridge of the roof – a later photograph depicts a new, longer White Horse Inn signboard running from chimney stack to chimney stack.

1927 George Reeves moved in with his family and became licensee – his daughter Hilda Primrose was then 12 years old. Beer was only 4d. and 6d. a pint and spirits 6d. a measure. The weekly takings had been advertised as being £25. It was a shock to find that the daily takings were only 19 shillings, he would add a shilling to make it up to £1 to keep the books straight. After six months there was no improvement so he sued the brewery for misrepresentation and claimed back £150 from the £500 he had invested.

1928 major alterations took place and electricity was installed and the street lamp changed, this was apparently unpopular with the local people - previously the pub had been lit by oil lamps which also gave out a certain amount of heat - the street lamp would have been gas-fired.

At a later date the inn took over the cricketers' teas and then began to cater for weddings, charabancs and cycling clubs – the cyclists would wash at the pump outside before eating as much as they could manage for 1s. 8d.

1939 during the war officers were billeted at the inn and soldiers in the national school, many of them found a 'home from home' at The White Horse and several letters were received from grateful relatives of the men.

After the war George Reeves became chairman of the Leighton Buzzard Licensed Victuallers' Association.

1955 Hilda, having worked in the inn since she left school at the age of sixteen, was given the tea room as a 40th birthday present from her father for her to manage as a grill room, with a new kitchen attached. The grill room was such a success that the kitchen had to be altered three times before it was large enough.

1971 George Reeves died after 44 years in the inn, Hilda then took over for the next five years, having spent nearly fifty years of her life at The White Horse.

1988 Mr. David Sparrow took over as licensee and was still trading in 2006.

Mr. David Sparrow, Landlord. 1810 BCR; 1898-1940 K.

**Flitton** 1719 ER; 1839 P; 1850 S; 1869, 1898 Ampthill - 1940 K; 1981 still trading.

**Flitwick, 101 Station Road** The original house was thought to have been built in 1868 when deeds show the land passed from John Dowdeswell, a farmer and corn merchant of Maulden to William John Giddings, a builder from London. It was initially let to Wells Hogge and Lindsell, brewers of Biggleswade.

1876, 8th June – a sale catalogue describes the property as abutting on to the village street and about three minutes walk from the station. It is described as a 'brick and slate, sashed and spouted public house with cellar, taproom, parlour, back kitchen, pantry, three bedrooms'; there was a timber and tiled stable and chaisehouse, a timber and tiled corn barn, also piggeries, a corn barn with a floor, a pig yard and a small garden with a well. A plan of the site shows a passage between two other parcels of land leading from the White Horse yard to market garden land which had been worked for gravel except for a small piece of grass in the north west corner which had gravel beneath it.

The property was bought for £680 by Francis Allfrey, a brewer of Newport Pagnell, whose brewery eventually became part of Charles Wells & Company who continued as owners. BLAR – Ref: WL 1000/1/FWLK 2/3/7 1940 K.

**Harrold** 1839 P.

**Haynes, Deadmans Cross** 1754 recorded as an alehouse.
1785 listed as trading under White Horse. BCR. 1850 S; 1869-1940 K.

**Hockliffe** 1712 the house was recorded as trading.
1894 listed as a Commercial Hotel, Posting House and a Railway Sub-Office[3].
BCR. 1850 S; 1869 K; 1914 K.

**Houghton Regis, Cumberland Street** 1914 K; 1940 K.

**Husborne Crawley, Turnpike Road / 1 Mill Road**
Situated on the corner with School Lane.

1819 listed as 'The Windmill', later to become 'The Magpie'.

1826 listed as White Horse by the time Joseph Morris, owner and brewer, sold out to J.W. Green Limited.

1831, 28th December – a mortgage for £30,000 of various properties from John Morris to Mary Jane and Sophia Morris, which included the White Horse with three closes of meadow – 12 acres in total. John Roote was licensee.

1840, 3rd April and 16th June 1882 - a mortgage from John Green to the Morris family of several properties which included the White Horse and a number of neighbouring cottages bought by John Morris.

1881 listed as trading with William Bunker, a cattle dealer, as licensee.

1887, 18th March – an auction of property took place here.

1893 William died and his wife Lizzie took over for the next twelve years – she was also a cattle dealer.

1907 the public house was conveyed to Morris & Company with other properties and to J.W. Green limited in 1926.

1927 the house was valued for rates, it consisted of a private lounge, kitchen and scullery, a public bar, smokeroom and a cellar; there were four bedrooms. Outside there was an unlicensed slaughterhouse, a barn, coach house, store, shed and a stable for one horse. There was also an adjoining shop, thought to have once been the butcher. The average takings were £11 per week and the licence cost £11 per annum; about sixty five barrels of beer were consumed each year and about nine gallons of spirits. A small amount of tobacco and confectionary was sold. The landlord complained of poor trade; he did not use all the bedrooms, nor the shop which had been occupied by the Air Raid Warden during the Second World War.

By 2006 the house was owned by Enterprise Inns. BCR. 1850 S; 1869-1940 K.

**Keysoe Row, Kimbolton Road** 1869 listed as trading with Thomas Cope as licensee.

A traditional half-thatched building situated near to two airfields, the main customers during the Second World War were American airmen and the RAF.

2007 the White Horse was included in the Good Beer Guide.

Landlord and KSGH. 1814 ER; 1850 S; 1869-1940 K.

**Linslade, 9 New Road** c.1839 built by Thomas Garner, a Leighton Buzzard maltster who unfortunately became bankrupt; some of his papers have survived which include expenses incurred during the building of the pub:

*December 1839*
For glazing, painting, wirework and 1 square glass in tap room window - 1s.6d.
Walls and ceilings plastered, the roof slated and a well sunk and built.

*March 1840*
For two chimney pots delivered - 10s. 0d.
For fixing with cement - 2s. 0d.

*May 1840*
For building materials, over 1,000 bricks, 6 fire bricks, a sink stone, 2 oven lumps (sic.) and 12 plain tiles - £17 8s. 9d.
He also bought bottles and glasses.

*June 1840*
Bedsteads, 2 card tables, baize and glue purchased.

*October 1840*
The words 'TAP ROOM' painted on a door - 9d.
Mahogany sign board - £1 16s.0d.
Painting - *of a white horse perhaps!* - drawing, writing and gilding the same - £2 2s. 0d.
New board in blank window - 8s. 6d. - *1851 was the last year of the Window Tax.*
4lbs sugar - 4s. 0d.; 1lb tobacco - 4s. 8d.; ½lb coffee - 1s. 2d., also cheese, butter, candles, eggs, currants.
To Dr. Wagstaff for attending Mrs. Garner and prescribing various potions including powdered rhubarb and caster oil - £2 0s. 6d.

December 1841 Thomas Garner was charged with being an insolvent debtor but the White Horse continued to trade under new ownership. The present sign 'Good Stabling' recalls the days when the yard echoed to the sound of horses.
1853 a Public House license was granted to the White horse.
1872 listed in the Returns of Public Houses with William Field as landlord; the owner was George Franklin of Leighton.
By March 2006, the inn had closed to reopen in February 2007. MB. 1883-1939 K.

**Riseley** Situated opposite Lowsdon Lane. An early 19th century house built from Riseley brick. First licenced as a beerhouse with a six-day licence in1822 when John Ekins, a farmer and brewer, was landlord.

By 1850 it was listed as White Horse and hosted the annual feast of the Tradesmen's Club, also teas for the bowls and cricket teams; the bowling green and cricket pitch were close by the pub. The Harvest Supper and Sale were also held here.

1855, June – the Bedfordshire Times reported that the White Horse had been burgled – a sovereign, two half-sovereigns, some half crowns and four penny pieces were stolen – a local man was apprehended.

1884, June – During the afternoon nuts had been given to the village children by the Tradesmen's Club, these were thrown into the air and the children scrambled for them. The seventeen members had a 'most excellent dinner', after which they had played quoits and other games.

Before the First World War auctions were held here for villagers to coppice and clear of sections of woodland; the wood they collected was used for their fires, stoves and fencing. Small top branches and undergrowth were bundled into faggots and used in faggot ovens[4]. One method of sale at the auction was to stick a pin in a lighted candle, people would bid until the candle burned down to the pin, when it fell out the sale went to the highest bidder.

In 1944 the final landlord and landlady, Mr. and Mrs. Felce, took over and remained until the pub ceased trading. During the Second World War U.S. airman were regular customers being stationed at nearby Thurleigh Airfield, the pub often ran out of beer by lunchtime! In fact the airmen had the choice of two White Horse public houses within walking distance – Riseley and Keysoe.

1996 the pub was sold to become a private house, but is still named The White Horse. LHS Riseley and AGe. 1839 P; 1850 S; 1869-1940 K.

**Shillington** 1822 the inn is listed in the Licensing Records, but almost certainly was trading at an earlier date.

Hertfordshire County Record Office hold the Simpson's Brewery deeds. 1898 K.

**Silsoe** 1839 P.

**Southill, High Street** The inn was built on land owned by Samuel Whitbread.

c.1826 the inn was trading according to the Licensing Return for Biggleswade Division dated 1876.

1919 an inn sign was painted by the artist Michael Dignam which was taken from Sawrey Gilpin's (1733-1807) oil painting of 1759.

c.2000 internal refurbishment was carried out. The house has huge fireplaces and oak beams.

A small local holistic group meet here every two weeks. They tell of a little girl called Ellen and a man who sits at the end of the bar, together with several other odd happenings in the house. 1850 S; 1869-1940 K.

**Stagsden** 1822 listed in the Licensing Records.

1869-1914 listed as trading in Kelly's Directory.

By the end of the 20ᵗʰ century it had become a thatched, private house. AA.

**Stotfold, Turnpike Road** 1728 recorded as a beerhouse.

By 1854 a new house had been built on the site. The Porter family ran the pub until 1903.

The Stotfold Silver Band practiced in the adjacent barn.

1958 there was a flourishing darts team.

c.1971 the pub was demolished to make way for housing.

**Wilstead, 82 Bedford Road** 1846 the White Horse was first granted a licence.

1940 listed as trading in Kelly's Directory under Wilhamstead.

By 2008 it had become a private house named Elephant Castle.

**Woburn, Bedford Street / Leighton Street** 1661 Sir Jonas Moore's survey of the Duke of Bedford's property in Woburn records Edward Gale as tenant of the White Horse, which is shown on the south side of Leighton Street; it is described as 'a large tyled howse five bay with back gowse brew howse, barnes stables, corne rooms, facing the market, betwixt the last tenement and Richard Houghton's howse'. The property was copyhold, held from the Manor of Woburn Abbotts.

1693 John Gale, a yeoman, became the tenant, but Robert Geeves took over from him during the year.

1712-1751 the house was recorded in the parochial dues.

1752 alterations were made to the building.

1788 it is thought the building was no longer an inn because of the following entry:

> 'which said part of the said messuage shop and chamber was used as
> part and parcel of the said messuage or tenement heretofore called
> the Horse Inn East and fronteth North into the market place there.'

There must have been a second White Horse as between 1802 and 1833 as a White Horse is recorded in the parochial assessment register. In 1817 a survey of the Duke of Bedford's property by Thomas Evans describes it as standing a long way out of the centre of the town, just south of Timber Lane on the north west side of the road, on the site of the modern street numbers 31-36.

1822 listed as trading in the Licensing Records.

1853 Kelly's Directory listed Jesse Lewis, a shoemaker, as licensee until 1869.

1877 the house was no longer listed. LHS Luton and District. 1839 P; 1850 S; 1869 K.

**Wymington, 17 High Street** 1869 first granted a licence as a beer house with Edwin Goosey as owner and William Lewis as occupier.

1938 the house was rebuilt and was still trading in 2007. BCR.

**Yieldon** 1803 the White Horse was listed in the Licensing Return but by 1876 it was no longer trading.

**Notes**

1. A 'last' is the wooden or metal form on which a shoemaker fashions shoes.

2. A shoemaker. Named after the Spanish town of Cordova where leather was made out of goat skins tanned and dressed, or later of split horse hides. It was much used for shoes, etc. by the wealthy during the Middle Ages. SOE.

3. Railway Sub-Office. See R.S.O.

4. Faggots were burnt to embers in the oven, then food would be cooked. LHS Riseley.

# BERKSHIRE

## Slough

*'The most fertile district is known by the name of the vale of the White Horse,
which is bounded by a range of chalk hills, and receives its name from the figure
of a gigantic horse cut on the side of a hill so as to expose the white chalk below.'*
JB.

**Binfield** 1877-1935 K.

**Cippenham, 359 Bath Road** The name comes from the old English Cippan-
ham, or Cippa's homestead. King Henry III had a palace here, the site is known
as Cippenham Moat. Villagers grazed their cattle on the Green until the end
of the 19th century – it is the only original village green surviving within the
boundaries of Slough.

From 1827 listed as trading in the Licensed Victuallers records with John
Lawrence as licensee, the sureties were provided by James Darby of Cookam.

The White Horse was not listed by Pigot or the Post Office until 1869 when
Arthur Downes was licensee.

1877 listed in Kelly's Directory as situated at Two-mile Brook – according to
a local historian the pub was one of four each a mile apart; from 1887 Kelly does
not list the White Horse for the next thirty nine years. If it was an alehouse, it
might have been listed under the name of the beer seller.

1883 the White Horse public house is shown on the Ordnance Survey map,
also for 1897, 1899 and 1924.

1885, 1st August – the *Slough Observer* reported that Mr. Buchanan of the
Greyhound, Beaconsfield had sued Mr. Hine formerly of the White Horse who
had said he was moving away with his family to a large town. Mr. Buchanan
had seen an advertisement for the sale of the inn in a daily paper; he agreed
to pay £120 which included the furniture, fixtures and goodwill and paid an
installment of £12. It was then discovered that Mr. Hine had moved only a mile
away to the Windmill public house, presumably taking his customers with him.
Mr. Buchanan declined to take the White Horse and Mr. Hine refused to repay
the £12 - the Court ordered it to be repaid with costs.

1928 listed as trading in Kelly's Directory with T. Jn. Matthews as licensee; the
pub is then listed consistently.

1932 the Ordnance Survey map shows the public house as 'One Mile House'
although Kelly's Directory lists it as White Horse; it appears as a different shape,
so alterations to the building must have taken place.

c.1950 still shown on the Ordnance Survey map until at least 1995.

1979 listed as an hotel in the Telephone Directory.

2000 the Summer Business Register listed the 'White Horse Big Steak Wacky Warehouse'.

2001, September - listed in Thompson's Directory, but does not appear in the May 2002 telephone directory.

By 2005 the building had been demolished and the site had become a car sales business. Landlady of The Pheasant at Burnham. 1883 Burnham K; 1886 Slough PO; 1915 Burnham - 1939 K. SLSL.


**East Ilsley** 1620 a charter was granted for a market, corn and sheep fairs were held regularly. Over the years the numbers built up to some 80,000 sheep penned in one day and with some 50,000 being sold.

1724 Daniel Defoe (c.1660-1731) described the fair as 'the greatest for sheep in the kingdom'.

1776 the opening of the Oxford and Newbury turnpike created a staging post for a major north/south route along the ancient Wheat Road. This small village with a population of 750 had twelve public houses, all close to the sheep pens in or near the High Street. Accommodation, food and fresh horses for the passing trade were essential; the White Horse with a back yard and garden was small with limited sleeping space for visitors.

1777 The Reading Mercury and Oxford Gazette reported a break-in at The White Horse bar in the dwelling house of John Goodall, thought to have been the landlord; he was a widower with three children, his wife Sarah having died in 1753.

1782 the creditors of the corn dealer, John Goodall (junior), were requested to meet here.

1799 Smithfield Market opened in London with two classes for cows and two for sheep!

1815, 29th September – William Seymour, a victualler was authorised to 'keep a common Ale-house or Victualling-house, at the Sign of the Whitehorse in East Ilsely . . . keep good Order and Government, suffer no disorder to be committed, or unlawful Games to be used in the said House, Yard, Garden, or Backside thereunto belonging, during the Continuance of the said Licence . . . ' William died in 1835.

c.1836 Thomas Hearne/Hearnsay, was recorded as 'occupier' of the inn. At this time John Satchell was the owner, who also owned The Star and The Crown and Horns; by 1872 these three pubs were owned by W.H. & F Blatch of Theale.

1839 George Matthews, a carpenter, was landlord with his second wife, Sarah and three children; she had been the widow of William Cooper, the landlord of the White Hart.

By 1844 Pigot's Directory lists Joseph Cooper, who had eight children as landlord; he was Sarah Cooper's brother-in-law.

1851 Frederick Stillman, a victualler and builder, was landlord. He had a wife and two children; there was also a domestic servant living with them. By 1851 they were not listed in the census.

1852 Joseph Stanmore Ralph was landlord, his son also Joseph, became landlord in 1860.

1854 Billings Directory records John Lock as landlord, he had previously run the White Hart for many years.

1860 Joseph Stanmore Ralph, a farmer from Wootton in Oxfordshire, took over the pub with his wife and six children. The family stayed here for eleven years, by 1871 the census lists only three children. Joseph was thought to be a very successful landlord as unlike previous landlords he had no secondary occupation; he later moved to the Star and then to the Crown & Horns, here he was fined £1 9s 6d with 10s 6d costs for adulterating his whisky.

Between 1872 and 1881 there were a number of landlords before George Tuson, a shoemaker, took over.

1891 John Evans, a publican and groom, followed three landlords between 1885 and 1890. He lived in the pub with his wife and a domestic servant.

1891 Samuel Somerville, innkeeper and policeman, took over and remained here for the next ten years.

c.1910 a photograph taken during a Sheep Fair shows the White Horse as one of four adjacent pubs in the High Street.

1911 the pub closed.

1965 a photograph shows the new front of Wheatsheaf House – formerly White Horse. The front having been rebuilt after an accident caused by a milk tanker which demolished the front elevations of both The Lamb and the former White Horse. LVR. Q/RLV/3/5, SGB and S. Burnay LHS – East Ilsley. 1830 P.

**Emmer Green, 9 Kidmore End Road** Situated in the heart of the village on the busy main route from Reading to the north, it is thought a coaching inn has stood on this site since the 16th century. The stabling and adjacent smithy remained until the early 20th century. Throughout the 19th century the blacksmith was also the publican. The inn baked its own bread and sold groceries up to c. 1920, local  people were able to take their Sunday joints to be cooked in the bakehouse ovens.

Food and lodging were always available, guests slept in an open dormitory but there was no bathroom until c.1950.

1938 the darts team won the Reading darts trophy.

Over the years various internal alterations took place. The inn was comprised of several small games rooms – one cribbage, one for darts, one for skittles and one for singing and talking.

During the 1950s a new inn sign was designed based on the Tincommius[1] coin found near St. Barnabas Church which depicted a warrior mounted on a white horse, the sign was replaced some years later by a horse's head on a green background.

1973 internal walls were demolished and old beams exposed to create an open plan area which included a dining room.

1983 the main entrance was moved to a side annexe.

1993 the name was changed to the 'Pickled Newt' however, due to local opposition, it reverted to the 'White Horse'.

Both in 1998 and 2002 the inn was closed for two months while being refurbished.

2005, September – the house re-opened after further refurbishment and was advertised as 'a top quality community local, attractively built in brick'.

Emmer Green Village Website. 1915-1984 K.

**Finchampstead** 1907 listed as a Posting house in Kelly's Directory. 1877-1899 K.

**Hermitage, Newbury Road** The building is thought to date from the 18th century.

2006 the house was closed, but reopened in 2007.

1877-1939 K.

**Lambourne, Oxford Street** 1830 P; 1840 R; 1844 P; 1891 *R.S.O.* - 1915 K.

**Maidenhead, 105 High Street** In 1782 the traveller, Herr Moritz, commented:

*'Maidenhead itself is a place of little note. I asked them to make
me a mulled ale and had to pay ninepence for it.'*

Sometime the name was changed to The Brewers' Tea House, in 2006 the name returned to The White Horse. CPM. 1823-1844 P; 1877-1939 K.

**Newbury, 40 Northbrook Street** Listed under 'White Horse Cellar'.

1830 P; 1840 R; 1844 P.

## Reading

*92 Silver Street* By 1883 Kelly's Directory lists the White Horse at No. 39.

1823-1830 P; 1840 R; 1844 P; 1877 K.

*74 Caversham Road* The signboard was painted by G.E. Mackenney in 1972.

1877-1939 K.

**Twyford, Shurlock Street** Originally a cottage looking out over open fields.

1750 it appeared in the Court Rolls for the Manor of Hurst and belonged to Moses

Sadgrave, the village baker. Moses and his wife lived there for fifty seven years and had six children. He died in 1758 requesting in his will that his wife should be given firewood in 'a reasonable and sufficient quantity to keep a good and constant fire'.

The property was left to his grandson, Richard Terry, who sold it in c.1762.

1774, October – the cottage, then known as Lambournes, with a barn and half an acre of land in an area known as 'the Gravel Pitts in Twiford Butts', was bought by a brewer from Reading, Stephen Flory, who built The White Horse, the first public house away from the main street.

1809 by this date it was owned by William Garrard and leased to Thomas Phillips, a mealman. William Garrard died in 1828, leaving his daughter, Elizabeth, as trustee. Shortly afterwards the landlord, Thomas Sowden, died and the lease passed to Henry Mason for £60 - he was listed as a victualler, innkeeper, livery stable keeper and landowner of several plots in the area.

In November 1831, once the disputes over William Garrard's will had been settled, a Manorial Court was held at The King's Arms when Elizabeth surrendered The White Horse to her brother Thomas, manager of a brewery in Friar Street.

By 1843 the brewery had been auctioned together with its public houses, including the White Horse. In 1846 Thomas Garrard died, leaving the property to his eldest son, The Rev. Thomas Garrard of Uffington who two years later sold it to the Wokingham Brewery, one of the three owners being John Rodgers Wheeler, the landlord, along with two business associates.

1857 The White Horse was sold for £403 to brewers at Theale.

c.1885 George Priest, a journeyman bricklayer, was listed as landlord – his wife Annie ran the bar; they were there for ten years.

From c.1900 records show that the house became rowdy and was described as the roughest pub in Twyford. On Saturday nights it was necessary for the village policemen to be in attendance.

1909, March – the house, with an annual value of £12 15s. 0d., was referred to the Compensation Board[2] and by December it had been closed.

1910 the property became a private house renamed Hornbeam Cottage.

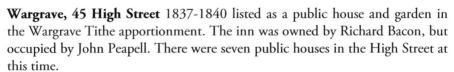

JSR. 1830-1844 P; 1877, 1883 R.S.O. - 1907 K.

**Wargrave, 45 High Street** 1837-1840 listed as a public house and garden in the Wargrave Tithe apportionment. The inn was owned by Richard Bacon, but occupied by John Peapell. There were seven public houses in the High Street at this time.

The 1841 census lists a Thomas Cooper – an agricultural labourer, but gives no address. By 1861 the census lists the publican as Thomas Cooper, a victualler, his wife, a dress maker and three agricultural merchants.

1871 Henry James Miller is listed as innkeeper.

1877 listed as trading, but by 1881 the census does not list the White Horse, however the house appeared to be occupied by William French, a jobmaster – he is described as a fly[3] proprietor in the 1891 census.

1901 the house is thought to have been renamed Orchard Cottage and in the occupation of Alfred Priest, a domestic gardener. It was later named Tudor Cottage and is one of two houses in the village to have a fire insurance mark on the wall. P.M. Delaney of LHS – Wargrave. 1877-1883 K.

**Winkfield Row** The village originally sprawled along the Forest Road but by the 20th century it was gathered round the White Horse pub, the school and the Working Men's Club.

By 2007 the pub had become an Italian Restaurant named Don Beni. JH&GS. 1877-1939 K.

**Wokingham, Easthampstead Road** The cottage next door – now 'White Horse Cottage' was the original inn which was trading in 1849.

1860 the inn moved next door. 1877-c.1897 Hones Green; 1899-1939 K.

### Notes

1. Son of Commius and king of the British Atrebates, c.25BC-AD1.

2. This stipulated that a licence could be withdrawn if the premises were structurally unsuitable or if the licence holder was considered unfit. JSR.

3. A light one-horse covered carriage formerly let out on hire. Col.

# BUCKINGHAMSHIRE

*'A large portion of this county is laid out in dairy and grazing
farms which supply the London market with butter, fat oxen,
lambs, calves, hogs, early ducklings, and the like.'*
JB.

**Aylesbury, 60 High Street / Market Street** Thought to have been trading before
1823 when the White Horse was listed in Pigot's Directory.

1872 listed as trading in the Returns of Public Houses and Beerhouses with
John Ward as landlord, the owners were Messrs. Parrott, who also owned the
White Lion and the White Swan in Aylesbury.

1823/4-1844 P; 1851 S; 1883, 1899 2 listed - 1915 K.

**Beaconsfield, London End** Thought to have been trading before
1823 when it was listed under White Horse in Pigot's Directory.

1872, 29th September - listed in the Returns of Public
Houses and Grocers with out-door licences under White
Horse; with George Hare was the landlord and Neville,
Read & Co. of Windsor were the owners. 1851 S; 1883-1939 K.

**Bletchley** 1898 K.

**Burnham, Lent Rise** The house is thought to have been trading before 1820.

1872, 29th September – the Returns of Public Houses and Grocers lists the
White Horse with out-door licences; James Cox was the landlord and the owners
were Nevill, Read & Co. of Windsor. 1883 K; 1886 Slough PO; 1915 K.

**Chearsley** Thought to have been trading before 1820.

1872, 29th September – listed as trading under White horse in the Return
of Public Houses and Grocers with out-door licences; Benjamin Wilson was
landlord and J. H. Peel of Watlington was the owner.

After the pub closed the building was occupied by a bakery until c.1977 when
it became a private house. 1883 K; 1899 PO; 1915 K.

**Chesham, 2 Amersham Road** Situated by the old village pound. 1838 the house
was granted a license for the first time, the owner and leaseholder was T. and J.
Nash of Chesham and the occupier was Reuben Geary. At one time it was known
as The Waggoner's Rest.

1872 the Return of Public and Beer Houses in Buckinghamshire lists two
White Horses under Chesham. The authors are unable to verify the location of
this other public house, whose license was first granted in 1830 when the house

was owned by Jonathan Bunker, the leaseholder was Messrs. Darvell of Chesham and the occupier was John Clare.

A photograph shows the White Horse inn in Amersham Road situated close to 'the early fuel facility on the London road'.

By September 1994 the name had been changed to 'The Wild Rover' and an extension had been built. B&A and Landlady. 1883-1939 K.

**Dinton** Thought to have been trading from at least 1810.

1851 the census lists Edwin Horwood aged 39, a butcher and victualler, as landlord. He lived here with his wife Hannah aged 41, their nephew James Ware a journeyman butcher aged 22 and their niece Jane Kirby a barmaid aged 16.

1872 the Return of Public and Beer Houses lists Spencer Smith as licensee and the owners as Messrs. Weller of Amersham. LHS Buckinghamshire. 1883-1939 K.

**Farnham Royal** 1915 K.

**Great Kingshill, Heath End Road** 1836 a beerhouse license was granted to the White Horse.

1872 listed under Great Missenden as a beer house with William Blunt as landlord and Robert Douglas as owner. 1899 Gt. Missenden PO; 1915 High Wycombe - 1939 K.

**Hedgerley, Village Lane** The inn is thought to date from the mid-seventeenth century, however the Licensed Victuallers records list it for the first time in 1753; it was owned by Wethered's Brewery of Marlow for 250 years.

1992 it became a free house.

In 1997 the beer was still served straight from the barrel by gravity.

SLSL. 1883 K; 1899 Gerrards Cross PO; 1939.

**High Wycombe, 95 West Wycombe Road** 2004 the pub 'has strippers every lunch time except Sunday and on a Wednesday evening. Usually stunners, nice stage and atmosphere, bar service a little slow: but they do lap dances (for £20) which makes up for it.' There are also Rock bands. 1844 P; 1883-1939 K.

**Longwick, Thame Road** 1834 the house was shown on the Ordnance Survey map, it was probably a significant hostelry at this time and is thought to have been a drovers' inn.

2007 the building was demolished.

LHS – Dr. A.J. Macfarlane. 1851 S; 1899 Princes Risborough PO; 1915-1939 K.

**Padbury, Main Street** A mens' drinking pub, which sold beer and wine only.

1832 listed as trading in the Return of Public and Beerhouses with John Kirtland as licensee; the owner was Joseph Groom of Great College Street, London.

1872, Monday, 24th June – the White Horse, an old established free public house was put up for auction. The house included a taproom, bar, bar parlour, club room, good bedrooms, scullery, cellarage, a blacksmith's shop and shoeing shed, stable and outbuildings, large yard, garden and premises; 'eligibly situate abutting on the Main Street and in the centre of the Village; in the occupation of Mr. John Kirtland, at the very moderate rent of £14 per annum'. Joseph Groom was still the owner.

c.1904 a photograph shows the Padbury Friendly Society at the front of the house with their flag and musical instruments; they used the club room in the pub for many years.

1957 the pub closed to become a private house. CB&CA and CBS. 1939 K.

**Stewkley, 155A High Street North** c.1994 the pub became a private house. 1883-1939 K.

**Stony Stratford, 49 High Street** Situated on Watling Street, a major thoroughfare from London to north Wales used by Roman and medieval armies. Between the 15th and 17th centuries there was a significant increase in the number of travellers, which meant a rapid growth in the numbers of inns, this included The White Horse which was listed as trading in 1540. At the height of the coaching trade thirty or more stage coaches passed through the town daily - the travellers all requiring food, accommodation and stabling.

1872 listed in the Returns of Public Houses with Samuel Vaughan as landlord and Mary Bottoms of Passenham as owner. The leaseholder was Phillips and Brothers of Northampton. CBS. 1830-1844 P; 1851 S; 1883-1939 K.

**Whitchurch, 60 High Street** A 17th century inn, the date 1837 can be seen on the chimney stack along with three sets of initials - the Master Bricklayer and his two brickies. 'White Horse Lane' runs beside the building, here there are three cottages which were once the local gaol.

1872 listed in the Return of Public Houses with Albert Holloway as landlord.

2006 clay pigeon shooting was organised here by Guy Bond, former manager of the Great Britain shooting team. There is a conference centre and restaurant, also accommodation. Licensee. 1844 P; 1851 S; 1883-1899 K; 1939 K.

# CAMBRIDGESHIRE

*'This county, at the time of the Roman invasion, formed part
of the kingdom of the Iceni and, according to Whitaker, was
inhabited by a tribe of that people called the Cenomanni.*

*Over this vast expanse [of fen land], the towns and villages,
built upon little elevations, through the moist and foggy air rise
upon the view like so many islands, and the turrets and spires
can be seen at the distance of many miles.*

*There are still about 150,000 acres of unimproved fen land in
this county'.*
JB.

**Alconbury** The author was unable to locate this house; in 1898 Kelly listed a
White Horse Inn some four miles south at Northend, Huntingdon.

**Barton, 118 High Street** In 2006 this old coaching
inn offered accommodation, a restaurant seating 40
people, a lounge bar, the Village Bar and a large garden.
2008 Greene King advertised the inn 'To Let' with
several other public houses.

KSGH. 1850 S; 1875 PO; 1900-1937 K.

**Bluntisham** The inn is situated in
the centre of the village overlooking the square.

1795 listed in the Licensed Victuallers in Hurstingstone
Hundred with Thomas Watts as licensee, at this time
there were 10 pubs in the village.

1801-1807 the local farmers met here to discuss
parish affairs.

1819, 5th November – a meeting of the Commoners of
Bluntisham was recorded as being held at the White Horse for the first time, when
William Asplan Jnr. and Thomas White were elected Fen Reeves for the ensuing year.
Meetings for the Fen Reeves were held here in 1821, 1827, 1842, 1861 and 1868,
those in intervening years were held in one of the other public houses in Bluntisham.

1889 listed as trading.

1909, 5th April – Huntingdonshire Petty Sessions granted the house an alehouse licence
for a year, the licensee was Harry White and the owner was Philip Hudson of Cambridge.

1914 Kelly's Directory no longer lists the White Horse.

CCC, CFT, SHM and KSGH. 1830-1839 P; 1850 S; 1885-1910 K.

**Buckden, Silver Street** Situated on the west side of the street close to the corner with Church Street.

1881 the census gives the landlord as Mr. John Jeakins, a blacksmith and publican; the smithy was attached to the pub. The premises were owned by Marshall's Brewery of Huntingdon.

c.1910 a photograph shows the White Horse with the new bakery and coal merchants opposite, these replaced a row of thatched cottages and bakery which had burned down in 1908.

c.1930 listed as a beerhouse.

Before the pub closed in c.1950 some local people referred to the pub as 'Jenny Sears' – she was the last landlady; the building, with the smithy, became a private house with no trace of the old front door leading on to the street. Buckden LHS and CCC. 1898 K.

**Burwell, High Street** c.1760 recorded as trading under White Horse. A two-storey building in white brick of 18th century design, although the building is thought to be earlier because a 17th century ovolo-moulded[1] cross beam exists in one room.

1846 both Pigot and the Post Office list a White Horse Inn in their directories.

1849 The White Horse Hotel was extended towards the street in white brick with central double doors and a parapet; soon afterwards an addition was added on one side in the same style to provide a first-floor assembly room; a bowling green was created about this time. At the opening of the new building in December there was a tea party and evening entertainment by the Newmarket and Burwell brass bands. These bands accompanied the Odd Fellows Society through the village to the White Horse to celebrate their first anniversary in April 1851.

1862 Mr. G. H. Clifton in his 'Burwell Alphabet' records that 'John Carter, brewer, maltster, and proprietor of the White Horse Inn was considered a good judge of horseflesh, and was known far and wide by dealers and others who came and stayed at his house for Reach Fair', which was under two miles away.

This large building seems to have been used for aspects of village life, for instance the club room had at one time been used for a fish and chip shop, also for fruit and vegetables; there was also Miss Ellis's barber's shop in the same building and the White Horse Bowls Club nearby. At the back of the White Horse Inn, there was a Variety Hall named Rialto which was also used by travelling shows and a cinema group.

1908 in June, the Swimming Club held their first annual dinner and smoking concert here. *Shock horror in 2007!*

1929 recorded as trading.

c.1930 it was listed as an hotel.

1975 the hotel closed.

By 1990 the club room had been demolished and replaced by a hairdressers.

2007 alterations were made to the main building.

HMR, CL, CLu, HMR, CCC, VCH Vol. X and K.S.G. H. 1850 S; 1875 PO; 1900-1916 K.

## Cambridge

No addresses are given for two White Horses listed in the 1830 Pigot's directory, one in the 1850 Slater's Directory and one in the 1916 Kelly's Directory; they are probably Castle Street and Trumpington Street.

**95 Castle Street** Originally built as a farmhouse in 1423.

c.1900 Jenkin's fish shop occupied part of the premises and there was a wood yard on the Northampton Street side of the inn.

1921 A.B. Gray wrote about the inn in his 'Cambridge Revisted':

It is 'interesting on account of a hiding hole used, probably in earlier days, by highwaymen and ingeniously contrived in the exceptionally wide central chimney breast. The little hiding place was entered from a sitting room through an aperture at the back of the fireplace, now walled up'.

By 1926 the White Horse was listed at 2 Castle Street.
c.1935 the inn was no longer licensed.
By 2007 the building had become the Cambridge and County Folk Museum.
1823/4-1839 P; 1850 S; 1875 PO; 1900 K.

**36 Coronation Street** 1875 listed as trading in the Post Office Directory.
By 1913 it had ceased to trade and had become a shop.

**Kings Parade** 2002 according to Zach Kincaid of Trinity International University in Illinois, USA, a 'White Horse Tavern was located on King's Parade on the University of Cambridge campus, part of King's College, located in St. Edward's Parish. Dating from the 16th century, the tavern was demolished in 1830 with the reconstruction of the west side of the street and Porter's Lodge. There is some stained glass with the words 'White Horse Inn' over the entrance to the 'café style meeting rooms below the chapel at Trinity'. *However, the author has been told there are no meeting rooms below the chapel.*

**Trumpington Street (on the corner with King's Lane)** Built probably in the early part of the 16th century.

The Cambridge Reformers met here, they were engaged in the compilation of the Liturgy. The leader of the White Horse Inn group was Robert Barnes – martyr – who was an Austin friar born in 1495, many students went to the monastery where he was Prior to hear him preach. Thomas Cranmer, the first Archbishop of Canterbury in the reformed Church of England, was a member of the group that met at the inn to study and debate the writings of Luther which were smuggled into the East coast ports of England from Antwerp.

An elaborately carved settle in one of the lower rooms was known as 'Miles Coverdale's Seat' - Bishop Coverdale (1488-1568) made the first printed English translation of the Bible in 1535. The seat was later presented to the Cambridge Antiquarian Society; by 1876 it was in the care of the Fitzwilliam Museum.

1823 Pigot's Trade Directory listed 'Ye Whyte Horse'.

While fitting a new stove Mr. Cory[2], the owner, discovered a faded fresco of Kings College Chapel behind some panelling.

1871 the inn known as 'Cory's House' was pulled down to make way for the new building of Kings College. WBR, KSGH.

**Chesterton / Old Chesterton, 98 High Street** A large inn which was demolished in c.1910 - it was no longer shown on the street map. Terraced houses were built on the site, whilst more houses were built over the yard at the back to become Thrift's Walk, the name of the landowner. LHS Chesterton. 1846-1875 PO; 1900 K.

**Comberton** 1846 listed as trading under White Horse in the Post Office Directory, but by 1925 it was no longer listed. 1875 PO; 1900 K; 1916 K.

**Cottenham, 215 High Street** Once situated on the site of 'The Limes' and thought to have been the largest inn in old Cottenham, dating from the 17th century; it belonged to the Dowsing family but occupied by Thomas Branscombe. 1730 Thomas Branscombe was granted a licence to sell liquor.

1753 Alice Dowsing was the licensee, she was followed by her son, William. The inn had been named the Grey Horse and the Yorkshire Grey.

c.1786 the inn became the White Horse.

1792 William's daughter Ann and her husband Robert Norman took over the inn and brewery.

Before the Enclosure Act of 1842, the annual Ordermakers', or Parish Officers', Feast was held here; they supplied beef and mutton, the meal for bread, and flour for the puddings and two buckets of coal to warm the room all bought from tradesmen in the village. Ann Norman cooked the meal – in 1811 she received £2.10s. 0d. for beer and ordermaking and 14s. 0d. for doing the cooking and making sauce.

1829 Robert Norman died and his son, Richard took over the business.

1850, April – a great fire destroyed The White Horse. By this time Richard's son, also Richard, was managing the business aged 22. His grandmother, Ann Norman gave part of the burnt site to another grandson, Norman Smith who built the house named The Limes at 219 High Street. Richard built the large Cottenham Brewery on the remainder of the site, which became No. 217 High Street.

1855 Richard sold the brewery to his brother-in-law, Charles Male.

1885 Charles Male died and the brewery had closed by 17th October 1890 when it was sold by auction. 1846 PO.

**104 Rooks Street** c.1852 the house was licensed as The White Horse, was owned by William Smith and kept by William Porter.

1854, 1st October – William Porter and the landlord of the Waggon and Horses were drowned while returning home with two friends from a musical evening at Burwell when their cart overturned into a large ditch. William's widow then ran the house until c.1859.

1880, December – the landlord, Jonathan Piggott sold his utensils and stock in trade by auction, which included a bagatelle table with massive turned legs.

1888 John Burgess, a painter and glazier became the licensee.

1889 a man from Peterborough fraudulently told Mrs. Mary Burgess that he was employed by the Post Office, she had given him credit of 4d. in money, 2d. of whiskey and 1d. of beer.

1898, 16th October – according to the diary of a local lady, Jonathan Pauley the landlord of the White Horse did a moonlight flit.

1900, 3rd November – the landlord Robert Lindley Booth died, his widow took over the business until 1903.

1908 Herbert Hulton, the landlord, was sued by the previous licensee, Walter Player, for money due to him on the take-over. Mr. Hulton was then declared bankrupt and his furniture was sold by auction on 2nd July which included a Turner and Price billiard table of 10ft. 6ins. x 5ft. 6ins. and a John Bull cigarette delivery machine.

1910, 13th October – Lacons, the brewery advertised The White Horse Hotel for sale. It was described as:

> 'Brick built with stone facing, slated roof, bar, commercial and smoke rooms. Private sitting room, tap room, kitchen, pantry, etc. 5 bedrooms, spacious club and billiard room, large yard, 4 loose boxes, chaise house, harness room, corn store, chaff house and stabling for 6 horses.'

1911 bought by Dales, brewers, who advertised The White Horse as a 'family and commercial Hotel, motor buses passing the door, trap to meet train by appointment. Dewar's and Buchanan's special Scotch whiskies and Dunville's special Irish'.

1884 the inn became a Free House.

By c.2004 it had ceased to be a public house and had become The Curry Palace.

Cottenham Village Society and KSGH. 1875 PO; 1900-1937 K; 1956 Hotel; 1964 C&L.

**Eaton Socon, 103 Great North Road** Dating from the 13th century, 'Ye Old White Horse' was situated on the Great North Road and was a Royal Mail staging post for coaches travelling between London and York. The building was re-fronted during the 18th century and many other alterations have taken place, though the old

English fireplace where gentlemen of the road are rumoured to have hidden or made good their escape remains.

The novelist Tobias Smollett (1721-1771) used this inn as a model for The Black Lion in his book 'Sir Launcelote Greaves'; Oliver Goldsmith (1728-1774) also had the White Horse in mind when writing 'The Deserted Village'.

Prior to 1837 Charles Fox, a mail coach guard, held the licence; he had seven children and five servants.

1838 Charles Dickens and his artist friend, Phiz (Hablot Knight Browne 1815-1832) stayed here – he used the inn in 'Nicholas Nickleby', published in 1838-9, calling it The Cock.

1869 Mrs. Jane Taylor, a farmer and brewer was licensee.

BHS, KSGH and HoW. 1955. 1823-1839 P; 1850 S; 1869-1898 St. Neots K; 1964 C&L.

**Ely, Back Hill (South Side)** Trading from the 17th century, this inn closed in 1893. It is not listed under White Horse in Kelly's Directory of 1875, but could possibly be the beer retailer, William Buckle, on Back Hill. RHo.

**Fenstanton** 1855 listed as trading at 'Fen Stanton' with William Bloom as licensee.

1898 recorded as an hotel in Kelly's Directory. 1885 K.

**Foxton, 45 High Street** 'The White Horses'. Situated on a mediaeval site.

c.1552 the house was rebuilt by Henry Sturmyn – apparently a lowly thatched cottage, but a salubrious one!

1633 Will Sturmyn died here; he had ten children born between 1608 and 1628, his wife outlived him by eleven years. One of their daughters, Mary Muncey who ran a private school, was summoned to the Archdeacon's court in 1686 because she never attended church.

During the 18th century the village blacksmith lived here.

1855 the house was granted a licence; Joseph Cooper was the landlord, the owner was Fordham's Brewery.

c.1880 it was destroyed by fire along with several other houses.

1882 rebuilt as a public house, but is no longer thatched.

RP and Landlord. 1875 PO; 1900-1937 K.

**Gamlingay** 1871 listed as trading with David Lancaster as licensee. 1916-1937 K.

**Girton** c.1760 the White Horse started trading.

A great annual feast was held here in late June and for some years took place in the garden of the house opposite the White Horse. The event was planned months in advance and a ham was sent to Cambridge to be smoked. It lasted for at least three days – the men celebrating on the Monday and Tuesday; the Wednesday was Mothers' Day when the women had a holiday and dressed up in their best clothes.

Most Cambridgeshire villages celebrated the bringing in of the last, or horkey, load of the harvest with a parade and events on the green; in the evening horkey suppers were given by the farmers. Mr. T. Osbourne, who was born in 1854, remembered that the Horkey 'was always on a Friday night as the farmers knew no one would be fit for work the next day. On Saturday there was a procession headed by a wagon with three horses driven tandem. On the first sat the Lord of the Harvest, on the second the Lady of the Harvest suitably dressed up. The third wagon was loaded with hay and all the children sat on top. The procession went to every farmer in the village who gave money to the men for their dinner, afterwards the men walked in to Cambridge. The Lord and the men visited butchers and grocers who contributed to their dinner fund which they had at a public house'.

At sometime the small thatched cottage with a chimney stack at each end was re-thatched and enlarged by adding an extra storey with two dormer windows and a central chimney stack.

c.1880 the inn was used as the village clubroom.

c.1910 the pub ceased to trade.

VCH Vol.IX, The Girton Times – Summer 2006 and K.S.G.H. 1846-1875 PO.

**Haddenham** 1847 listed as trading with William Hobson as licensee.

1906 the house had a six day beer licence, the owner was William Cutlack of Littleport and the licensee was Charles Crane who had been here for 14 years; there was stabling for one horse only.

1912 the house was closed. IOE and KSGH. 1846-1853 PO.

**Huntingdon, Ermine Street** The inn was Situated on the Roman road which originally ran from a point near Pevensey in East Sussex, to York; later it ran from London to Lincoln. The name Ermine was derived from a group of Anglo-Saxons who settled in Cambridgeshire near to the road called Earningas, people of Earna.

B2. 1823/4 North End - 1839 P; 1850 Northend S; 1885-1936 K.

**Ickleton** 1937 K.

**Isleham, 13 Church Street** The inn opened in c.1800.

1829 the building, probably timber-framed, was destroyed by fire but was rebuilt as a grey brick house.

By 1857 a society of Ancient Shepherds were meeting in the clubroom, by 1873 they had a membership of 120.

Arthur Haughton, born in 1911, mentions in his book 'Memories of Isleham' that in his youth the inn was kept by Arthur Collen who was a smallholder; behind the building there were stables, piggeries and a cowshed.

The inn closed sometime after 1930 and became a private house named Claremont House. VCH and Mr. V.R. Place. 1875 PO; 1900-1937 K.

**Keyston** 1885-1936 K.

**Kimbolton, Stow Road, Newtown** The house was situated next to cottages built c.1820 called Valentine's Row which became Valentine Gardens.

By 2007 it had been enlarged and refurbished. 1885-1936 K.

**Linton** 1859 S. *This is incorrect – The 'White Hart' was trading here between 1780 and 1911.* LHS - Linton.

**Little Chishill** 1848 the landlord was listed as a blacksmith and victualler.

1895 listed as trading with Josiah Jude as licensee. 1863 W; 1886-1937 K.

**Little Downham / Downham in the Isle, Cannon Street** 1871 The White Horse is recorded for the first time in the census; Kelly's Directory of this date records the landlord's name only, he was described as a beer retailer.

c.1900 the Post Office changed the name of the village to Little Downham so that its mail was not sent to Downham Market some fifteen miles away.

1902 the Justice Returns listed John Herbert Frost, a labourer, as landlord; the owners were A & B Halls, brewers of Fore Hill, Ely. The pub had very little land, unlike the twenty others in the parish who had a few acres to augment their incomes. Mr. Frost was followed by Mr. Shelton as licensee who had four sons and three daughters living in the house.

1906 the house still only had a beer licence; the licensee was John Frost, who had been here for 3 years. There was stabling for one horse only.

1932 the pub was closed - Ernest Culpin, a mole catcher, was the last licensee, he then took over a nearby pub in the village called The Spade and Beckett[3]. The White Horse was then let to various people until 1964 when it was demolished and a bungalow built on the site. KSGH and Mr. R. Martin.

**Littleport** 1875 the Post Office Directory lists a White Horse, allegedly in Granby Street. There was a James Harrison, a licensed victualler, who was possibly a retailer at this inn. He had been married in Littleport in 1850 but by 1881 he had moved to Spalding, Lincolnshire; the 1881 Census records James and his wife Eliza – both licensed victuallers - as living at the Vine Inn, 20 Commercial Road, Spalding. LAO and LHS – Littleport.

**Manea, West Field** 1861 Recorded as trading with Thomas Padgett, a shoemaker, as licensee.

1906 licensed as a beerhouse; there were frequent complaints of drunkenness, the pub being difficult to supervise as it was situated on a road leading to a fen. The owner was Morgan's Brewery of Norwich and the licensee was John Livett who had been here for 12 years. There was stabling for 3 horses. IOE and KSGH.

**March, Church Street / West End / Whittle End**
Situated beside the River Nene, the thatched house dates from c.1800 and was built for the bargees who would have stopped for refreshment while selling their wares up and down the river. The riverside footpath is one of the oldest parts of the town and was once known as Whittle End, many of the cottages are thought to date from the 16th and 17th centuries.

1875 recorded as trading with Joseph Housden as licensee.

1906 the owner was recorded as C.S. Lindsell of Chatteris and the licensee, who had been here for 4 years, was Alfred Harwood. The premises included one acre of land, stabling for 3 horses and accommodation for 3 vehicles.

c.2002 the pub closed after the landlord suddenly left – apparently not wishing to pay the new business rate – the brewery sold the property to developers who built on the land to the rear and the pub was restored to become a private house.

IOE and KSGH. 1823-1839 Whittle End P; 1830; 1850 West End S; 1875 West End Whittle End PO; 1900-1937 K; 1960 JM.

**Milton, High Street** Originally built as three cottages and converted into a pub with a cellar sometime before 1853 with stables and a horse pasture at the back – to become the present large garden; two of the front doors can be seen, the third is bricked up. The passage beside the building, once used by the horses, leads to the back where an old door gives entrance to the flat above the pub with large open attic rooms throughout. Landlady. 1853-1875 PO; 1900-1937 K; 1960 CCC.

**Oakington, 28 Longstanton Road** The pub was still trading under White Horse in 2008 when it was taken over by the landlords of the Duke of Wellington at nearby Willingham, they ran both houses. 1846 PO; 1900-1937 K.

**Pampisford, London Road (on the corner with Brewery Road)** 1688, October – the deeds described the site as a 'messuage, tenement and croft or orchard wherein Charles Parish now dwells'.

Sometime during the 18th century, the area on which the White Horse stands was awarded to John Whestone, a yeoman of Pampisford under the 'Inclosure Act' when the property was described as 'a messuage with appurtenances or croft, orchard and pightle[4] adjoining the street'.

c.1750 cottages on the site were occupied by families named Turner and Rollinson.

By 1831 the land and property had been sold to William Parker-Hamond.

1841 it is believed that beer retailing was taking place on the site. Pampisford's Brewery was being established at this time, the water below the chalk in this area having great purity and being ideal for brewing ales.

By 1866 Charles Smith had a grocers shop and bakehouse here; the brewers, Bathe & Co. owned part of the property when Sam Parsons was a beer retailer.

1889 The White Horse Inn was included in the sale of the property to Philip Llewelyn Hudson, who also bought the brewery.

c.1900 the house would have probably been open all day. There was a long tap-room, lounge bar and snug bar, there was also a billiards room in the building to the rear. Charabancs stopped at the Inn, some carrying hop-pickers on their way to Kent. One elderly resident remembered when she was a child playing with her friends outside the inn on summer evenings, waiting for the coaches to leave as the occupants would throw coppers to the children.

c.1937 the building was demolished and a new White Horse was built near by.

OCM. 1900-1937 K.

**Ramsey, High Street** 1823/4-1839 P; 1850 S.

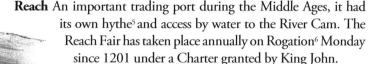

**Reach** An important trading port during the Middle Ages, it had its own hythe[5] and access by water to the River Cam. The Reach Fair has taken place annually on Rogation[6] Monday since 1201 under a Charter granted by King John.

This small 18th century thatched alehouse standing at the end of a row of cottages with only one small tap-room and no bar was too small to offer accommodation; it became a brick building of two storeys with a tiled roof, gabled parapets and kneelers[7].

1875 listed as trading.

1920 William King took over as landlord. He died in 1954 and was succeeded by his wife.

1960 Wilfred King became landlord until 1967 when it closed to become a private house. KSGH, RCH. 1937 K.

**St. Ives, Bridge End** 1823/4-1839 P; 1850 S.

**Soham, High Street** 1846 PO.

**Stetchworth** 1900-1916 K.

**Stretham, Chapel Street / Square** 1871 listed as trading with Edward Jackson as licensee.

1906 listed as a beerhouse, the brewers Greene King were the owners and the licensee was Mary Morden who had kept the house for 9 years. There was stabling for 3 horses and accommodation for 2 vehicles.

c.1915 listed as trading in Kelly's Directory under the name of the landlord who was a beer retailer.

By 1989 the house had closed. It was mentioned in a book of village reminiscences published in this same year. It was thought to have traded under White Horse prior to 1915. KSGH and IOE.

**Swavesey, 1 Market Street** The principal inn situated in the market place at the corner of the High Street is thought to have been trading from c.1500, although the earliest recorded date is 1668. The main dock by the market was connected to a number of watercourses and at one time was used by barges carrying heavy goods; however after drainage the water-table lowered and the dock became the Town Pond which has been filled in to become an area of mown grass.

Dating from c.1700, the traditional Cambridgeshire timber framed and brick building with a tiled roof is thought to have once been thatched; there are three rooms upstairs and three rooms down. It has been altered at least four times, the marks of which are shown on the east gable. The final alteration would have been to raise the eaves to give more space in the three upstairs rooms. The windows and door are of 19th century design; the front door is a third of the way along the façade with a chimney directly above it; the chimneys at each gable end of the building date from the late 18th or early 19th century.

By c.1800 it was one of twenty one inns and taverns in the village.

By 1820 it was one of seven public houses to be fully licensed and considered to be the most important one the village.

1828 John Dodson is listed as licensee.

1839 tithe records show the locations of the inns with names of the proprietors only – there are no sign names.

1858 P. Dodson is still listed as licensee.

Over the years it has been used as an auction house, a public health office and a meeting place for many local societies which included several football teams, a golfing society, an MG owners club and a Cambridge motorcycle scrambling club. Charity events and musical evenings were also held here.

2006 the landlord Will Wright, who had been a professional drummer and sound engineer, was the third generation of his family to live in the village.

There are exposed oak beams, open log fires and old photographs of the village hanging on the walls, along with prints and drawings of Cambridge colleges.

2006 JS, LHS Swavesey and District and BBC. 1858 K; 1875 PO; 1900-1937 K; 1996 WA.

**Tilbrook, High Street** Recorded as trading from 1738, for many years the only public house in the village.

1822 is the earliest surviving licence.

1869 listed as trading with Charles Storey as licensee.

1876 listed in the Licensing Return with Thomas Smith as owner and Eliza Storey as occupier.

1986 the pub closed, to re-open after refurbishment on 12th August 1989.

2004 the house was again refurbished but has retained the wooden beams and fireplaces. KSGH. 1839 P; 1850 S; 1869-1936 K.

**Waterbeach, 12 Greenside** The house consists of two buildings of different of different early dates.

1797 listed as trading with Jane Harrison as licensee.

1803, Wednesday 4th May – an auction sale of various properties took place at the White Horse, including one cottage called 'The Brewhouse'.

1813 listed in the Enclosure Award.

1837, 25th August – the White Horse was included in the sale particulars of Steward Cotton & Co., brewers. It was described as having a club-room, small parlour, tap-room, cellar and five bed-chambers. A back kitchen led on to a large yard, where there was a coal-house, range of stables and piggeries.

1911 the northern part of the house was rebuilt, this was thought to have been the club-room.

c.1950 shown on the village road plan. KSGH and DC. 1846-1875 PO; 1900-1937 K.

**West Wickham, 3 High Street** It is thought the house may date from c.1790 when the 2nd Earl of Hardwicke of Wimpole Hall bought the West Wickham estate and was Lord of the Manor; the White Horse was part of the estate and was held on a copyhold[8] tenure.

1795, 25th January – the *Cambridge Chronicle* reported an inquest held at the inn, the coroner having most likely ridden the twelve miles from Cambridge due to the roads being in such a bad state.

The case concerned the seven-day-old child of Sarah Pluck who was unmarried. The verdict was wilful murder by strangling. A warrant was issued for the mother's arrest, but she was never caught.

1808 Jonathan Atkins, the parish clerk for West Wickham and nearby Horseheath became the licensee – he had just been married for the first time at the age of fifty one, but died the following year. The license was transferred to his wife, Mary; as she was his widow, she was permitted to hold the licence.

1810 Mary married a widower Walter Price, who took over the licence.

1826 Walter died and the licence reverted to Mary, who died two years later, the tenure passing to William Price, thought to be her stepson.

1839, 4th May – William advertised the copyhold for sale in the Cambridge Chronicle.

1842, 7th May – the White Horse was again advertised for sale, along with other properties owned by Nutter & Elliston, brewers, who had become bankrupt, as did many small breweries in this era.

Between 1842 and 1892 the history of the pub is a little unclear. The tenants of the adjacent buildings included John Simkin, a blacksmith and brewer, whilst Martin Jeffery was the innkeeper and a butcher. His wife Mary Ann Hayward had been married to William Walker, tenant of the White Horse from 1848 – the Hayward family were then associated with the pub until 1894.

1894 the Manor Court Book recorded that Frederick Christmas, a brewer of Haverhill became the copyhold tenant. Christmas & Co. then bought the property freehold after the break up and sale of the Hardwicke estate.

1917 Greene, King took over Christmas & Co. and the pub remained in their ownership until 2003/4.

The White Horse had only one bar and served an agricultural community; between c.1930 and c.1980 one room in the house was a small shop until the building was refurbished.

2004 the property was sold to a pub chain who in turn sold it to be converted into a private house, in 2007 the village was still campaigning to save it and re-open it as a pub. LHS – West Wickham. 1875 PO; 1900-1937 K.

**Whittlesey** 1904 listed as a beerhouse with J.W. Carter as licensee.

1906 the owner was J. & H. Phillip of Stamford; there was stabling for 2 horses and accommodation for 2 vehicles.

1925 Charles Bingham, a pork butcher, was listed as licensee.

1971 the house was closed to become a private house named The White Horse.

2007 there was an iron silhouette of a white horse in the window.

IOE and KSGH. 1937 K.

**Willingham, Main Street** 1851 the Directory of Cambridgeshire listed the White Horse kept by Charles Underwood who was an auctioneer.

By 1871 Alfred Underwood had taken over, he was also an auctioneer as well as a publican.

By 1875 the name had been changed to the George and Dragon, Alfred was still the licensee.

1883 Henry Simmons was the licensee, he was also a monumental mason.

1916 the house was no longer listed. KSGH. 1853 PO.

**Wisbech** 1892-1900; 1914 Newton K.

**Witcham, 7 Silver Street** The White Horse was the only house in the village to sell spirits, the others being ale houses; there was an off licence and children would come to a window to buy their drinks and crisps. At one time pigs were kept at the back of the building on what became the car park.

1871 Thomas Wallis was listed as licensee, he was also a blacksmith.

1899-1904 Arthur Giddens, a saddler, was listed as licensee.

1904 Ely Petty Sessions Division records that the house was fully licensed and owned by A & B Hall of Ely. There was stabling for a horse and space for a vehicle; one of the three bedrooms was available for travellers; the sanitary conveniences were described as 'good'. There was one door at the front of the building and two at the back. The gross annual value was £13 17s. 0d. and the rateable value was £11 3s. 0d.

During the Second World War the pub would have been patronised by the New Zealand Bomber Command who were based at nearby Nepal Air Field.

c.1990 an extension was added to the restaurant.
2005 the house was voted the Ely and District CAMRA Pub of the Year.
Morris dancers perform at the inn during the summer.
J. Wells. 1875 PO; 1900-1937 K; 1960 JM.

## Notes

1. A convex moulding of which the section is a quarter-circle or a quarter-ellipse, receding from the vertical downwards. SOE.
2. May have been related to Isaac Preston Cory (1801/2-1842) who matriculated at Clare College in 1820; a writer and Hebrew lecturer at Caius College where in 1824 he was elected to a fellowship. Five of his brothers were educated at Cambridge.
3. A drainage tool.
4. A small field or enclosure; a close or croft. SOE.
5. A landing place. OxN.
6. Rogation Monday, Tuesday and Wednesday before Ascension Day were observed by Christians with processions, special prayers and blessing of the crops.
7. Knee-timber; timber having a natural angular bend, suitable for making knees in shipbuilding or carpentry – also a bent piece of timber used in carpentry. SOE.
8. A tenure less than freehold of land in England evidenced by a copy of the Court roll. Col.

# CHESHIRE

## Wirral

*'In some parts of this county potatoes are cultivated to a great extent. In a great part of the county, however, the dairy is the great object of the attention of the farmer, and cheese has long been one of its staple productions, not less than 12,000 tons – the produce, according to a late survey, of 32,000 cows – being made annually'.*
JB.

**Chester, 66 Handbridge** In 1778 there were 141 alehouses in Chester, two of which carried the sign of the 'White Horse'.

c.1912 a photograph shows Javelin Men[1] outside the inn prior to escorting the Judge to the Assizes at Chester Castle.

2007, February – the inn was closed.

AJM. 1822-1834 P; 1849 NCD; 1855-1869 S; 1896-1939 K.

**Churton-by-Aldford, Chester Road** Originally a neatly thatched building thought to have been built between 1822 and 1850 which had become an alehouse by 1850 when a book records Moses Baker as supplying ale, this could have been from the White Horse.

1891 listed as trading under White Horse in the Inns Register, it was owned by the local Farndon Brewery.

1900 the thatched house was demolished, possibly by fire.

1901/2 a larger Edwardian-style public house was built set back from the road to allow room for customers' horses and traps. There were several outbuildings at the back, including a stable attached to the east building which housed shire horses, presumably used for pulling the beer barrel wagons. At the time of re-opening it was a tied house owned by Bent's Brewery of Liverpool.

The roadside sign was changed several times over the years, usually to an iron-reinforced wooden sign, by 2008 it once again resembled the original 1902 road sign. Local Historians. 1939 K.

**Congleton, Prospect Street / 43 Astbury Street** One of three public houses near Prospect Mill, the workers in this silk throwing mill would have been the main customers.

1860 recorded as a beerhouse, it retained this licence until it closed.

1875 the Ordnance Survey town map shows the pub situated on the corner of Astbury Street and Prospect Street.

1906, March – the *Congleton Chronical* reported the request for licence renewal was being debated by the Congleton Adjourned Brewster Sessions. The White Horse had been transferred seven times in as many years and six times in the last five years. The property consisted of two bedrooms – used by the tenant's family - an unused clubroom and a rarely used stable. It was owned by the North Cheshire Brewery of Macclesfield, John Birtles was licensee.

In this same year Joseph Hey, who had been tenant from December 1903 to May 1904, sued the previous tenant, Mary Drinkwater, for misrepresentation and had recovered damages - he had paid £40 for the in-going, the weekly income had been about 25 shillings per week and he had lost £30 on the house.

During 1906 the average daily customer numbers were 51; 98 barrels of beer had been sold in 1903 -1904 and 45 barrels in 1906. In spite of the reasonable sales the licence was referred to the Compensation Board.

1907, 12th January – the premises, still only a beer house, closed due to the licence renewal being refused – there were five public houses within a radius of 200 yards.

By 2000 the site was occupied by private garages. LMu.

**Disley, 18 Buxton Old Road** The brewers Frederic Robinson hold deeds which record that this site would be held for 999 years from 3rd January 1770; there is no indication that there was a public house here at that date.

The authors have not checked the year 2769!

1887 the deeds record the house was trading as the White Horse.

1889, 10th December – Sarah Moore sold the house to Bell & Co. Ltd., brewers of Stockport.

1892 listed as trading.

1949 Bell & Co. Ltd. was taken over by Frederic Robinson.

1982 this old hotel underwent alterations.

FR. 1834-1842 P; 1849 NCD; 1855-1869 S; 1896-1939 K.

**Great Barrow** 1822-1828 John and Samuel Dodd listed as alehouse keepers. The early photograph from which the line drawing was taken shows a large thatched building with a well-dressed man in a stovepipe hat, a lady in a full length dress sits outside the door. AJM.

**Great Boughton** This township was one mile east of Chester on a Roman road – now the A51.

1822 listed under White Horse in Pigot's Directory.

1823 the recognizance gives The White Horse as the alternative sign of The Farriers Arms. AJM and JB.

**Knutsford** 1849 listed as trading in the National Commercial Directory – this is thought to be a misprint for either the White Bear or the White Lion as there is no further record of a White Horse. LHS Knutsford.

**Macclesfield, Churchside** Originally situated next door to the Old Sun Inn, the site of which became the police offices.

1659 is the earliest record for the deeds of this site, but there is no mention of a licenced house until 1818 when Hannah Halton was licensee, she remained here until 1825.

1827, 12ᵗʰ May – the local paper published the following advertisement:

'To be sold, on premises of late Hannah Halton,
White horse, Church Yard Side – furniture and brew vessels etc.'

1828 William Lawton was listed as licensee.

By 1838 it is thought that the house had been rebuilt and named The Guildhall.

1857 the local council acquired the property for £600 and converted it into extra police offices and a residence for the Chief Constable.

PW and WRG. 1822/3 P; 1855 S.

**Middlewich, Lewin Street** 1767-1828 the Holford family held the license for over sixty years; prior to 1779 this alehouse appears to have been known as 'The Horse and Jockey'. AJM. 1822-1834 P; 1855-1869 S; 1896-1939 K.

**Nantwich, 22 Pillory Street** An old coaching inn listed in the Licensing Records as trading from 1774.

A 1920 postcard shows the inn sign 'White Horse Inn, Good Stabling, Cover provided for Traps, Carts and Cycles'.

NaL and Landlord. 1774-1828 AJM; 1822-1834 P; 1855 S; 1896-1939 K.

**Neston**, or **Great Neston** The inn was trading by 1778; a whitewashed, thatched building with two external chimneys situated at Neston Cross.

c.1870 a photograph shows a wheel-operated well outside the inn which had been sunk five years earlier; it was replaced by an ornate drinking fountain probably in 1877 when the picturesque building was demolished and replaced with a new, larger public house.

1892 listed as trading.

c.1930 when the public house closed, it is thought it became a betting shop; later it was turned into offices.

2007 the house became vacant for letting.

Burton and Neston LHS. 1778-1828 AJM; 1822/3-1834 P; 1855 S; 1896 K; 1914-1939 K.

**Northwich, 5 Crown Street** 1816 was the earliest recorded date of this inn with John Roscoe as landlord, it was known to have been situated very close to the old Crown inn.

c.1830 it lost the alehouse licence, but was granted a licence as a beerhouse.

1870 the landlord, Hugh Mares, applied for a full licence but was turned down; the house closed at his death in 1880.

AJM and Northwich Library. 1816-1828 AJM; 1822/3-1834 P; 1849 NCD; 1855 S.

**Rostherne** 1770-1828 AJM; 1834 P.

**Sutton, Bridge Street** This is an area within Macclesfield.

Listed as a beerhouse from c.1850. By c.1870 it had been renamed 'New Bridge Inn' and later became the 'Sceptre' which closed in 1925.

LHS Macclesfield and Mr. P. Wreglesworth.

**Tarporley, 46 High Street** 1828 listed as trading with Charles Rogerson as licensee.

The original building was burnt down and replaced by the White Horse Café – the wrought iron bracket which carried the café sign can still be seen.

1971 Bedford's Toy, Gift and China shop was trading here, later to become the White Horse Gift Shop until the late 1990s; a shop still trades on the site.

RCT and Tarporley LHS. 1834 P; 1855-1869 S.

**Waverton** Situated near the centre of the old village near to the modern pre-school building.

A 1737 map shows the building called 'Sign of the Lyon' – not White Horse.

1784-1787 listed as trading under the licensee's name, Joseph Wright (1).

1811 owned by the Rev. R. Massie with a number of other buildings and some nearby small fields; the tenant was still Joseph Wright – who died in 1825.

The 1822 Vestry Minutes record a Joseph Wright as licensing the White Horse for a fee of £29; the Land Tax in 1824 was 3s. 5d., this was presumably his son, also Joseph (2) who continued as tenant until 1828.

1832, October - the *Chester Courant* advertised a sale of land to be held in the White Horse on 1st November.

1837/8 the Tithe Redemption still gives the Rev. R. Massie as owner with Joseph Wright (2) as tenant.

1842 the public house and over four acres of land was purchased as part of a larger transaction by the Marquess of Westminster.

1844 Joseph Wright (2) died and his Wife, Anne took over as alehouse keeper.

1848, 2nd February – the Eaton Estate transferred the alehouse from Anne to her son, also Joseph (3), at a reduced rent of £18 as the licence had been terminated at the request of the local inhabitants.

1851 the census records Joseph as a tailor and parish clerk, living at the White Horse Inn with his wife, his mother and two young children. Also in the house were an aunt, who was a blind pauper, a niece Ann Price aged four years, a lodger aged fifty seven and a thirteen year old servant.

By 1856 the family had moved to Avenue Farm nearby.

By 1865 the inn had closed and had been demolished – for many years the parish bell ringers and choir frequented the White Horse Inn and there had been 'occasional problems'. LHS Waverton, AJM and WPCC.

**Willaston, Wirral** A house plate in the gable reads 'R.V. 1733' which is thought to be Richard Vernon and the building date.

1749 Mr. Vernon was constable and burley man at the manor court; records show that from time to time he was ordered to dig out his ditches and lop his trees.

1822 Quarter Sessions list the inn as White Horse with Thomas Lloyd as licensee.

1823 listed as either 'Grey Horse' or 'White Horse'.

By 1834 the name had been changed to the Nag's Head with William Delamore, a farmer, as licensee.

By 1871 Mr. Delamore's son-in-law, Thomas Jones had taken over.

c.1980 the name was changed to the Songbird and the brewery who owned the premises had a new inn sign made. This was unpopular with the villagers who called it the Twittering Tit – within a year the name resumed to the Nag's Head. ECB and AJM.

**Winwick** 1834 P; 1848-1855 S; 1858 Warrington PO.

**Notes**

1. One of a body of men in the retinue of a sheriff who carried spears or pikes and escorted the judges at assizes. They were last provided at Chester in c.1924. SOE and CCA.

# CLEVELAND

### Redcar and Stockton-on-Tees

*The name means "cliff-land" and was historically located
entirely to the south of the River Tees and is now part of the
North Riding of Yorkshire. It was created as a county in 1974
from its predecessor Teesside.*

*In 1996 the county was abolished but the white horse supporter
to the dexter side of the county's coat of arms is a reminder of the
importance of a 'Yorkshire Grey'.*

**Loftus, High Street / Market Place** The 1891 census
records Peter W. Dalling aged 31, a veterinary surgeon
from Edinburgh, as landlord of the White Horse Inn.
He lived here with his wife, Elizabeth, and two children,
Muriel aged 2 and Veronica aged 1. There were three
servant girls, Birtha Bottomly aged 16, Mary Loverin
aged 17 and Theresa Crocker aged 13.

   The 1901 census lists him aged 41, veterinary
surgeon and innkeeper, his wife Lizzie and five children,
the youngest being one year. Charles Towsdale aged 26,
servant and barman was also living here. 1876 K; 1897-1937 K.

**Redcar** 1857 PO – *this is possibly a mistake for White 'House' which was a well-
known large building at this date.* Mrs. V. Robinson, M.B.E. – Local Historian.

**Stockton-on-Tees, Church Street / Church Row /
Church Road** By 1894 it is listed in Whellan's Trade
Directory as the Castle and Anchor.

   LHS – Stockton-on-Tees. 1828/9-1834 P; 1876 S.

# CORNWALL

*'The packhorse was the general beast of burden in this county until the 19th century due to the steep hills and narrow roads. These sturdy horses, travelling one behind the other in a string, set out from assembly points which were often inns – hence a number of houses in Cornwall carried the sign of The Packhorse.'*
HLD.

*'From its vicinity to the Atlantic, the climate is extremely damp; but, except to persons of scorbutic habits, it is highly salubrious, the inhabitants being in general healthy and vigorous, and attaining many of them a remarkable longevity.*

*Oxen are generally used for the draught, and in the mining districts many mules are bred, and are in general request. Goats are more numerous here than perhaps in any other division of the kingdom.'*
JB.

**Breage** 1786/7 in the Breage Overseer's Accounts there is a reference to '32 lace of road leading from Redalan to the White Horse'; a 'lace' was historically 324 square feet or a Cornish perch.

**Kilkhampton** 'The White Pack-horse'. 1770, 15th October, a survey was held here – Thomas Jolliff was landlord.

By 1777, 8th September - the house had been renamed The White Horse Inn when the landlord, Thomas Heard, moved to The Crown at Bideford in Devon – where there was also a White Pack-horse.

The mule trains which plodded between mines, stamps, smelting houses and the ports do not seem to have a similar commemoration. ShM and HLD.

**Lanreath, Churchtown** 1857, 22nd May – recorded as a beerhouse adjoining the Calvinist Chapel.

1860, 26th October – a Quarter Sessions appeal by Richard Davey, a Lanreath auctioneer, who was granted a full public house licence after three or four years of applying, in spite of having improved the premises.

1867, 15th February – a notice was published in the *West Briton* newspaper that the brewery adjoining The White Horse was to be let. HLD and RCG.

**Launceston, 14 Newport Square** 1690 originally built as a farmhouse; by 1714, when King George I came to the throne, a licence had been granted and it was named The White Horse Inn after the Hanoverian Dynasty. Its outbuildings and stables stood around the old farmyard.

1750, 17th December - a sale was held at The White Horse. 1755 Officials of the annual Launceston Races were provided for at the inn, the race course being at nearby St. Stephen's Down. Only inhabitants of Launceston, St. Stephens and Lifton were allowed to erect booths or tents to sell wares on or near the race course – they were obliged to pay half a guinea. Every morning before the race meeting hunting took place with fifty couples of foxhounds – cock fighting also took place at both the White Horse and the White Hart.

1758 John Pearce, the licensee moved to the King's Arms.

1787 the inn 'furnished neat carriages, able horses and very good accommodation'. During the 19th century the inn was kept by the Burt family who were also wool merchants, saddlers and ironmongers, their business was carried out on the premises.

1871 a sale notice described the inn as a substantial construction of stone and slate, 'containing Parlour, Tap Room, Bar, Bar Parlour, Kitchen, Dairy and six Bedrooms with Warehouse and room over, also a Brewhouse, Stabling for eleven horses, with Lofts over, Chaise House, Cart Shed in Stones and Thatch and a Yard'.

1910 George Burt, whose family had kept the inn for most of the 19th century, sold the property to Walter Longman, who bottled sherry, whisky, Guinness and Plymouth gin on the premises which had been delivered to him in five gallon jars. The Guinness labels were sent from Dublin until the firm cancelled the licence to bottle their stout; he was a follower of the local hunt and kept orphaned fox cubs in his stable. His wife Polly was a good cook and never turned anyone away, she took over after his death in 1936.

1941 Walter's nephew, Edward Longman, became the licensee; he developed a wholesale and retail wine business – the wine was stored in the old brewhouse. Large stocks of it were sold when he retired, including an 1893 bottle of hock with a White Horse label.

1972 Edward retired and Mr. S. Fields took over; it was then leased by the Hairs brothers.

1996 a fight broke out at the inn between several councillors after one of their meetings. The landlord, Mr. Nick Hairs, wrote the following in a letter to the Cornish and Devon Post:

> 'I don't expect members at the Town Council to use my public house as a battle ground in their pursuit of their own ideology'.

1998 once again Mr. and Mrs. Fields ran the inn.

2001 Mrs. Fields sold the premises to Mr. and Mrs. Howard who sold the freehold to the St. Austell Brewery in c.2003.

The original cellar, which was not deep, became a bar.

The nearby St. Stephen's Down, once the race course, became the present day golf course.

It is thought there may have been another White Horse inn within the town walls of the borough of Launceston; this may have led to confusion with the well established White Hart Inn.

HLD, MM, RHL, ShM, J. Hairs – owner and Mr. Clive Tallent – licensee. 1830-1848 P; 1914-1939 K.

**Liskeard, The Parade** The original White Horse Inn was described as a large house, centrally situated near the junction of roads leading through the town; overlooking ground where the local fairs were held.

1757, 18th April - the well-accustomed premises were to be let due to the death of Thomas Clemens the landlord; his widow Mary had a stock of cider already laid in which she kept on the premises until they were let.

1784, 2nd February - John Hick advertised that he 'hath now fitted up his house on the hill in a neat and commodious manner as the White Horse Inn and Tavern'.

1805, February - as the inn was conveniently situated, Russell's and other stage-wagons put up here.

c.1811 The White Horse was demolished to make room for improvements to the town; its name was taken by another inn situated on The Parade.

HLD, ShM and AHL. 1848 P; 1914-1939 K.

**St. Columb Major, Fore Street** From 1747 to 1811 the inn traded as The Ship; it was situated at the junction of Fore Street and Broad Street.

Towards the end of the 18th century William Tom/Thom was the landlord; he married Charity Bray and on 10th November 1799 their son John Nicholls (sic.) Tom was born.

1814-1818 the inn was listed as the White Horse Inn.

William Tom and his family had moved to The Joiner's Arms by c.1820 when this recently insured building was destroyed by fire which caused his wife, 'Cracked Charity', to lose her mind; she died in the County Lunatic Asylum some four years later.

Young John was an unpopular and vain boy, after leaving school he was given a job in an attorney's office where he did well but rebelled and left when his father re-married the local school mistress. He was then taken on as a cellarman in Truro by Plumer & Turner, wine merchants and maltsters, whose business closed five years later.

By this time John had married Catherine Philpot and had moved into his father-in-law's house, which was old and dilapidated, to rebuild it into a 'commodious mansion' with premises at the back to carry on his maltster business.

A sudden fire burnt the whole of this working area which had recently been insured for £3,000; after due enquiries this sum was refunded in full.

Locally John had become known as 'Mad Tom', looking dignified in strange clothes, he had also become an orator.

Having rebuilt his business, a consignment of malt was sent by ship to Liverpool from where he wrote to his wife on 3rd May 1832 telling her that he had 'discharged the vessel of the malt which has given every satisfaction to the purchasers. The measurement has exceeded my expectations by twenty-four Winchesters[1] . . . I have paid the captain of the vessel all the freight.' Signed John Nichols (sic.) Tom.

During the next fourteen months he embarked upon a life of fraud, masquerading as Sir William Percy Honeywood Courtenay, Earl of Devon, Knight of Malta and King of Jerusalem.

By July 1833 he was involved in smuggling activities near the Goodwin Sands off Dover, to be prosecuted for perjury and indicted at Maidstone Assizes in Kent on 15th July.

When his relations in Cornwall heard of his exploits, they made representations to the Home Secretary of insanity; as a result he was held for four years in Barming[2] Heath Lunatic Asylum near Maidstone in Kent. Some reports gave him as 'having disappeared' but he was weaving fantastical stories – some of which were believed by his followers – such as having visited the eccentric Lady Hester Stanhope in the Levant.

SBG, IR, HLD and Mrs. Glanville – Recorder of the St. Columb Old Cornwall Society. 1747-1811 'The Ship'.

*Dear Reader, should you wish to follow Mad Tom's saga upon his release in 1838, please turn to The White Horse Inn at Boughton-under-Blean in Kent.*

**St. Ives, Gabriel Street** The main entrance of this large building is in Royal Square, the rest is in Gabriel Street.

Originally the house was the town residence of the Banfield family.

By 1853 the name had been changed to The Royal Western Commercial Hotel.

1864 and 1873 listed as trading with Richard Hodge as landlord.

1891 John Tremayne was listed as licensee, he ran the house for the next thirty five years.

2000 the hotel still continued to flourish. HLD.

**Truro, Lemon Street** A beerhouse, situated on the corner with Charles Street, this area was developed between 1800 and 1820.

1866, 18th October - the landlord Richard Brewer was fined for harbouring prostitutes. Six months later the premises a 'dwelling-house and shop lately occupied as an inn' was put up for sale. It is perhaps worth noting that the King's Head was on the opposite corner, 'just as lively a house as the White Horse'; in 1927 this too it had been 'for a considerable time well-known to the police as the resort of women of a very suspicious character'. RCG and HLD.

**Notes**

1. The shortened form of a dry or liquid measurement applying to a bushel, gallon or quart; the standards of which were originally deposited in Winchester.

2. It is thought that the word 'barmy' (insane) could have been derived from Barming Heath Lunatic Asylum in Kent.

# COUNTY DURHAM

*'Cattle are abundant, and of a superior kind, particularly the Durham short-horns, as they are called. Teeswater was long celebrated for its breed of sheep, but these have been replaced by the Leicestershire breed.*

*The S.E. part of the county is famous for draught horses, known by the name of Cleveland bays'.*

JB.

**Darlington, Durham Road / North Road / Harrowgate Hill (in the parish of Whessoe)** It is thought that a public house has stood on this site for centuries. Described as a traditional hotel which has catered for travellers on the Great North Road for many generations, it stands at the end of 'the dry road' so called because the landowners – the Barningham ironworking family – prohibited the sale of beer between North Road Station and Burtree Lane. This lane, an ancient 'cole street[1]', was a coal packhorse route from the south Durham pits, skirting around Darlington to avoid the tax collectors – Harrowgate Hill was once independent of Darlington and until 1900 the rates in this area were lower than the rest of Darlington.

The house was of plain Victorian brick, it was known as 'Glebe Hotel' and the customers were mainly Irish labourers.

1888 the landlord, James Lynch, emigrated to the United States.

The local inhabitants showed their independence by electing their own 'Mayor of Harrowgate Hill'; the election always took place here, the qualification for this job being that he had to be able to stand a round of drinks at the bar.

According to reports from local newspapers, by 1922 the hotel had become The White Horse.

1922, 25th March – an obituary reads:

> 'On the 25th inst. at the White Horse Hotel, Harrowgate Hill, Darlington, Alfred Richmond, beloved husband of Jane. Internment, North Cemetery, Tuesday, cortege leaving residence at 2.45. Friends kindly accept this intimation.'

1925 the owner, Plews & Son sold the hotel to the brewers J.W. Cameron & Co. Ltd.

c.1938 the owner, J.W. Cameron of Hartlepool applied for planning permission to demolish the building and rebuild a new one behind it, however the Second

World War broke out and the hotel closed being left boarded up and derelict. Known locally as 'the Ghost Hotel', it became a target for vandals.

c.1948 planning permission was given and a Tudor-style White Horse Hotel was built just behind the old site. Stones from the old hotel were used to build the new attached cottages and the ground floor fireplaces.

1951, June – the opening of the new hotel was reported in the *Darlington and Stockton Times* – 'a feature of the bars are two electric glass-washing machines which sterilise glasses when used in conjunction with a germicide'. A photograph of the hotel dated c.1951 is held by Darlington Library.

1973, 2nd October – according to the *Darlington and Stockton Times,* the announcement on Friday of the closure of The White Horse came as a shock to the customers as well as the manager, who was told to be out by the following Friday. It was closed for six months while extensions which included a dance hall were built.

1974 plans for a discotheque were accepted by the Borough Council - workmen arrived one day to find vandals had broken in and had turned on the taps of the water containers, flooding a room.

1975, 26th April – the hotel was again closed for structural alterations for a new games room and carvery.

1975, 18th June – angry local residents forced the hotel - owned by a Greek catering firm - to drop the idea of the discotheque and to modify its conversion plans.

1979, 22nd May – the hotel applied for planning permission for an extension of forty new bedrooms and bathrooms. Darlington Council had received three objections, one being Whessoe Parish Council who were 'absolutely flabbergasted' when three days later permission was granted for the extension and eighty four car parking spaces.

1980, 29th November – the opening of the new wing provided guests with first class accommodation and food; all the beef was delivered fresh from Scotland. The function room could cater for 70 people.

1981 and 1985, 9th November - the hotel was listed in the Egon Ronay Lucas Guide which praised the charming and helpful staff and awarded a good rating for its up-to-date accommodation.

By 1982 the premises were leased to the Reo Stakis Organisation of Glasgow and had changed from an hotel to a public house.

2003, October – the White Horse was sold as an hotel for £900,000.

2005, 27th April – the Planning Committee of Darlington Borough Council decided they would grant Planning Permission to demolish the hotel and replace it with 64 flats. Local residents met in Whessoe village hall to discuss how they could prevent this.

2006, January – a referendum was held and 98 per cent of the local residents voted to keep the 3 star hotel.

2008, 17th September – the White Horse closed despite local residents campaigning to save it; flats were to be built on the site after all.

BPe, JWC, M. Cartwright – local campaigner and North East History. 1938 Whessoe K.

**Ferryhill, 23 Market Square** 1876 S; 1897 PO; 1914-1938 K.

**Great Aycliffe** 1834 P; 1848 Aycliffe S.

**Pelton** 1876 S.

**Sedgefield** 1848 S.

**Notes**

1. Money; also slang for coal. SOE.

# CUMBRIA

*At one time the people here were known as the Cumbri,*
*meaning fellow-countrymen or compatriots. It is from here*
*which the words Cumbrian and Cumberland originate.*
CAd.

*'While her soil, her rivers, and her shores are thus beautiful,*
*even the most bleak and barren of her moors and mountains are*
*not less so, almost all of them teeming with valuable minerals'.*
JB.

**Appleby-in-Westmoreland, Bongate** The name of this street stems from the Norman period when the villeins, or bondmen[1] lived here, their lives were governed by the lord in the castle.

1869 Slater's Directory lists Thomas Warwick as landlord – he was also listed as innkeeper at the Royal Oak in Kelly's Directory of 1873 and at the Butchers' Arms from 1881-1885. It is thought that there may have been a beerhouse named White Horse close to the Royal Oak. LHS – Appleby-in-Westmoreland. 1842 P.

**Beckermet (St. John's)** 'The White Mare Hotel'. The building dates from 1863 but there is no reference to a 'White Mare' or 'Horse' in the local Trade Directories of 1882/3 or 1901. 1849-1860 Haile S.

**Carlisle, 44 English Street, 2 White Horse Lane, Blackfriars Street** Map references of c.1865 show that these two addresses are for the same site in White Horse Lane; the earlier buildings extending from Blackfriars Street to English Street.

1884, 19th September – the *Carlisle Journal* reported:

> 'In September 1884 Superintendent Russell, inspector under the Food and Drugs Act summoned an innkeeper for selling him a half-pint of gin diluted with water. The public analyst proved that the gin was 17% below the lowest legal standard, i.e. 52% below proof. The defendant's wife said that she only put a pint of water to a gallon of gin because it had come from the spirit merchant. The fact that the defendant was new in business and probably not aware of the liability, a fine of 10s and costs was made by The Board.'

Map Sheet XX III 3.25 from 50" Series c.1865. 1828/29 2 listed P; 1834 P; 1848-1884 S; 1914 K.

**Cleator Moor, Jacktrees Road** One of about thirty public houses in the area at the turn of the 20th century.

1925 Miss Rose Richardson was recorded as providing beer which is thought to have been delivered from the local brewery in barrels. The following year Mrs. Mary Watson took over as beer retailer.

1941 due to a coal mine running underneath, the pub with the road and several cottages subsided; the area was levelled and by 2007 there was a children's play area on the site. LHS – Distington. 1938 K.

**Dalton-in-Furness, 6 Market Street** 1798 The Universal British Directory listed the White Horse.

1809 a poster from the Cumbria Soulby Collection (Ref. ZS 346) advertised 'All that commodious and well accustomed public house, commonly known by the Sign of the White Horse' to be sold on 'Friday the 15th day of December at 6 o'clock in the evening'. It was described as a large house consisting of three parlours, a kitchen and brew house, a large dining room, convenient bedrooms and excellent cellars. There was stabling for ten horses and an adjoining garden. Everything was in 'complete repair, and may be entered upon at May-day next'. There was also 'a very good garden situate at the head of Dowker Lane, near Dalton' and 'a close of excellent arable ground called Mary Bank Parrock, by estimation 1 acre statute measure'. The owner, Mr. John Gardner would show the premises and provide further particulars.

The premises were 'held of the Manor of Dalton, by payment of the Yearly Rent of 2s. 9½d. and a fine of only 3s. 4d. upon change of tenant'.

1996 the hotel is a listed 19th century building, it replaced the earlier inn.

CRO. 1829 PWD; 1855 S; 1858 PO and 1881-1924 Ulverston K.

**Eamont Bridge** In this town bordering Westmorland, it is said that at one time there were two White Horse Inns, also an entrenched amphitheatre called 'King Arthur's tilting ground'; this legendry King was usually depicted upon a white horse.

JB. 1828/29-1834 P; 1848-1884 S; 1895 PO; 1901 BUD.

**High Crosby** 1829 the Parson and White Directory for Crosby-on-Eden Parish lists John Goodfellow, a victualler, as licensee of the White Horse.

1842 the High Crosby Baptism Records list Margaret, daughter of innkeeper John Noble and his wife Mary. By 1844 their son, George, had been baptised.

1843 John Noble is still listed as licensee in the alehouse recognizance, but by the following year the house had closed and John Noble had moved to the Old Boot Inn at Whiteclosegate.

1847 the White Horse is not listed in the Mannix and Whellan Directory.

By 1860 the house had been demolished. CAS. 1828/29 P.

**Kendal, 41 Strickland Gate** 1820, 1st November – a borough notice regulated the placing of stalls at the quarterly fairs – the cloth stalls were to be placed from the White Horse Yard. 'These quarterly fairs and marts were the events of the year. Peripatetic traders visited the town, and country folk, for miles around, flocked in and bought clothing and goods to last them till the next similar occasion. With side shows they were entertained – wild beasts or bearded women, six-legged calves or merry-go-rounds. The shopkeepers had great roasts of beef in their back rooms, and plenty of beer for the free use of customers to feast at will.'

1821, 2nd March – the inn was advertised for sale by the Executors of the Late Adam Walker. It was described as a 'commodious and well-accustomed inn' with a 'good brew house, granary, and excellent stabling for 33 horses, with other outbuildings, now in the occupation of John Walker as tenant'.

1832 William Wilson was recorded as publican.

The Philharmonic Society held its meetings here.

Extracts from *The Westmorland Gazette* include the following:

> 1825 a local dancing master, Mr. Winder, held a ball at the White Horse.

> 1833, 6th April – from the Sporting Intelligence section concerning Pigeon shooting[2]:

> 'Another match came off on Monday last in the fields at Aikrigg End, for dinner and wine, headquarters the White Horse. The day was very unfavourable, being excessively cold, with a strong east wind, so that when the birds went away from it, their flight was rapid enough to baffle the most experienced shot. Each man liberated his own bird from the trap; the bird had to be picked up by the umpire within bounds. After a good dinner at the White Horse inn, and when wine had made the heart glad and the party gay another match was made for Ten Sovereigns to come off in a week.'

> 1842, 19th February:

> 'Mr. William Wilson's hunt, of the White Horse Inn, came off on Monday last. A fox was let go at the back of the Helm *(a local hill)* which made a short but excellent run, and was finally killed in full view. There was a good muster of horsemen as well as pedestrians in the field. After the sports, a large company retired to the White Horse where they partook of a very substantial dinner. After the cloth was removed, Mr. Holmes of the King's Arms, was called to the chair whose attentions greatly contributed to the harmony of the meeting. It would be injustice not to remark that the wines and spirits of Mr. Wilson were of the first-rate character, and proved an auxiliary in bringing forth many a tale of tall stone walls and hairbreadth scapes. The company finally departed in high glee and good humour.'

Two months later on 9th April the following advertisement was published:

> William Lickley 'respectfully returns thanks to his Friends and the Public for the extensive patronage he has enjoyed for Eleven Years in the Union Tavern and embraces the present opportunity to inform them that he has removed to the large and commodious inn called the White Horse, situate in the Town of Kendal where he hopes, by hospitality to the comfort of his Guests, to merit a share of Public suppers.'

1854, November – the publican, John Kelly sold all the domestic effects, brewing vessels and innkeeper's stock-in-trade.

1864, April – the Rate Book records the Executors of John Kirkbride as owners of the inn, a photographer's shop at No. 43 and a stone mason's yard at No. 39 - William Clark was the publican of the White Horse until September when William Todhunter took over.

1873 Mrs. Ann Todhunter was listed as publican – the inn closed the following year and was converted into a grocery shop.

1874 there was a total of over fifty five people living in White Horse Yard, which was much longer at this date – some twenty five lived in the cottages, the rest were workers.

1882, February – an article published in the *Westmorland Gazette* suggested that the original Pack Horse Inn had been on the site of another White Horse Inn; the building was renamed White Horse when the Pack Horse moved to its present site.

1917, 27th October – the following article appeared in the *Westmorland Gazette*:

> 'At 7.45 am on Wednesday *(24th)* there was a fire in a wooden building at the top of White Horse Yard in Stricklandgate, in a portion used by William Carlisle as a stable. The fire originated in that part of the building. The horse was out at the time. With plenty of water from the main the fire was quickly extinguished. The building was insured.'

c.1960 White Horse Yard was severely truncated.

A.R. Nicholls, Local Historian and JFC. 1828/9-1834 P; 1858 K.

**Kingsbridge, Kingsbridge Ford** The bridge spans the River King – 'Kingwater'; the origin of this name is lost in the mists of time but it could have referred to King Edward I (1239-1307) who it is known stayed at the nearby Lanercost Priory which was founded in the 12th century and became not only the favourite of kings, it was also taken under the aegis of several popes.

1828 listed in Pigot's Directory under The White Horse, the following year the landlord was recorded as a victualler and blacksmith.

By 1847 the house was renamed The Three Horse Shoes – the arms of the Worshipful Company of Farriers. The sign can still be seen in strong sunlight, on a cement panel high on the gable end of the building. At one time there was also a cobbler working in the premises.

July 1909. *The Carlisle Journal* published the following article 'A Burglar at Work':

> 1908 Martinmas – 11th November. Irving Dand of no fixed abode stayed one or two days at the inn; the following June he called again having visited four other local inns. He was later charged with 'stealing 18 shillings from a cup in the bar of the Kingsbridge Ford Inn on June 26th. Sarah Little, landlady of the inn, said prisoner went into the house in the early morning and was in and out until the evening. The night before a sovereign was left as usual in a cup in the bar for change, made up of a half-sovereign, three two-shilling pieces, and four separate shillings. After prisoner had left eighteen shillings were missed from the cup, and some sixpenny pieces that had been wrapped in paper. Dorothy Harding, servant with Mrs. Little, said prisoner had a drink and she left him in the house when she went out to get some coals. When she returned the bar door was standing partly open. It was shut when she went out'.

> 1909, 30th June – Mr. Dand was charged on remand at Brampton Police Court with intent to commit a felony. The prisoner pleaded guilty to one burglary 'but not to that at the Kingsbridge Ford Inn. He was committed for trial to the Quarter Sessions on both charges and was sentenced to six months hard labour'.

c.1911 the inn closed to become a private house. In 1945 the tenancy was taken over with the blacksmith still working – he retired in 1950.

1984 the house was sold to the tenant, who renewed the windows and porch.

TFB; PWD; BSM(a) and Owner.

**Kings Meaburn** An early 19th century building shown on the 1823 Hodgson map. 1829 it is not listed as trading in Parson and White's Directory.

1847 William Dent and his wife Sarah are the earliest recorded publicans at The White Horse Inn.

c.1850 a tailor, Robert Parkin was landlord, he employed two journeymen tailors.

1880 listed as trading with Anthony and Ruth Carlton as publicans.

1890 John Walker became licensee with his wife Annie.

c.1900 James McCune ran the pub.

Amongst the travelling tradesmen who sold their wares in the village was Bill McGuinness from Shap who sold herrings, he rang a bell and shouted "fresh herrings, fresh herrings, 20 for a shilling"; Tommy Lowther of Penrith sold pots and pans from his cart pulled by his piebald pony.

By 1929 Jonty and Elizabeth Eggleston had taken over, they had three daughters Isabel, Dolly and Violet. Jonty rented a three acre field on the right hand side of Haggs Lane at the top of the hill, this became known as Jonty Field. The Egglestons continued to run the pub until 1945 when Jonty died, his daughter and son in law Dolly and Sid Longstaff took over. They later ran the George Hotels at Orton and Morecambe.

The inn continued to trade for about the next twenty years with four changes of landlords.

1973 the village shop opened at one end of the inn with Mike and Kath Rowan in charge.

1977 Bob Gate became licensee and shopkeeper with his wife Mary. In 1982 after the death of the postmistress and her husband Maisie and Robert Gate, the shop at the White Horse was specially altered to incorporate both the Post Office and the Victorian wall letter box which was moved from the old Post Office.

1996 alterations again took place when John and Helen Wylie moved in to run both businesses.

2000, September – the Wylies left and the pub was run by the village for two months until Mr and Mrs J Hamilton moved in with their three children.

The current owners of The White Horse Inn are the Addison family who have been land owners in the village for many years – Mrs. Caroline Addison has written a history of the village. CAd and J. Hamilton, landlord. 1855-1921 Morland K.

**Mainsgate, Mainsgate Road (near Millom)** The Irish Sea washes the parish of Millom on the west and south, and the Duddon bounds it on the east, forming a bay celebrated for the excellence of its cockles, muscles, salmon and sand eels.

1867, May - John and Hannah Sawrey purchased the site from William Eagers, it was bounded on the south by Thurston Street, on the north by a street forty feet wide and on the east by the estate of Thomas Woodburne.

1869 the building of the White Horse Inn was completed and the tenant, William Jones, moved in.

1874 John Sawyer died and the trustees, Redhead and J.W. Robson, a brewer from Ulverston, leased the property to William Jones.

1881, 2nd April – William Robson agreed to sell the property to Thomas Woodburne, but instead the premises were sold to the Hodbarrow Mining Company on 6th May for £2,200, they continued to run the inn, collecting a weekly rent of 3s. 0d.

1883, August – the licence was not renewed. By this time the land surrounding Mainsgate was subsiding due to underground iron ore mining operations and the inn was demolished, along with Hospital Row, Hunter Row and other cottages; these were rebuilt further inland at Steel Green, but not the White Horse Inn. Historically, this became one of the most important and productive haematite mines in the world. Eventually the area filled with water creating a lagoon, which for a time was controlled by pumps; a defence wall was built near the sea which cracked and fell into the lagoon. A more substantial wall was built which still stands.

JB and C.J. Gregg – local historian of the Millom Folk Museum and M. Scott – local historian.

**Marton, Silver Street** The surrounding area was once involved in iron ore mining. At the end of the nineteenth century this small wayside alehouse was one of five local hostelries. It probably mainly catered for the miners, as it was not on a direct route to anywhere, had no stabling or outbuildings and only a small garden.

From April 1880 to 1882 there was a succession of landlords, one being Charles Wilson who was fined ten shillings (50p) and costs for opening during prohibited hours. He was followed by Matthew Askew who adulterated the gin - he had to pay £2 and costs; he lost his job and was followed by Margaret Craghill.

1949 the house, with two front rooms and no cellar, situated at the end of a small row of adjoining houses was sold for £1,100 to Mr. Lewney as a private residence. It is likely that the house was still trading, as immediately prior to the sale a quantity of small barrels of liquor were discovered under the stairs, which Mr. Lewney shared with his work colleagues at the shipyard in Barrow.

2006 the Lewney family, who had never made any structural alterations, sold White Horse House for £145,000.

The new owners completely renovated the building, adding a larger front porch with a new entrance at the side. Mr. Clive Richardson and LiF.

**Penrith, 1 Friar Street** The original headquarters for the Penrith Cricket Club and the Friars Bowling Club; when the public house closed, the Cricket Club held a meeting there to toast the memory of the Club's founders. The site is now occupied by Merlin Court flats.

1834 P; 1848 S; 1855 White Horse Cellar S; 1869 S; 1895 PO; 1914-1938 K.

**Great Dockray** In August 1988 the name was changed to 'Bootleggers' but by May 1996 the sign was once again 'The White Horse'. 1884 S; 1895 PO; 1938 K.

**Sandwith** Originally one of a pair of cottages which had a long leasehold for 1,000 years from St. Bees Free Grammar School; in 1771 the reserved rent was one halfpenny.

The adjacent property was a blacksmith, when people brought their horses to be shod or farm implements to be repaired, they stood outside talking – someone had the idea to sell them beer.

1820 Joshua Tyson bought the cottage from Henry Fletcher and Ann Kirby for £72 and converted it into a public house; it remained in the Tyson family until 1895 when Ann Tyson died, she left it to her son the Rev. Joshua Tyson and her son-in-law Edward Telfer who sold it for £500 to William Jackson Armstrong. The house was run by tenants for many years.

One of three public houses in the village, The White Horse was known locally as the 'Bottom House' – the others were known as 'Middle House' and 'Top House'.

1966 the pub closed and was bought back by the Telfer family to become the village Post Office which had previously been housed in Middle House; when the Post Office was closed, the building became the Telfer family home.

Whitehaven LHS and Mr. John Telfer - descendant of Edward Telfer and owner. 1848 S; 1882 licensed victualler PD; 1895 Whitehaven PO; 1914-1938 K.

**Scales** c.1610 a coaching inn situated on the road between Penrith and Keswick.

Named after the white horses stabled here, which when harnessed to the passing coaches or waggons as lead or cock horses, helped them up the steep hills - west over Burns Brow a 1 in 7 incline, or east over Troutbeck Moor. It is said that these grey horses, once uncoupled, returned unaccompanied to the inn.

There was an old Bible, dated c.1800, which was always kept in a spice cupboard; it is reputed that the inn will fall down, should the Bible ever be taken from the building.

The renowned yeoman farmer and huntsman, John Peel (1776-1854), who stood 'six feet high and more', used to call in here to stable his horse Dunny before leading the fell hounds on foot over the Skiddaw Forest and the Blencathra country, this would have included the moorland behind the Inn known as 'White Horse Bent[3]'. He acquired fame from the Cumberland hunting song 'D'ye ken John Peel' which maintains that he 'lived at Troutbeck once on a day' - in fact though born and buried at Caldbeck, he spent most of his life at Ruthwaite from where he hunted hounds for some 46 seasons.

The building has been extended at least three times; the present bar stands on the site of the old stables. The original salt cupboard is still in the hall; the old court cupboard was sold to an American in 1936.

A. Charlton – licensee and MJB. 1884 S; 1895 PO; 1901 Threlkeld BUD; 1914-1938 K.

**The Hill (near Millom)** 1848 Listed as trading with John Hodgson as licensee.

1861 the census does not list the White Horse, but does give John Hodgson as an innkeeper and waller.

1871 the census lists the White Horse Inn with Henry Hodgson as innkeeper, his wife Jane and two sons aged 13 and 10.

1873, 23rd August – Elizabeth Frearson, a butcher in Broughton-in-Furness, took on the ownership with James Frearson.

1875, 10th July – John Brocklebank took over as licensee from Henry Hodgson.

1876 the owner was recorded as James Wilson Brockbank of Bank Spring Brewery, Kirksanton; his manager, Henry Hodgson was also recorded as the licensee.

1881 the census shows John Brocklebank as landlord, his wife Ann, three children and two servants.

1882, 22nd April - Adam Thackeray is listed as innkeeper.

1891 the census does not list the White Horse by name but the innkeeper is Adam Thackeray, he lived here with his wife Elizabeth, his daughter, grandson and a niece.

1896, 20th August – two leases concerning the inn were drawn up as a Condition of Sale, the first from Elizabeth Frearson to Messrs. R. and P. Hartley for a term of fifteen years from 12th May 1897 at a rent of £29; the second concerned a close of land called Redlands adjoining the public highway leading from the Green to Millom Castle from Elizabeth Frearson and John Garner to the Hartleys for a term of fifteen years at a rent of £22.

1901 the census lists the inn with James Simpson as innkeeper, his wife Anne and three children. On 16th February the premises were sold to J. W. Brockbank and Sons, one of the partners was Arthur Hodgson Fox-Brockbank.

1912, Tuesday 23rd April at 4.30 pm the white Horse was sold by auction. The premises consisted of a bar, two smoke rooms, sitting room, kitchen, beer cellar, spirit cellar, wash house and four bedrooms. Outside there was a stable with three stalls and a hay loft over, other out buildings, a garden, orchard and about two acres of pasture in front of the house.

1962, April – the licence expired and the pub closed to become a private house.

CAS and Ms. C.J. Gregg – local historian at Millom Folk Museum. 1834 BG; 1855 Boothe - 1869 S; 1895 PO; 1914-1938 K.

**Waverton** Originally a small hamlet situated on the main route from Wigton to Maryport, the inn is thought to have been built c.1830 when traffic was increasing through the village.

By 1834 it was listed as trading in Pigot's Directory with William Shepherd as licensee, he was a victualler and also a farmer, as were the three following landlords.

It is thought to have been a posting house until the arrival of the railway in 1845.

1865 the Ordnance Survey map shows the inn situated in the centre of the small hamlet.

1898 listed as trading.

By c.1900 the building had become a guest house named Green Lea.

Between 1906 and 1925 it was listed as Refreshment Rooms.
By 2008 it was a private residence – Greenlea House.
Mrs. Shirley Thornhill – local historian. 1848-1869 S.

**Notes**

1. A serf or slave. SOE.

2. Pigeon shooting or Pigeon match shooting. There were two kinds of match shooting, one for the wealthy and influential usually supported by private subscriptions, the other, funded by public contribution, was generally organised by innkeepers who might offer three pieces of plate, or purses of sovereigns to be given to the three best shots.

A popular sport throughout England, it was considered 'the perfection of the art of shooting flying' because the pigeon glided silently from the trap without showing any alarm. The trap was a shallow box of about a foot long and eight to ten inches wide which was sunk into the ground, parallel with the surface twenty-one yards from the footmark. It had a sliding lid with string attached which was pulled at the command of the gunner, who was not permitted to raise his gun until the bird was on the wing; the bird must fall within one hundred yards of the trap. DPB.

3. A rough moorland grass – Agrostid Canina.

# DERBYSHIRE

*'The mines and miners are governed by certain ancient
customs and regulations; one of the most remarkable is
that by which an adventurer discovering a vein of lead
unoccupied in the king's field, has the right of working it on
the land of any person without compensating the proprietor.*

*This custom is still in force, but gardens, orchards, and
highways are exempted.'*
JB.

**Ashbourne, Buxton Road / The Piggery** Situated
on the road leading to the Pig Market at the top of
the Market Place; there was a clear view from the
inn of the Old Derby Road, this made it an ideal
'look out' for Highwaymen. When they saw travellers
setting out to Long Lane they would take a short cut
down Dobbin Horse Lane and lie in wait for their
unsuspecting victims.

There was a grocery shop behind a stable door underneath the
first floor of the inn. There were ten or twelve houses in White Horse Yard, some
were private dwellings and the others were for ostlers and grooms.

1936 the inn closed and was demolished to become two private dwellings; a
hairdressers shop opened on the site of the grocery.

Mr. and Mrs. G. Shaw – LHS Ashbourne. 1822/3-1842 P; 1876-1916 K.

**Bakewell, The Square** 'The Rutland Arms'.

A well-known 18th century tavern and posting house.

In 1789 The Hon. John Byng, who became Viscount Torrington (1740-1811),
author of the Torrington Diaries, visited the inn and recorded that he paid one shilling
for a quarter of cold lamb, a cold drink, salad, tarts and jellies, the wine cost one
shilling and three half pence, the brandy and rum three pence
and the hay and corn for his horse was five pence – totalling
the equivalent of 15p. He visited again the following year.

c.1795 Jane Austen (1775-1817) is reputed to have
stayed here while writing 'Pride and Prejudice', which
was not published until 1813.

In 1796 the White Horse was the scene of a riot against
the Militia Ballot when the mob burnt the list of men
liable to serve, in front of the inn. However they
dispersed peacefully, having paid for their drinks.

1804 the young Duke of Rutland (1778-1857) demolished the tavern along with other buildings to form Rutland Square; the inn was rebuilt and enlarged to cater for coach travellers – the stable hands slept in a small building adjacent to the hotel, which by 1980 had been named the Stable Bar; later to become the Starboard Quarter and by 2007 it was The Ridgeway Gallery, still owned by the hotel.

1818 as many as 600 travellers passed through the town.

1828/9 Pigot's Directory still lists the inn under White Horse.

1861 Mrs Beeton, in her 'Modern Household Cookery' mentions that a cook at the White Horse made a mistake with some ingredients which resulted in the traditional Bakewell Pudding.

1862 Slater's Directory records William Greaves as running coaches daily to Rowsley, Buxton, Sheffield and Whaley Bridge. The Manchester to London coaches also stopped here for refreshment. Mr Greaves, whose family was associated with the inn throughout the 19th century, married a sister-in-law of Sir Joseph Paxton, who designed the Crystal Palace and he was also head gardener and man of affairs to the sixth Duke of Devonshire at nearby Chatsworth.

The Hotel brochure 'A Brief History of The Rutland Arms Hotel'; TN 15th April 1996 and Mr. D. Parkinson.

**Birch Vale, New Mills Road** Rumour has it that the Waltzing Weasel at Birch Vale was at one time known as the White Horse. Trade Directories from Derbyshire Record Office between 1800 and 1900 were unable to confirm this, but in 1940 there was a White Hart in the nearby village of New Mills, which may account for any confusion.

**Chesterfield, West Bars** c. 1786 William Launt owned the inn and other buildings in the yard; there was also a steam cornmill which was sold in 1803.

1898 the inn was demolished. The Portland Hotel was built on the same site.

DCA. 1822, 1828/9 3 listed - 1842 P; 1876 K.

## Derby

*Cornmarket* Originally the 'Crown and Cushion', one of twenty two inns in this 16th century market.

1737, 20th October – renamed the White Horse by a new landlord who had lately kept the White Horse in Rotten Row.

1739, Thursday, 13th March – *The Derby Mercury* published the following advertisement:

> 'To be Lett and Enter'd upon at Lady-Day next. The White Horse Inn, standing in the Rotten Row. Enquire of Thomas Trimer in Derby who is now selling off his Woollen Drapery, with a design to leave off the business.'

1742, 16th December – *The Derby Mercury* published the following notice:

'That William Sherrin, who lately kept the White Horse in the Rotten Row, now keeps the same sign at the Old Crown and Cushion Inn in the Corn Market; where all Gentlemen Travellers, etc. may be assur'd of good Entertainment. Also all Carriers Waggons, Pack Horses, etc. for which there is great Conveniences.'

**96 Friar Gate** 'Old White Horse'. A medieval thatched building thought to have been cruck-framed[1]. It was one of the many inns built to serve the 14th century cattle markets at the upper end of Friar Gate. It was also the home for a short time of the evil Ellen Beare, a murderer, abortionist and adulteress.

1732 the inn was mentioned in *The Derby Mercury*.

c.1870 a water colour of this building was painted by A.H. Keene, now in the Derby Museum and Art Gallery.

In 1876 it was photographed by R. Keene before its demolition, it had been acquired by the Great Northern Railway to make way for a railway bridge; on 13th October this ancient inn was closed and later demolished. DCA, MAC.

**High Street** 1843, 19th July - a malthouse in the yard of the 'White Horse Tavern' was advertised for sale. DM.

**25-29 The Morledge** Situated on the corner with Thorntree Lane. Originally one of a line of three small inns - the Noah's Ark, the Cossack and the White Horse – named after the white horse of the Hanovarians in 1714 when George I landed at Greenwich. Later the inn became the 'Bishop Blaze' for a few years.

By 1842 the inn had been granted a full licence. 1874 the name was changed to White Horse and by 1900 was listed as No. 27.

By 1903 the owner was Pountain Girardot & Forman of Derby; the licensee was George Walker who paid a rent of £80; there was a Poor Rate Assessment of £82. The inn had seven bedrooms, nine beds, two stables and a coach house.

1920 the house was demolished together with the Cossack.

By 1923 a fine, well proportioned terracotta faced building had been built incorporating the site of the Cossack as well as an adjacent shop, and it opened as the White Horse Hotel.

c.1994 the name was changed to The Foal and Firkin.

By c.2000 its name had once again reverted to the White Horse, but by 2007 it had become The Court House. MAC. 1941 K.

**85 Regent Street** 1822 listed as trading in Pigot's Directory, it was one of two inns in this street.

By 1865 the inn had been granted a full licence.

1903 Sarah Ann Eyre was listed as licensee, she paid a rent of £60; the owner was A. W. Thies. There was a Poor Rate Assessment of £47, accommodation for four people and stabling for three horses and their vehicles.

By 1937 it was owned by Pountain's Brewery.

1958 the inn was closed. MAC. 1828/9 P; 1900-1941 K.

**Rotten Row** One of six taverns which stood in this 16th century street.

1737, 20th October - the landlord moved to take over the 'Crown and Cushion' in Cornmarket, taking the sign of the 'White Horse' with him.

**Sadler Gate Bridge** 1765, 1st February – advertised to let by *The Derby Mercury* as 'White House' – *a misprint* – 'near to Sadler Gate Bridge'.

**Horwich End, Macclesfield Road** During the 19th century many dwellings in this area were built at about 600 feet above sea level due to the height of available water; there was a well behind No. 54 and a stone trough outside No. 50 which provided a welcome drink for horses travelling along this road.

The inn was situated on the corner of the road to Whaley Bridge.

WBL. 'White Hors'. 1828/9-1842 Chapel-en-le-Frith P; 1876 Chapel-en-le-Frith K - 1941 K.

**Kelstedge** 1680 this date appeared on a stone set in the wall, but it is believed that it had been brought from a nearby farmhouse when extensions were carried out.

1889, 29th June – the house was offered for sale by auction in the *Weekly Independent*.

The house was demolished during the 20th century.

DCA. 1822/3-1842 P; 1876-1941 K.

**Melbourne** Developed into a small manufacturing town during the 19th century, the population doubled and there were at least seven public houses; the Ring of Bells which closed in c.1816 is thought to have been named after packhorses as the front packhorse always wore a collar of bells.

Illustrated in 'More Melbourne Memories' p.39 – published in 1987 by the Melbourne Civic Society.

1941 the White Horse was listed as trading in Kelly's Directory.

**Old Whittington, 143 High Street** The original building must have dated from the 18th century as the date 1780 can be seen on the fireplace. The inn was thought to have been trading by the early 19th century and probably belonged to the Elmwood House estate. A blacksmith's shop was attached to the side of the inn and records show that the blacksmith lived on the premises.

1891 Harry Parker, aged 27 was listed as landlord, his wife Florence was 24. Their son later became the village coal man.

1900 a photograph of the inn taken from the front shows a white block on the left of the building, this is thought to be a gents' urinal.

1909 Charles Henry Judson was listed as licensee and blacksmith.

1962 the inn was sold to Whitbread.

During the 20th century the house was rebuilt, one of the original buildings can still be seen at the rear. OWL, TNu and Brian F. Bakel. 1828-1842 P; 1876-1941 K.

**Stanton** The White Horse, thought to date from the early 17th century, is the older of the two public houses situated in the village on the eastern side of the A444 road. The original oak beams and part of the early structure can still be seen.

1786, 4th and 5th December – an Indenture of Lease and Release recorded Edward and Elizabeth Lovedon as releasing approximately 251 acres of the Stanton Manor lands to five people, which included the Eagle and Sun, thought to be the original name of this inn. The tenant was Thomas Briggs; it had a barn, stable, outhouses and gardens. The field behind the house was the home ground of the Stanton Cricket Club, the Stanton Football Club and the Quoit Club who were undefeated champions in 1901.

1820 listed as trading with John Orgill as licensee, he was succeeded by Anthony Orgill in 1835.

1928 William Insley was licensee and a keen member of the Quoit Club.

1999 the ownership passed from Bass Brewery at Burton on Trent to Burtonwood Brewery at Warrington.

2001 the pub closed while refurbishment was carried out at a cost of £118,000 to reopen in March 2002 with Neil McCormick as licensee.

KG, DFe and Chesterfield Library. 1876-1941 K.

**Woolley Moor, Badger Lane** Situated on the original packhorse route and toll road from Stretton to Woolley Moor; a 'badger' was a measure of salt. Traders who had collected salt from the deep Derbyshire mines to sell from their wagons would have refreshed themselves here. Miners living in the village worked in the mine at nearby Morton.

The front room was used as the local magistrate's Court; offenders were taken to a cell in the Greyhound at Milltown, if found guilty.

Marked by paving stones along the tops of the hills opposite the inn is the old coffin route to Ashover from the neighbouring hamlets of Lea and Dethick, which did not have burial grounds.

There used to be cottages behind the inn; the Old Toll Bar Cottage still stands at the end of the lane.

The inn sign depicts a horse and cart carrying a 'badger'. EB and Licensee.

**Notes**

1. A pair of curved wooden timbers supporting the end of the roof in certain types of building. Col.

# DEVON

'... the vocation of postmaster was inseparably connected with the innkeeping trade, and by the end of the century all towns and many villages on main roads had one inn among the other inns known as the Post House, and displaying, besides its own peculiar sign ... a post-horn in token of its special function'.
AD.


'The valleys are remarkably rich, abounding in orchards, and producing the earliest grass and finest hay. There is also very rich meadow-land in the vales of Exeter and Otter.

The principal manufacture is that of woollen-cloth and kersey[1], more than two-thirds of which is purchased by the East India Company'.
JB.

**Bampton, 7 Fore Street** At one time known as Bathampton. James Bell in his Gazetteer of 1834 describes the town with stone houses 'irregularly scattered over a space'.

c.1664 it is thought the inn was known as 'Sanders'.

1706 a mortgage relates to the White Horse Inn.

There is some confusion over the street numbers between No.5 and No.9 which is a handsome and substantial Grade II listed site overlooking the busy market place. Bampton is especially well known for its Horse sales. The White Horse was an important commercial and posting inn and hotel, its front entrance opening straight on to the street with no space for a carriage, hence an archway leading into a large courtyard to the rear of the building was essential for the carts and carriages attending the auction and market. Between 1743 and 1785 The White Horse was the venue for many property sales advertised in the *Sherborne Mercury*.

1772, 29th June – 'To be Lett, for any term that may be agreed on, and entered upon immediately, that old and well accustomed Inn, known by the name of The White Horse Inn, situate in the market place, in the town of Bampton; together with the little meadow called Cookhays, and Cookhay's Stables and Garden. Any person inclined to take the same, may have an opportunity of buying the stock of liquors, vessels, brewing utensils, etc . . . '

1827 an abstract of title, a lease dated 1863 and a survey of dilapidations dated 1873 referring to the White Horse are held by the Devon Record Office.

1830-1848 Pigot's Directory lists the house as a Commercial and Posting Inn and in 1883 Kelly's Directory describes it as a Family, Commercial and Posting Inn.

c.1900 a caption to a photograph reads 'the White Horse tap (cider bar)' – a cider house was at the rear of the building. Major alterations to the hotel and surrounding buildings took place as a later photograph shows a taller building and a porch over the pavement with a statue of a white horse on its roof; the archway which led into the courtyard had become a shop front with two floors above and the adjoining cottages had also been refurbished with an added third storey.

1998 Mr. Derek Aldridge carved a new horse from wood which took the place of the old one on the porch. Inside the horse he put a book about local people and memorabilia from Bampton. Dev. 2147Badd/M/T9-10; 1044B, Martin Upton – Landlord and Exeter Central Library. 1866 2 listed PO; 1939 K.

**Barnstaple, 29 Boutport Street** 1837-47 listed by the Magistrates' Clerk office as trading with W. White as landlord.

By 1939 it was listed as an hotel.

By 2006 the premises had become 'Shamus O'Donnells'.

1823-1848 P; 1866 PO; 1883-1939 K.

**Bideford, Union Street** In 1857 the inn was listed as The Pack Horse and by 1866 it had become The White Pack-Horse. In Cornwall, the packhorse was the general beast of burden until the 19th century. Dev. 1805 WCO; 1879 BD; 1883 K.

**Churston Ferrers, Dartmouth Road** 1860 built on Lord Churston's estate as Richmond Villa, perhaps for the use of guests arriving by train at the new railway station nearby.

c.1890 a local school master, John Williams-Bennett, moved into the house. He had thirteen children; once they had grown up and moved out he took in paying guests and by 1901 it had become 'The White House Hotel'.

During the Second World War it is thought that General Eisenhower stayed here while visiting U.S. troops who were stationed nearby; he was probably boosting morale after the loss of almost 1,000 troops during Exercise Tiger at Slapton Sands – about eight miles away – or giving encouragement before the D-Day landings.

1950 restoration took place when much of the original decoration, such as the marble fireplaces, were removed.

1960 the house was renamed The White Horse – the owners had connections with horse racing.

From 1991 the hotel underwent restoration of some of the original Victorian workmanship.

2006 it was still a hotel. Landlord.

**Devonport** Formerly called Plymouth-Dock, from the foundation of the dockyard in the reign of William III, from 1824 the port was known as Devonport.

JB. 1848 Plymouth P; 1866 Plymouth PO; 1939 Devonport K.

**Exeter, Goldsmith Street** 1715, 24th June – a mortgage records that The White Horse Inn was at sometime used as a bridewell[2]. *The mother of the artist who drew the many illustrations in this book was born in a bridewell in 1913 at Prescott Road in Liverpool. Her father was the police Sergeant in charge, her mother had the title of 'Female Searcher'. The bridewell consisted of two cells; she would boast that she had been born in a jail, and had been christened Ivy Prescott Warburton.*

1946, 4th June – Henry William Hayes, a prisoner for debt in Exeter Gaol was recorded as occasionally residing at The White Horse Inn.

Dev. D1/27/38 and 53/6 Box 75.

**St. David's, High Street** 1838 listed in the Exeter City Improvement Commissioners Valuation of the 141 inns in Exeter.

**St. Sidwell's Street** 1782, 30th August – the Exeter Flying Post published a notice about a Public Survey to be held at the White Horse Inn 'between the hours of two and six in the afternoon (unless disposed of in the interim by Private Contract . . . )'. The survey was concerned with 21 acres of land called Heathfields 'in the parish of Heavitree near Broad-Clyst Turnpike[3] Road, about one mile and a half from the city'.

1844 Pigot's Directory records Robert Drake Beavis as licensee.

By 1870 the inn had closed. Dev. and WCSL. 1823/4-1848 P.

**St. Martin's Lane / Southernhay Recorded** as trading during the reign of Henry VIII (1509-1547) when James Gatemaker, the landlord, paid a rent of 13s. 4d.

1575 recorded as trading with a rent of £4 10s. per annum.

1787, 27th September – the resident landlord 'who had kept the White Horse in Southernhay many years, takes the Valiant Soldier in Exeter'. EFP and DRO.

**Farringdon Hill** 1796, February – a sketch of the house and a short article was published in the Gentleman's Magazine.

**Heavitree** A parish about half a mile north of Exeter city centre.

Mr. J. Young in his notes on Exeter inns records that in 1797 William Vallance, the landlord of the White Horse, moved opposite to the Horse and Jockey, a flourishing coaching inn, which belonged to the Baring Estate. Dev.

**Honiton, High Street** 1823/4-1848 P; 1866 PO; 1883-1919 K.

**Ivy Bridge, 43 Fore Street** 1897 listed as trading in Kelly's Directory, by 1939 it was listed at No. 45.

**Lifton 'The Arundell Arms'** Between the 7th and 9th centuries Lifton was a military outpost of the Saxons. King Alfred (849-901), was Lord of the Manor of Lifton – a Royal Manor. It is believed that there was an alehouse on this site, probably well before the 16th century – only the stone entrance still exists. In those early days the inn was known as The White Horse, which could link to the Saxons and their white horse emblem.

1555 Queen Elizabeth I sold the Royal Manor of Lifton to William Harris of Hayne whose family was linked by marriage to the Royalist family of Arundell.

1644, July – during the Civil War, King Charles I spent the night at the Manor House – which became The Old Rectory – on his way to rally the Royalists of Cornwall.

The present building dates from c.1730; the cockpit in the hotel garden dates from the late 18th century and still has the original walls and is used as the rod and tackle room. The circular earth mound inside, where cocks were set to fight, remained until c.1970 – a pair of the cockfighting spurs are displayed in the cocktail bar.

1783, 24th July – the *Exeter Flying Post* advertised the inn To Let; again the paper mentioned the White Horse on 15th February, 1786.

1815 William Arundell bought The White Horse Inn, which was not part of the estate, for £1,680; he thought this was too expensive but worth it because 'it is the best and most respectable inn in the village' – one of about fifteen inns in Lifton who all brewed their own beer.

It had seven bedrooms, three parlours, a bar and stables; it was renamed The Arundell Arms.

Between 1842 and 1850 William Arundell lost his inheritance through, it is said, gambling away the whole of his estate – including the inn – as a stake in a snail race.

During the early 19th century the landlady Mrs. Elizabeth Ball, took bookings for the Exeter to Falmouth mail coaches, which was a two-day journey on the turnpike roads – about a week to London – until c.1860 when the Great Western Railway Station opened at Lifton on the Exeter to Launceston line, making easy access to London until its closure in 1962.

There was a tailor's shop opposite which became the hotel annexe; the old school building in the car park became the conference centre; the old petty sessional court building, with its barred cell windows, adjoins the hotel.

By 1900 there were eight bedrooms and stabling for ten horses, also additional stabling for visitors. The hotel provided a pony and trap service to and from the station, to take customers to the fishing beats and commercial travellers on their rounds – often covering twenty miles a day.

1926 Kelly's Directory records 'There is excellent family and tourist accommodation at the Arundell Arms and Devonshire Fishing Hotel'.

1932 the hotel was bought by Major Oscar Morris, a keen fisherman, who closed it for four months during extensive alterations; it was reopened with a celebration dinner for all the men who had done the work. The old Tinhay Limestone quarry was acquired and stocked with fish, it had been flooded by the accidental broaching of a deep spring in Victorian times; this, together with the beats on the local rivers, contributed to the hotel becoming a centre of organised fishing.

1961 Gerald and Anne Fox-Edwards became the owners.

1970 Roy Buckingham, the hotel's ghillie won the Welsh Open Fly-Casting Championship.

1972 Gerald Fox-Edwards died, but Anne, who became Mrs. Voss-Bark, continued to run the business and by 2005 she had spent 45 years at the inn.

The Arundell Arms website. 1826 L.

**Moretonhampstead, 7 The Square** During the 16th -18th centuries Moreton was an important cloth-making town, the largest employer being the woollen industry.

The White Horse was recorded as trading by the mid 17th century; it became an important posting inn with its own malt house alongside.

The first known landlord was William Soper who died in 1795, he was succeeded by his wife Mary who was a very merry widow and one of the most popular landladies in the town.

1799, April - Mary was to marry a Sergeant of Marines, who jilted her by saying he had been ordered back to barracks in Newton Abbot – however her son was born in October!

1803 Mary's three year old pig weighed 705lb, 'the fattest ever seen in Moreton since the memory of the oldest living person'. This would be 320.45 kg. live weight!

Sometime before 1821 the house was owned by Samuel Cann.

1822 the Gray family took over, they were farmers from Addiscott, sometime later the house was renamed Gray's Hotel which the family continued to run as an hotel until the 1930s.

1838 there was a major fire at the hotel, the landlord had allowed a customer to store rock powder[4] in an outhouse, the fire fighters were unable to do anything until it exploded. Neighbouring properties were also damaged.

There is supposed to be a ghost who switches on the lights and slams doors in peoples' faces. 1823-1848 P; 1866 PO; 1883-1914 K. LHS Moretonhampstead.

**Okehampton** 1775, 6th October – the White Horse was advertised To Let, as a result a new landlord moved in on 5th January the following year.

1796, 16th June – once again the inn was advertised To Let and a new landlord moved in.

1805 W.C. Oulton described the house as the 'principal inn' in his 'The Traveller's Guide'. EFP. 1826 L; 1866 PO; 1883 K.

**Tiverton**

Originally the town grew from the important ford across the River Exe used from prehistoric times up to the Roman occupation.

***12 Gold Street*** 1780, 12th May – the White Horse was mentioned in the *Exeter Flying Post*.

1823 recorded as trading by the Excise Office. 1848 P; 1883-1939 K.

***Globe Street*** 1780, 29th December – the *Exeter Flying Post* advertised the 'New White Horse' To Let.

**West Teignmouth** 1866 PO; 1883 K.

**Woodbury Salterton, Sidmouth Road** Dating from c.1890 this house once traded as the White Cross. Landlord and HB.

## Notes

1. Coarse, narrow wool cloth and usually ribbed.
2. 16th century from St. Bride's Well in London near to a royal lodging given by Edward VI for a hospital, and later converted into a house of correction. There were several hundred bridewells throughout the country. Col., SOE and LE.
3. An advertisement was published in the same edition of the Exeter Flying Post:

> 'Notice is hereby given that the Tolls arising on the several Toll-Gates erected on the Exeter Turnpike-Roads, called or known by the Names of the . . . Heavitree . . . Gates, will be let, by Auction, for One Year from Michaelmas next, to the best Bidder, at the Guildhall of the City of Exeter, on the Third Day of September next, between the Hours of Eleven in the Forenoon and One in the Afternoon, in the Manner directed by the Act passed in the Thirteenth Year of the Reign of His Majesty King George III' (1773).

4. Gunpowder for blasting work.

# DORSET

*'There are numerous manufactories in this county. Among
these are several for a particular kind of baize*[1]*, and for the
manufacture of shirt-buttons. The latter affords employment to
a great many women and children.*

*There are a few silk-spinning establishments and about 9,000
people are employed in the neighbourhood of Bridport and
Beaminster, in the making of twine, netting, ropes, sail-cloth,
and other hemp goods, for the supply of which a great deal of
hemp and flax is grown in this county.'*
JB.

**Beaminster** The *Sherborne Mercury* published the following events:
   1756, 5th April - a sale took place at the White Horse.
   1756, 15th May - the landlords of the Kings Arms and the White Horse offered
a silver bowl for a bell ringing match.

**Blandford Forum, 21 Orchard Street** c.1770 two White Horse inns were listed
as trading, one may have been at Stourpaine.
   1855 Kelly's Directory lists the address as 'New Buildings' with W. Long as licensee.
   1939 Kelly's Directory gives the address as 15 Orchard Street.
   c.1950 the premises were was sold as a private house.
   DAS, DHC and H&W. 1842 P; 1855-1875 PO; 1898 K.

**Briantspuddle** c.1753 listed as trading with Richard Roberts as landlord.

**Bridport** c.1770 listed in the register of alehouse recognisances.

**Dorchester, Pease Lane (to become Colliton Street)** Trading from c.1890 with
William Henry and Ida Crocker as licensees; the pub was situated below Grey
School Passage on the north side of this narrow back street where there was also a
blacksmith, a church hall, an abattoir's yard, a sweet and grocery shop, a marine
store, a garage and the Boys' School. DHC and ISt. 1875 PO; 1898 - 1931 K.

**East Stour / Stower** 1861 the census recorded James Combes as innkeeper, aged
50, his wife Ann was 60 and they had a daughter, Lavinia, aged 26.
   1889 Kelly's Directory lists Albert Trowbridge as landlord, and in 1939 he
records Robert S. Perry as licensee.
   By 1982 it had become a private house.
   H&W and DHC. 1898-1939 K.

**Evershot** 1755, 15th September and 1756, 13th September – 'sword and dagger' competitions took place at the White Horse; the combatants carried a weapon in either hand.

1787, 12th February – a cockfight was organised here. Cockfighting was made illegal in 1847.

**Hinton St. Mary** 1753, 9th July – the *Sherborne Mercury* reported that a cudgels match was held here – the weapons used were short stout sticks.

1855 listed as trading in Kelly's directory with D. Mitchell as licensee.

1960, September – the lease ran out and the pub reverted to a free house and was part of the Pitt-Rivers Estate. H&W and DHC. 1855-1898 PO; 1915-1939 K.

**Litton Cheney** There are ancient standing stones situated in the nearby parish of Gorwell named 'The Grey Mare and her Colts'.

The inn, with its garden, stands at one end of this small quiet village alongside a stream.

1920 the inn was burnt down and rebuilt.

2007 the bar still had some of the original flagstone flooring and an old fashioned wood-burning stove. The small dining room displayed paintings by local artists. NQ 3rd S Vol. VI Oct 29 1864. 1855-1898 PO; 1915 K.

**Lyme Regis, Brown Street** 1842 Kelly's Directory lists Thomas Brown as landlord of the White Horse Inn, but by 1848 it was no longer listed.

**Maiden Newton** A fine old stone hostelry dating from the 17th century. Thomas Hardy (1840-1928) mentions it in his novel 'Tess of the D'Urbervilles' published in 1891, when he refers to 'the inn at Chalk-Newton' where Tess breakfasted on the way to Flintcomb Ash.

1740, 16th September – The *Sherborne Mercury* advertised the White Horse To Let as an ancient and well-accustomed inn.

1751, 23rd September – once again the inn was advertised To Let – 'with handsome gateway, etc.'

1812, 23rd March – a sale took place at the White Horse.

1855 listed as trading in Kelly's Directory with J. Peckham as licensee.

c.1900 it was destroyed by fire and rebuilt in red brick, continuing to trade.

The stump of the old Market Cross could still be seen in front of the house.

c.2000 the White Horse Inn Hotel was converted into flats.

ShM, DHC, CGH and DWB. 1898 PO; 1915 K. Hotel.

**Melcombe Regis, 33 St. Mary's Street** 1855 listed in Kelly's Directory with John Farwell as licensee.

By 1980 the site had been converted into shops.

DWB and DHC. 1823/4-1842 P Weymouth; 1875 PO; 1898-1939 Weymouth K.

**Middlemarsh** c.1770 listed in the register of alehouse recognizances.

1788, 9th June – the *Sherborne Mercury* reported that there had been a fire at the White Horse.

1855 listed as trading in Kelly's Directory with Mrs. E. Rogers as licensee.

1887 the inn appears as 'The Horse on Hintock Green' in 'The Woodlanders' by Thomas Hardy.

By 1928 it was described as a 'picturesque building of weatherworn brick; the tiled roof is laid to a pattern and the tiles themselves are moss-grown, the chimneys are massive and elaborated with dentil courses under the copings'.

c.1965 the inn was sold as a free house.

By 1985 the coach house and brew house had been modernized.

HL, H&W, PE and DHC. 1855-1875 PO; 1898 Sherborne - 1939 K.

**Minterne Magna** 1746, 9th June – the *Sherborne Mercury* mentioned William Mullet as landlord of the White Horse.

**Pokesdown, 947 Christchurch Road** By 2004 the White Horse, next to the railway bridge, had been demolished for redevelopment. Apparently nobody regretted its closure in what once was considered a poor area of Bournemouth.

1895-1915 Bournemouth K; 1939 Boscombe K.

**Poole, 31 West Quay Road** 1939 listed as trading in Kelly's Directory with G.T. Arnold as landlord.

**Stourpaine, Shaston Road** An old coaching inn once named 'The Black Horse', but by 1855 it had become the 'White Horse' with S. Seager as licensee.

It is said to be haunted, and that there was a tunnel to either the nearby church or village from one of the original three cellars, two of these are now filled in; one had a blocked-up archway thought to have led to the tunnel. The stables have been converted into a bar.

Landlord, DHC and H&W. 1855-1898 PO; 1915-1939 K.

**Swanage, 11 High Street** Originally trading as Burt's Restaurant, it was situated in a row of terraced houses; an illustration showed horses and carts outside and the building with a pitched roof which was altered to a flat roof.

c.1933 there was a fire in the top storey which would account for the flat roof.

c.1950 it was named the White Horse. Landlady in 2006, Mrs. Ross Shakespeare.

**Notes**

1. Woollen fabric resembling felt, usually green. Col.

# ESSEX

*'No less than thirty-six times in this county do we meet with the*
*sign of the White Horse.'*
MCh.

*Over one hundred years later some fifty White Horse Inns were*
*known to have traded in Essex.*
*'The annual quantity of grain sent from hence to London is*
*estimated at 25,000 quarters of wheat, and 150,000 quarters*
*of malt, besides large quantities of beans, peas, etc. . . . Its*
*principal manufactures are baizes and other coarse woollens.*
*Calico-printing is also established here; and on the banks of the*
*Lea are some mills for making sheet-lead'.*
JB.

**Ashdon, Crown Hill** Situated opposite the village shop, the White Horse, a brick building, was one of four public houses in the village.

During the 20th century the pub adjoined the Conservative Club.

1895 listed as trading with Walter Nation as licensee.

1989, March – the Conservative Club building was donated to the village to be used as the Village Hall, at the same time the pub was purchased on behalf of the village to house the caretaker.

By 2008 the former pub had been converted into offices.

1839 P; 1848-1863 WDG; 1870-1886 K.

**Brentwood, 10 High Street** 1895 listed as trading with Henry Fox as licensee.

The White Horse was a fare stage for the many bus and coach routes which ran through the town; it was always a popular public house but never a spectacular one.

For some years a postman used to wait outside the pub for one of the buses which brought a special lockable box containing overspill mail from nearby Billericay; the postman would unlock the box, remove the sacks of mail and take them to the Brentwood sorting office.

c.1990 the pub closed to become Kentucky Fried Chicken.

Brentwood Museum Society. 1839 P; 1845-1859 PO; 1863 WDG; 1870-1937 K.

## Chelmsford

1845 the Post Office listed two White Horse inns, one probably being Great Baddow. 1895 listed as trading with Mrs. Susannah Everett as licensee.

*1 High Street* 1654, April – A 'Horrid Murder, committed upon the person of Thomas Kidderminster of Tupsley in the County of Hereford, Gent., at the White Horse Inn in Chelmsford . . .' Mrs Kidderminster had heard nothing of her husband for nine years; she knew he had planned to visit Cambridge, Chelmsford and London. Nine years later in 1663 she was shown an announcement that human bones had been discovered in a back yard in Chelmsford. She walked from London, stopping for the night at the Black Bull in Romford where she talked to a woman who knew the White Horse inn at Chelmsford the landlord of which was highly respected, but his unpopular predecessor, Sewell, could well have been a murderer. The woman introduced Mrs. Kidderminster to Moses Drayne an old ostler of the White Horse; reluctantly he told her that he could remember a man of Thomas Kidderminster's description and that there had been a chambermaid, Mary Kendall of Brentwood who would remember. At this stage Mrs. Kidderminster called in the police.

On the night of the murder Mary had been moved to a remote bedroom in the White Horse and her door had been locked until the following morning; she believed Mr. Kidderminster had left but was suspicious when she found that the door of his room was kept locked for several weeks. When she eventually went into the room, she recognised garments belonging to him and took them to Mrs. Sewell, wife of the unpopular landlord; she beat Mary who realised during the ensuing row that he had been murdered when the £600 he was carrying was stolen. Moses Drayne, the ostler, was hanged; it was proved that he and Mrs. Sewell were accomplices in the murder.

1667 a ½d. token was issued by John Turner at The White Horse – *a horse*. GCW No. 68.

1770 plans were made to buy the White Horse and other land around it to build a new gaol near to the assize house, however this was strongly opposed by county landowners, farmers and graziers. If the inn was demolished they would be deprived of a convenient place for their accommodation and transacting their business on market day; a prison near the assize house and the market cross could spread infections to the public and it would mean tampering with the town's water supply. There was also a right of way which had to be considered from Greyhound Lane through the backyards and gardens of the White Horse to the rear of the Coffee House at No. 2 High Street which had no access for vehicles from the street. The row went on for two years between the county authorities, the M.P. and the local people.

1772 the plans for the new gaol on the site of the inn were shelved, it was finally built on the site of the old gaol across the river at Moulsham.

1776 the inn, having been sold to John Harriot, was demolished and two brick private houses were built on the site. One described as 'being fitted up in the most elegant manner' had a coach house and walled garden planted with fruit trees. The other was owned by Mr. and Mrs. Roffe who ran a boarding school for young ladies and gentlemen in the front part and let the back premises as genteel lodgings suitable for an elderly gentleman or lady with one servant, free from the noise of the school and with a view over the meadows and river.

MCh, GHM, HG, ERO and JB.

**25 Townfield Street** The Landlord believed the house was listed as trading in 1842. 1859 PO; 1863 WDG; 1870-1889 K.

**Chipping Ongar, High Street** 1789, September – recorded as trading amongst a list of inns in the Epping Division.

The Pubs, Inns and Taverns Index for England, 1801-1900 records the following:

1791 Thomas Hendry was listed as licensee, he was a victualler and Post Office keeper.

1822 – c.1832 Abraham Duvall was licensee.

The 1841 census lists the occupants of the inn as John Snow as publican and Post Office Keeper aged 40, his wife Frances aged 35, their six children – four sons and two daughters ranging in age from 4 to 12 years; Susanna Brown a servant aged 18; a visitor, Charles Cook aged 30 and three agricultural labourers, who were Joseph and James Hornsby aged 28 and 30 and a third aged 23.

1845 John Snow was still listed as the Post Office keeper until c.1848.

1848 and the 1851 census records Thomas Holt as innkeeper and Post Office keeper; he was aged 36 and his wife Susan 38; their three children – a son aged 3 and two daughters, one aged 1 year and the other 3 months; there was also a servant, Sarah Stow aged 33.

Between 1867 and c.1880 Mrs. Susan Holt was listed as licensee.

1881 the census shows the occupants as William Prentice, publican aged 46 and his wife, Ellen aged 47; they had two sons and two daughters aged between 10 and 22; also their son in law, George Day, a fishmonger. The lodgers were Charles Francis, a fishmonger, William Palmer a shoemaker, William W. Palmer a labourer and George Holt a drover.

1882 to 1891 at least three different landlords were listed – the census gives Walter Loven as publican, aged 27; he had three boarders, George Page, a general labourer, his wife Margaret both aged 50, with nine year old George Page.

The inn was listed as trading until 1899.

MCh and John Mead – local historian. 1824-1839 P; 1859 PO; 1863 WDG; 1870-1889 K.

**Coggeshall, West Street** 1895 listed as trading with Henry Warren as licensee. 1824 P; 1845-1859 PO; 1863 WDG; 1870-1889 K.

## Colchester

**4 East Street / Bridge** By 2006 the address was No. 67 and the pub had been converted into an office. 1824-1839 P; 1845-1870 PO; 1886-1937 K.

**The High Street** 1706 deeds of the property exist at this date.

1747, 17th June – the first landlord was buried in St. Mary's churchyard.

1755, Michaelmas – a freehold estate producing £40 a year was offered for sale which included 'The Three White Naggs', a good Inn, with lodging Rooms, Stabling, and all needful Accommodations', occupied by John Smith.

1756, April – described as 'a very commodious House which stands in the principal Street, where all Persons pass travelling from London to Harwich'. John Beets took the inn but only stayed until Christmas, the inn was again offered for letting at a moderate rent and the name was changed to the 'King of Prussia'. This referred to Frederick the Great, grandson of George I – who reigned from 1740 to 1786 – known as 'The Glorious Protestant Hero', Frederick was a most popular man in his day due to his victory at Rosbach on 5th November 1757; many innkeepers repainted their signs and renamed their houses The King of Prussia.

1763 listed as trading with Mr. Tetum as landlord.

1842 the deeds still record the name as The King of Prussia.

**Dengie** A coaching inn. The words '1702 the White Horse was established' were engraved on a mirror hung on the wall of the bar.

1841 the census lists William Grove, aged 33, as innkeeper, his wife Mary aged 31 and their two sons, William aged 2 and Thomas aged one month. There were 10 lodgers – Elizabeth Willis a servant aged 21; 9 were agricultural labourers; one 'independent' was aged 67.

1845 William Grove is recorded as running the Post Office.

By 1851 the census records William Grove as being a farmer and innkeeper; he had four more children; his nephew, an agricultural labourer, and his niece, the barmaid aged 15, were also living here along with the servant, Hannah Tyrell and George Payne an errand boy aged 9 along with eight other lodgers which included four agricultural labourers, a jobbing gardener and a sawyer[1].

By 1867 Mary Grove was running the inn and the Post Office; by 1870 she is also recorded in the census as a farmer. The Grove family ran the White Horse for almost forty years.

By 1881 Charles Brett was licensee, he and his wife had three children; a servant and five agricultural labourers were living with them. He remained licensee until 1917.

1895 listed as an *S.O.* for H.M. Post Office (See References).

c.2003 the inn closed to become a private house.

Landlord. 1839-1845 PO; 1848-1863 WDG; 1870-1937 K.

**Dovercourt, 489 Main Road** 1808 the Harwich Guide advertised the White Horse as 'a good and commodious inn . . . Which has lately been rebuilt by Thomas Cobbold Esq.' of Cobbold & Cox, brewers.

1817 the partnership agreement between Anthony Cox and Thomas Cobbold, who was a particularly good and knowledgeable brewer, was renewed and contained the following passage:

'Nothing herein contained shall prevent or hinder the said Thomas Cobbold from teaching or instructing at or upon the Brewhouse and Premises used for the time being in the said Partnership Trade and Business any person or persons in the act or mystery of brewing Beer Ale or Porter nor from receiving for his own use any Gratuity Premium or Emolument for so doing.'

Thomas, being manager, had a two-thirds share of the business and Anthony one-third. Anthony had a monthly allowance of £10 and Thomas £20 plus an annual salary of £100.

By 1847 Miller Christy in his 'The Trade-signs of Essex' recorded that the sign had been changed to the 'Great White Horse'.

1895 listed as trading at Upper Dovercourt with Paul Todd as licensee.

1824 P; 1848 WDG; 1859 PO; 1863 WDG; 1889-1937 K.

**Epping, High Street / Market Place** 1667 ¼d. token issued by Francis Furrill - *a horse passant*. GCW No. 169.

1789, September – listed as trading in Miller Christey's 'Trade Signs of Essex'.

c.1965 the site became a Barclay's Bank. 1823-1839 P; 1848 WDG.

**Great Baddow, 78 High Street** 17th century ½d. token issued by John Langston in Muchboddow - *a horse*. GCW No. 240.

1863 victualler, agent to Sovereign Life Office and to British Economical Manure Company.

WDG; 1824 P; 1845-1859 PO; 1870-1937 K.

**Great Dunmow** A tavern. 1824 listed as trading in Pigot's Directory.

1848 W. White's 'History, Gazetteer and Directory' listed the landlord as a timber merchant and victualler.

1839 P; 1845 PO; 1859 PO; 1863 WDG; 1870-1937 K.

**Great Chesterford** 1895 listed as a Sub Office for H.M. Post Office with Thomas Bareham as landlord.

1839 P; 1848 WDG; 1886-1937 K.

**Great Maplestead** Listed as the 'White Horse of Kent', but by the end of the 20th century it had become a private house. 1937 K.

**Great Sampford** 1937 K.

**Halstead, 12 Parsonage Street** A coaching inn. Two White Horse inns were listed in Pigot's Directory in 1859, also in Kelly's Directory in 1870 and 1886, one possibly being at Great Maplestead.

1895 listed as trading with Francis Hibble as licensee.

c. 1993 the inn closed to become a private house. 1848 WDG.

**Harlow, 160 Old Road** 1789, September – recorded in a list of inns in the Epping Division.

1841 a beer shop with Ann Prior aged seventy as licensee; she had two lodgers who worked in the nearby brickfields.

1851 a neighbour James Foster became the licensee. At this time J.G. Nash, known as a common brewer, supplied the premises with beer until 1877.

BHS and MCh. 1839 P; 1842-1859 PO; 1870-1937 K.

**Hatfield Broad Oak** 1789, September – Miller Christy listed the pub as trading in 'The Trade Signs of Essex'. 1839 P; 1845 PO; 1848-1863 WDG; 1870-1914 K.

**Hatfield Heath** The inn has always been popular with cricket enthusiasts as it overlooks the pitch on the village green.

1895 listed as trading with John Stevens as licensee. MCh. 1863 WDG.

**High Ongar** 1789, September – recorded as trading amongst a list of inns in the Epping Division.

1895 listed as trading at 'Ongar', *this could refer to Chipping Ongar,* with Albert Peachey as licensee; the house was also a Sub-Office for H.M. Post Office.

MCh. 1848-1863 WDG; 1886-1937 K.

**Kelvedon** 1859 PO; 1937 K.

**Layer Marney, 'White Horse Heath K.G.'** The letters 'K.G.' may refer to the Tolleshunt Knights who owned this land from 1207; the name Marney is taken from Marigny in Normandy where the family originated.

1863 the landlord was listed as a wheelwright and victualler in White's History, Gazetteer and Directory.

1874-1876 The Ordnance Survey map shows The White Horse at Smyth's Green where the inn had been re-sited as the local land owner, Mr Charles Round of Birch Hall, objected to seeing drunks lurching about when they left the inn, he therefore had it moved further along the road.

1895 listed as trading with Charles William Everitt as licensee.

1918 put up for sale, the catalogue is held by Essex County Archives Ref. A289.

c. 1994 it became a private house. 1839 Birch P; 1870 K; 1937 K.

**Little Totham** This Grade II listed timber-framed former coaching house dates from the late 16th century. It stands on the road between Goldhanger Creel and Totham Plains, a road which was used by smugglers who would have probably frequented the inn.

1768 the inn was granted a licence to sell liqueur with Stephen Rice as landlord; in 1774 he was followed by Thomas Crane who was there for ten years.

c.1803 Edward Quy became licensee and remained here for forty years, he was succeeded by his son Henry.

1806 Edward married Ann Mayne.

1845 the owner is recorded as Tolly Cobbold, brewers.

1859 Henry Quy retired as licensee; Richard Redgwell took over the licence, he also was a timber merchant and had his yard on the premises.

1870 Thomas Fordham became licensee, he was also a carpenter and was in business with Richard Redgwell. Thomas remained at the White Horse until c.1890.

The Lord of the Manor held his courts here up until the early 20th century when they were replaced by parish Councils.

1895 listed as trading with Arthur Fordham as licensee.

1902 the pub closed - it is thought that the landlord died of typhoid. The house was sold to Ernest Kempen from Cobbolds, maybe with the original settles which had remained in the bar for over two centuries, and the name was changed to White House. His wife opened a small shop selling sweets, cigarettes and soft drinks, she was the grandmother of Ruby McColm (née Kempen) and her sister, residents of Little Totham – aged 85 and 87 in 2008. Their grandmother also let rooms and it became a regular stop for cyclists. At the beginning of the Second World War their aunt sold the house, which had fallen into bad repair.

Over the years the property has been altered several times; in 1947 when Mr. and Mrs. Gregory moved here there was no electricity, no bathroom, no main drains and the garden often flooded. When the bathroom was installed, it was only the second house in the village to have one; electricity was installed in the house in c.1950. Later, an Irish family restored the house and by 2008 a couple with their young family had moved in.

Mrs. Lorna Key – local historian. 1839 P; 1845-1859 PO; 1870-1889 K.


**Maldon, 26 High Street** Originally a coaching inn.

1786 a White Horse Inn appears several times in advertisements in the *Chelmsford Chronical*.

1895 listed as trading with Frederick Handley as licensee; the address was given as 24 and 26 High Street.

2006 The inn still offered accommodation with food being served all day. MCh. 1805 WCO; 1823-1839

Commercial and Posting P; 1845 2 listed PO; 1859 3 listed PO; 1863 WDG; 1870 PO; 1886-1914 K.

**Mistley** The village is on the River Stour, where in 1834 there were good quays and 'commodious warehouses for corn, malt and coal'; Lord Nelson's (1758-1805) ship the 'Ampion' was built here. The quay, port and warehouses were the property of the proprietor of Mistley Hall.

The White Horse, a Georgian building, is one of several built by Richard Rigby of Mistley Hall, for his estate employees. It is situated opposite the Thorn public house where, in c.1644, Matthew Hopkins[2] the notorious witch finder[3] held court here.

Between 1848 and 1882 James Wilson was listed as landlord.

1891 the census listed Thomas Steward as licensee, he was also a carpenter.

1895 listed as trading with David Cooper as landlord.

By 1897 the owners were Charrington Nichol; by 1910 they had sold it to James Turner of London.

From c.1919 Saville Brothers of Stratford, London owned the house.

1932 the planning application for new toilets was approved.

1960, 2nd October – the pub closed to become a private house, named White Horse House – at one time lived in by Douglas Fisher, the BBC wildlife photographer. JB. and Michael Scurrell – local historian.

**Mundon, Main Road / Mundon Road** 1788 listed as trading in Stubbings Directory with James Thurlle as innkeeper, he ran the inn for over twenty years.

By 1839 Pigot's Directory lists John Deekes as licensee, he and his family were innkeepers here for the next sixty nine years.

1851 John Deeks, aged 69, was listed as a victualler and a bricklayer, he was also in charge of the Post Office; his wife Elizabeth was 63. Their son of 28, who was also a bricklayer, and their grandson John lived with them. There was a house servant and two lodgers, both agricultural labourers.

By 1861 the census records Elizabeth as innkeeper, she was also a butcher. She lived with her son, grandson and servant; there were four lodgers, two bricklayer's labourers and two agricultural labourers.

1862 Kelly's Directory lists Elizabeth, now aged 74, as Postmistress. By 1878 her grandson John, also a bricklayer had taken over. He was innkeeper for the next 30 years.

1912 Kelly's Directory lists George Cleveland as licensee and he was still here in 1937. 1839 P; 1848-1863 WDG; 1870-1937 K.

**Netteswell** 1789, September – Miller Christy recorded The White Horse in 'The Trade Signs of Essex'.

**Newport, Belmont Hill** Situated on the A11 road.

The building has a black and white beamed frontage and is thought to be over 300 years old; it is said to be haunted. A landlord thought it had been trading in 1913.

1978 the pub had its own football and cricket teams; in August there was a traditional match between the saloon and public bars. A-S and ERO. 1937 K.

**Nine Ashes** 1895 listed as trading with William Pleasant as licensee.

**Norton Heath** A Coaching inn dating from the 14th Century.

1895 listed as trading with John Willis as licensee.

**Paslow Common** c.1875 run by the same family for some thirty years until October 1905 when it was demolished. HoW 1955 Spring edition.

**Pilgrims Hatch, 173 Coxtie Green Road** c.1748 is the date of establishment given by the Landlord. 1870-1937 South Weald K.

**Pleshey, The Street / High Street** The village was known to have been occupied during the Bronze Age; it was named Tumblestown by the Saxon and Pleshey by the Normans from the word 'Plesseis' – an enclosed space. 'The village that boasts of knowing neither a teetotaller nor a drunkard relied entirely on its home-brewed liquors until 1900'.

The 15th century building and brewery still have the original roof timbers and are joined by a slate roof; more original timbers and flooring can be seen in the restaurant.

1987 – a new building added with kitchen and gallery.

1995 – a bar was added, 2000 – a function room.

The pub plays an important part in village life. There is a permanent exhibition of watercolour and oil paintings by local artists and a gift shop which used to sell Suffolk glass, and second hand books.

HPM and Pub Brochure. 1848-1863 WDG; 1870-1937 K.

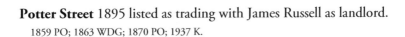

**Potter Street** 1895 listed as trading with James Russell as landlord.

1859 PO; 1863 WDG; 1870 PO; 1937 K.

**Ramsden Heath, Heath Road** A 17th century coaching inn with later additions.

During the 19th century the newspaper from Chelmsford would be delivered every week to the inn where it was be read aloud to those who could not read.

1895 listed as trading at Ramsden Bellhouse with Mrs Ann Wenmen as licensee.

Due to the railway cutting the parish was cut in half – one area became Ramsden Bellhouse and the other Ramsden Heath.

By 2006 the restaurant seated between 100 and 200 people.

2007 the inn was being renovated.

PE and Mrs. Isabel Johnson – Local Historian and Landlord.
1848-1863 WDG; 1870 Ingatestone PO; 1859 P; 1886-1937 K.

**Rayleigh** 1895 listed as an *S.O.* for H.M. Post Office with Samuel J. Bowen as landlord. 1839 P; 1845-1859 PO; 1863 WDG; 1870 PO; 1879-1937 K.

**Ridgewell, Mill Road** 2006 still trading as The White Horse Inn and Hotel. 1937 K.

**Rochford, 66 North Street** "Very old" – Landlord. 1859 PO; 1937 K.

**Roydon, 2 The High Street** The inn traded from c.1600, with its old beamed ceilings and walls it is the oldest inn in Roydon and thought to have been a coaching inn.

By 2005 it had become an Indian Restaurant known as White Horse India Cottage with accommodation available in an adjacent self contained cottage.

Landlord. 1937 K.

**Saffron Walden, Cats / Cate's Corner** The 1896 Ordinance Survey map shows this inn at the top of Hill Street with a smithy adjoining.

ECA. 1823/4 P; 1845-1859 PO; 1863 Hill Street WDG.

**Market Street** 1895 Listed as trading with Charles Watson as licensee.

1896 the Ordnance Survey map showed a smithy to the rear of the building.

ECA. 1839 P; 1848 WDG; 1859-1870 Market Place PO.

**Sible Hedingham, 39 Church Street** The White Horse stands in the centre of the village which is recorded in the Domesday Book as being part of the largest parish in England. Sir John Hawkwood, a medieval knight, was born here; his tomb is in the Parish church.

Half of this building – the restaurant and the rooms above – are dated from the 14th century, the rest of the house is of a later date. There are ancient beams, a large wood-burning stove set in a vast brick hearth, a collection of horse brasses – and a well-behaved ghost. Landlady. 1839P; 1848-1863 WDG; 1870 PO; 1889-1937 K.

**Starling Green** By 2008 the pub had become a private house named White House. 1886-1937 Clavering K.

**Southend-on-Sea, Southchurch Boulevard** Trading as a public house from c.1800 when it was a brick and weatherboard building set in three acres of its own land and standing beside a cart track.

1895 listed as trading at Southchurch with John Stebbing as licensee.

c.1900 the landlord was also the parish constable; the village stocks and a whipping post stood outside the pub and criminals were punished here.

At some time there was a silhouette of a galloping white horse on the west wall of the building.

The large present day building is situated on the corner of a main road junction someway from the centre of the town. The Boulevard was constructed for a tram service to run a circular route from the Minerva public house on the seafront, along Southchurch Road and Southchurch Boulevard – passing the White Horse – and back to the seafront.

2002 an application for a public entertainment licence was turned down.

2007 advertised as a 'bright modern pub'.

SCL, A-S, EvE 26/04/71, EvE 23/07/02 and IY. 1839 P; 1848-1863 Southchurch WDG; 1870-1937 K.

**Southminster, North Street** An Edwardian building, at one time named 'The Fox and Hounds'.

1895 listed under White Horse as a Sub-Office with Charles English as landlord.

Landlord. 1839 P; 1848-1863 WDG; 1870 PO; 1886-1937 K.

**Sun Street** 1789 recorded as trading by Miller Christy in 'The Trade Signs of Essex'. 1839 P; 1848-1863 WDG.

**Waltham Abbey, Market Place / Square** 1845 PO; 1870 PO; 1937 K.

**Widford** 1824 listed as trading in Pigot's Directory. Between 1848-1863 White's History, Gazetteer and Directory lists the licensee as a butcher and victualler. 1895 listed as trading with Mrs. Ann Thorn as licensee. By 2007 the site of the old public house had been converted into a restaurant seating 120 people named Masons Too. 1839 P; 1845 PO; 1870 PO; 1886-1937 K.

**Witham, Chipping Hill** 1839 P; 1845 PO; 1848 Newland Street WDG; 1859 2 listed. PO; 1863 WDG; 1870-1889 K; 1914-1937 K.

## Notes

1. One who saws timber for a living. Col.

2. Said to have been a lawyer at Ipswich and Manningtree; he made journeys for discovery of witches in eastern counties. In 1645 he procured a special judicial commission under John Godbolt to hang 60 women in Essex in one year, nearly 40 at Bury and many more at Norwich and in Huntingdonshire. He died of consumption in 1647. DNB and M. Scurrell – local historian.

3. One formerly employed to search for and obtain evidence against witches. SOE.

# GLOUCESTERSHIRE

## Bristol & South Gloucestershire

*'It is remarkable for its tide, which rolls in with great noise and
an impetuous elevation of three or four feet. Its usual rise at
Gloucester is 7½ feet, but it sometimes rises to 9 feet, and from
its violence has often done great damage to the surrounding
country. To guard against this, and the sudden inundations to
which the low grounds, especially below Gloucester, are liable,
drains, sea-walls, etc. have been constructed at great expense,
and are placed under the superintendence of a society called the
commissioners of sewers.*

*Other manufactures of the county are iron and brass ware, wire-
cards for the clothiers, pins, nails and writing paper.'*
JB.

**Almondsbury** 1891 the List of Licensed Houses in the County of Gloucester
records the White Horse as an alehouse.

**Bedminster, 166 West Street** 1775 was listed as trading with Dinah
Underhill as licensee.

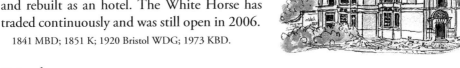

1880 the original building was pulled down
and rebuilt as an hotel. The White Horse has
traded continuously and was still open in 2006.

1841 MBD; 1851 K; 1920 Bristol WDG; 1973 KBD.

## Bristol

1518-29 'le Whightehorse' was recorded as trading.

1752, 15th June – 'Last Sunday night the dwelling house of Mr. Roads at the
'White Horse without Lawfords Gate' was robbed by a person who lodged there
of about twenty five pounds and made off undiscovered'.

Both these early references refer to unrecognised locations but will refer to one
of the following sites. D&W and BJ.

***Barrs Street and Horse Fair*** Situated on the corner with the Horsefair, the hotel
traded continuously between 1752 and 1953.

The census for 1841 records William Holder aged 60 as inn keeper with his
wife Eliza aged 40 and their son Thomas aged 5. There were six servants, two
travellers and Elizabeth Heath aged 40 of independent means.

The census for 1851 records Sarah Vickery a widow as landlady, aged 45 she was a licensed victualler, who was born in Canterbury; William aged 27 was a house proprietor. There was also a cook and a chambermaid.

1856 James Wild of the White Horse Inn and Commercial Hotel advertised the provision of wines and spirits of the first quality, good stabling and lock-up coach houses; he also provided omnibuses to and from every railway train. An addition to the advertisement reads:

'J.W., having become the Proprietor of the above Establishment avails himself of the medium – in conjunction with Mrs. Wild (late Vickery) – to return their grateful thanks for the liberal support hitherto afforded them, and respectfully solicits a continuance of such favours, feeling assured that the additional extensive arrangements now being made to the House, with an observance of moderate charges, will give general satisfaction.

Commercial gentlemen will find the above Inn to possess every comfort, combined with unremitting attention.'

1861 Richard Cowle aged 46, innkeeper, his wife Ann aged 45 and their daughter aged 14. Other residents included two servants, a corn dealer, a twine manufacturer and an ostler. Mr Cowle advertised 'Good Stabling, Beds and every Accommodation for Commercial Gentlemen'.

c.1872 an advertisement for the White Horse Hotel records Eliza Hember as landlady selling 'Wines & Spirits - Home Brew'd Beer - Superior Bottled ales and Stout' – there was also Good Stabling.

1881 William Jones aged 42, described as Head Licensed Victualler and his housekeeper, Elizabeth Matthews aged 24, also a barmaid, cook, a housemaid born in America, an ostler and an England Boarder mail cart driver.

1953 Barrs Street was closed and the hotel pulled down. The site was redeveloped by Debenhams department store. BRO and MW. 1752-1764 BRO; 1836 Commercial Tavern MBD; 1868 S; 1870 -1920 WDG.

**Fish Ponds** 1885 K.

**12 Hanover Place** An old cider house situated on the corner with Sidney Row.

1841 the census lists William Francis as landlord, he had a wife Grace, both aged 40 and five children; Ann Buss aged 50 was also living here.

1851 the census lists Ann Tregaskis, a widow aged 39 as publican; she had four children – Thomas aged 15 was a brass finisher and Rosina aged 7 was described as a scholar.

By 2007 it had been renamed The Orchard and could be reached from the city centre by the harbour ferry by day – alight at the SS Great Britain landing stage, walk down the street and take the first alley on the right. 1973 KBD.

**Horse Street** 1752-1764 BRO.

**Jacob's Wells** 1841-1853 Mary Cooper is recorded as licensee.

**49 Lower Ashley Road / No. 1 Lower Ashley Place, Lower Ashley Road** Recorded as trading between 1867 and 1953. 1973 KBD.

**Queen Street, Christmas Steps** Part of the old city wall, steep to the River Frome.
Listed as trading between 1855 and 1909.

**St. Mary Rexcliffe (Redcliffe)** 1751, 30th December                                                       –
*The Sherbourne* Mercury published a notice about a lost horse belonging to Mr. Barnstable at the White Horse in Redcliffe Street, Bristol.
1752 listed as trading.
1753, 26th March – The White Horse was advertised To Let – it was described as an inn and tavern, a well-accustomed house.
1775 Robert Powell is listed as licensee. ShM, BRO.

**St. Thomas Street** 1764 listed as trading under White Horse but by 1778 it had been renamed 'The Crown'. BRO.

**264 Two Mile Hill Road** 'A mining area where Charles Wesley (1707-1788) thrived and there are many non-conformist chapels, liberally dotted amongst the taverns'. In 1834 there was already a free school nearby for the sons of Methodist ministers. The Salvation Army building dating from 1879 would have been adjacent to The White Horse.
Listed as trading between 1853 and 1926.
1927, 8th December – an advertisement reads: 'The Popular Licensed Premises known as The White Horse'. Accommodation included two Bars, Bottle and Jug Department, Taproom, Skittle Alley, two Sitting Rooms, three Bedrooms and Boxroom and 'a good Cellar'. There was also 'a large piece of Garden Land adjoining, the whole containing by estimation about 30 perches.
This area was badly bombed during the Second World War and in the 1950s new houses were built; the White Horse was listed until 1953. JB, BRO Ref. 29595/2 and Mr. Thomas Gover.

**Waterloo Place** Recorded as trading between 1840 and 1904.

**166 West Street** 1868 S.

**Buckover** Situated on the A38 in the outskirts of Thornbury, it was originally a cider pub; a traditional public house with old fireplaces. 1868 S; 1914-1939 K.

**Cainscross, Cashes Green Road** c.1998 the pub was renamed D.B.'s.
By 2007 the house had ceased to trade and had become derelict.
1851 K; 1868 S; 1885 K; 1891 Stonehouse GLH; 1939 Stroud K.

**Charfield** 1885 listed as trading in Kelly's Directory.
By 1891 the alehouse had been renamed the 'Cromhall'.
During the 20th century it became a private house named 'White Horse Cottage'.
GLH.

## Cheltenham

*Regent Place* 1891 recorded as trading as a beerhouse. GLH. 1939 2 listed K.

*Townsend Place* 1891 beerhouse. GLH.

**Chipping Sodbury, 'Melbourne House', 29 Horse Street** By 1664 it is known that one of two buildings on this site was known as the 'White Horse Inn'.
1738 a clothier, John Edwards, may have been responsible for the present façade which in c.1760 was constructed across the two buildings.
The young Dr. Edward Jenner (1749-1823), an apprentice physician, lodged here. It was he who discovered that people working with cattle who contracted cowpox never caught smallpox. He coined the word *'vaccination'* from the Latin word *'vaccin'*.
1870 a Henry Watts is shown as living in Horse Street – a limeburner. According to Kelly's Directories, the Watts family occupied the premises at least until 1914. It is said that a Mr. Watts from Chipping Sodbury was the city architect of Melbourne in Australia and that he also gave his name to the Watts River. It is thought that the present name 'Melbourne House' referred to William Lamb, 2nd Viscount Melbourne (1779-1848), the British prime minister whose name was given to the Australian port in c.1835.
The original name 'White Horse' was revived when it became a charcoal grill; by 1983 it had been renamed 'Sultan'. GRO (Ref. Percy Couzens records); local historian, P.J. Elsworth, and the owner of the Sultan Restaurant. 1830 P.

**Coleford, Newland Street** 1849, 26th April – a surveyor of the town, George Gregg, was noting down information for a town map while standing in the Market Place on the corner of Newland Street. He recorded in his notes 'The White Horse Inn and Blacksmith Shop'.

c.1850 from about this date the house ceased to be an inn, but continued to trade as a blacksmith's shop and beer house.

By 2008 the premises had become a bistro and wine bar. LHS – Forest of Dean.

**East Dean** 1891 the White Horse was recorded as an alehouse. GLH.

**Frampton Mansell, Cirencester Road** 2004 The White Horse was the Les Routiers Dining Pub of the Year. 1868 Stroud S; 1870-1939 K; 1981 C&L.

**Hambrook, Bristol Road** A coaching inn on the London to Bristol road. The highwayman, Dick Turpin (1706-1739), is said to have called here.

1891 listed as trading as an alehouse under Winterbourne.

c.1920 a photograph taken outside this inn shows officials collecting evidence of a recent road traffic accident. GLH, GRO Ref. No. QY 6/4/1 p.43 C. 1868 S; 1897-1939 K.

**Hawkesbury** 1657 Token issued by Thomas Walker at the Horse in Hawkesbury – *a horse prancing.* GCW No.111.

1891 listed as a beerhouse. GLH.

**Henbury** 1891 listed as a beerhouse. GLH.

**Iron Acton** 1891 recorded as trading – beer, cider and perry were provided. GLH.

**Kiftsgate** Records dating from 1755 have been checked but only the names of licensees are given and not the names of individual public houses. It is understood that there was a White Horse Cottage in the parish, which may originally have been a small alehouse called White Horse. Local Post Office and GRO.

**Lydney** 1891 beerhouse. GLH.

**Mitcheldean, High Street** A sandstone brick building which has changed very little over the last hundred years, it is situated next to the Town Hall and close to the Mitcheldean Forest Brewery; at one time the inn could have been considered the brewery tap.

By 1674 the White Horse had been established in part of the building belonging to the Talbot inn – the Talbot had closed by 1696.

During the 18ᵗʰ century the White Horse became and important meeting place. On 5ᵗʰ October.

1761 the *Gloucester Journal* published the following article on the occasion of the coronation of King George III and Queen Charlotte:

> 'At Mitcheldean, in this County, on account of their Majesties' Coronation, the inhabitants expressed their joy in a very distinguished manner. The principal inhabitants met at The White Horse, and went in procession from there to the Church, when an excellent sermon was preached and two grand anthems were performed.
>
> The company afterwards returned to the White Horse to dinner, where a fine buck deer given by Maynard Colchester, Esq., who regaled the populace with a hogshead of liquor, and the day was spent in giving every proof of loyalty and joy which could become good subjects upon so happy an occasion.'

Due to the turnpike roads running through the town, William Price the landlord used this busy coaching era by hiring out 'neat Post Chaises, with careful Drivers, and Saddle-Horses'. It was not successful as a coaching inn but remained a popular meeting place.

From 1809 the Mitcheldean Friendly Society held their meetings here.

1830 listed in Pigot's directory with William Pearce as licensee.

c.1840 when Giles Gardener was landlord The White Horse Inn was listed with stables, yard, malthouse and garden.

1891 Frances Wintle of the Mitcheldean Forest Brewery bought the premises; listed as an alehouse with a rateable value of £27.

1894 Cornelius Baynham, the landlord, advertised the inn as a 'family and commercial hotel, the most replete in the neighborhood; every accommodation at moderate charges'.

1902 listed as an alehouse, the rateable value was still £27 and closing time was 10 pm. The Flooks family were licensees, they ran the house for 25 years. Mrs. Flooks was killed when she fell down the cellar steps.

c.1920 the public houses in the town were considered responsible for some of the street fighting, especially during elections when the 'public meetings were rowdy, with everyone sporting ribbons and favours'.

1923 the house was advertised for sale with freehold tenure, let to Mr. Harry Preece on quarterly tenancy at the low rent of £40 per annum. There was a chief rent of 2s. 0d. per annum payable to the Lord of the Manor. The compensation charge was £6.

The house was described as being built of stone with a slate roof. It had 10 bedrooms on the top two floors and half landings; there was a W.C. on the first floor.

The ground floor was comprised of the hotel entrance, serving bar, smoke room, commercial room, private sitting room, tap room, kitchen, and cellars. The yard was approached by a gateway from the main road, it had a stone built coach house or garage with a small room adjoining, stabling for two, a wash house, coal house, bottle store, etc. There was a range of stone built stabling for 10 horses, a small coach house with lofts over and a granary approached by a flight of stone steps. There was also a 'pig cot', public W.C. and urinal, a large kitchen garden and a skittle alley of timber with a corrugated iron roof.

1937 The house was once again put up for sale.

2006 the White Horse was still trading as a Free House.

GLH and LHS Forest of Dean. 1859 P; 1868 S; 1885K; 1939 K.


**Moreton-in-Marsh, Stow Road** Moreton means Moorland Settlement, 'in Marsh' was a corruption of 'henmarsh' – a boggy piece of land where there were wild birds.

1891 listed as a beerhouse owned by Henry Lardner & Sons with George Bumford as licensee.

1903 still listed as a beerhouse but owned by Flowers & Sons with John Tarplett as innkeeper, he was still licensee in 1939.

By 1997 the owners were Marston's Brewery.

2007 the house is still trading as the Inn on the Marsh. RN. 1891 GLH; 1939 K.


**Pilning, Northwick Road** The village grew from the arrival of the railway station.

An old village pub situated in a cul de sac due to the construction of the second Severn crossing. The author is unable to discover any history of this flourishing house.

**Painswick, Vicarage Street** 1891 listed as a beerhouse.

1957 recorded as trading.

By 1970 the pub had been demolished to make the entrance to the new White Horse Lane. GLH and LHS – Painswick. 1939 K.

**Sandhurst** 1891 listed as trading as an alehouse. GLH.

**Soudley, Church Road** The original White Horse was described as a modest mid 19th century beer house situated near the chapel.

1857 a local Friendly Society held its meetings here.

1891 licenced as an alehouse with a rateable value of £17.10s.0d.

c.1895 the premises were taken over by the Alton Court Brewery at Ross – some of their old beer bottles can be seen at the Dean Heritage Museum. The pub moved to larger premises at its present site; it was named the White Horse Inn and was licenced as an alehouse.

1962 the house and two parcels of land of 36 and 22 perches in Blakeney Walk were bought by West Country Breweries. LHS – Forest of Dean. 1870-1939 K.

**Staunton** Situated on the A4136 road. The original building dated from 1813, it was rebuilt during the 19th century to accommodate the turnpike road.

1845 listed as trading with William Morgan as owner.

1863 James Carver of the Royal Oak in Staunton also took over the tenancy of the White Horse, leaving Mrs. Carver in charge of the Royal Oak. James served behind the bar and shoed horses in the adjoining smithy, he was also a pigman and cider maker.

1891 the Alton Court Brewery at Ross bought the premises which had a rateable value of £7. 4s. 0d.. It was listed as an alehouse with John Hicks as landlord, he was also a plumber. He ran the house for at least 20 years.

1901 a collier, John Tye was summoned for assault at the White Horse and for 'being drunk and disorderly and refusing to quit the place'.

1962 'the inn in the Parish of Staunton known as the White Horse with the blacksmith's shop, outbuildings, garden and land thereto adjoining' was taken over by the Cheltenham Original Brewery.

2001, August – an application was made to convert the premises into a private dwelling.

2002 the pub was still thriving. GLH and LHS Forest of Dean. 1868 S; 1885 K; 1914 K; 1981 C&L.

**Tetbury** 1754, 18th March – the *Bath Journal* published and advertised for the house To Be Let:

> 'A very commodious and well accustomed inn known by the sign of the 'White Horse'. Situated in the Corn, Cheese and Bacon Markets. Having good stables, cellars and other conveniences all in good repair'.

**Thornbury** 1891, listed as trading as an alehouse. GLH.

**Westbury-on-Trym, 24 High Street** Dating from the 14th century, later to become a coaching inn; it has had many alterations over the years.

1706 documents dated 20th and 21st March relate to 'a close of land called Huntsgrove . . . near the White Horse Inn, Westbury-on-Trym'.

1891 listed as a beerhouse.

2007 the old serving hatch could still be seen on entering the house – *presumably the original 'jug and bottle' where people could bring their own vessels to have them filled.* The house has been known as the 'Hole in the Wall'. GLH, BRO and Landlord. 1973 KBD.

**Wickwar** 1851 listed as trading in Kelly's Directory.

c.1975 the pub closed and was converted into a private house named 'White Horse Lodge'. 1868 S; 1879K; 1891 GLH; 1914-1939 K.

**Winterbourne** 1891 GLH.

# GREATER MANCHESTER

*The first known recorded use of the term 'Greater Manchester' is in 1914 and it is described as one of the classic areas of industrial and urban growth in Britain; lying within the ancient county boundaries of Lancashire, Cheshire and Yorkshire.*

*Created as a metropolitan county in 1974 and then as a ceremonial county in 1997, it is the only urban area in the UK outside Greater London to officially bear the name 'Greater'.*

**Ainsworth, 12 Church Street** Situated at the corner with Bradley Fold Road, The Old White Horse was built in c.1880. 1881-1924 K.

**Ashton-under-Lyne, 58-60 Stamford Square** A cannon is shown on a plaque set in the wall giving the date 1876 and the words 'Stamford Square Nelson Square'. Landlord.

**Bury** 1822-1834 P; 1848-1855 [2 listed] S; 1858 PO; 1881-1898 K. *These dates may refer to Tottington, Walshaw, Harwood, or Ainsworth.*

**Eccles, 110 Gilda Brook Road** Situated on the Manchester-Warrington Road where a toll bar was in operation outside the inn. 1803 known as 'The Shovel and Broom', to become 'The Trafford Volunteer' six years later, having taken the name of the local volunteer forces' company raised to defend England against a Napoleonic invasion. In 1810 the inn was renamed 'The Volunteer'.

1825 listed as The White Horse in the licensing records.

1914 the house was damaged by fire.

1919 purchased by the brewers, Frederick Robinson Ltd. who rebuilt it in c.1935.

1969 the site was compulsorily purchased to make way for the 'South Lancashire Motorway' which became the M60.

In 1973 a new White Horse inn was opened some 200 yards away from the original site. TF and FR. 1834 P; 1848-1858 S; 1881-1898 K.

**Farnworth, 12 Egerton Street, Moses Gate** The hotel was demolished and a supermarket and offices built on the site.

**Harwood, 177 Stitch-mi-Lane** 1701 a stone with this date was once built into the wall of the house, which had been built by the Bromley family.

c.1900 the house, built of stone, was refurbished in part but retained some of its original features; there are stained glass windows and plaster freezes depicting the Canterbury Tales.

2001 the celebration of 300 years trading.

Landlord. 1822-1834 P; 1848-1855 S; 1858 PO; 1881-1898 K.

**Hollinwood** This was possibly mistaken for the White Hart. 1848 S. OLS.

**Hulme, 245 Stretford Road, Hulme Walk** 1937 listed as trading, but by c.1997 due to major housing redevelopment this White Horse no longer listed. PO.

**Irlam, 575 Liverpool Road** c.1845 the building is thought to be of this date.

In 1860 Charles Sherlock is given in the rate book as occupying a blacksmith's shop and cottage on the highway, employing two men and two boys; an agreement dated 1869 refers to a blacksmith's shop and a beerhouse called 'The Old Grey Mare Inn'; a plot of land was included bounded on one side by the turnpike and on the other by the tramway on to Chat Moss.

1872 listed in the licencing records.

1874 the inn was sold to a brewer for £350 under the name 'Grey Mare' but exclusive of the land, which by 1877 was in the possession of the Mayor and Aldermen of the City of Salford.

1890 A sale took place at the Boars Head; Lot 11 was 'a beerhouse The Old Grey Mare Inn with blacksmith's shop adjoining and near to the Railway Station' – it sold for £700. Papers relating to a second sale refer to the public house or beerhouse known as White Horse Inn – the average annual sale of beer for the previous three years being 326 barrels.

1962 the premises were owned by brewers.

There are five televisions including a big screen for major matches, pool table, darts board and juke box. CWh.

**Kearsley, 217 Bolton Road** Originally three cottages; a stone in the wall once gave the date 1824. During the Second World War RAF personnel were billeted here whilst training at Bolton Technical College.

**Leigh, 2 Railway Road** One of Leigh's main coaching inns with a regular run for Manchester via Astley Green. 1854 the house was rebuilt in the same late Gregorian style. By 1924 it had become the White Horse Hotel. The name changed several times over the years and by c.1990 it was known as 'Strutz'. Trade listed 1834 P; 1848-1855 S; 1858 PO; 1881-1881 K.

**Little Lever, 30 High Street** According to the landlord this is an 'old building'. By 2008 the pub had closed.

**City of Manchester, 21 Hanging Ditch** The original building dated from 1740; this old hostelry was next to 'The Manchester Corn, Grocery, and Produce Exchange Limited' – on the left, west side, looking towards Corporation Street.

1772 listed in Raffald's Directory.

1896 'White Horse, William Robinson, 21 Hanging Ditch' was listed in the Manchester, Salford and Suburban Directory for the last time.

By 1904 this hotel had been demolished for street improvements.

Fred L. Tavaré, NQ and MCN.

**Middleton, 78 Long Street** The inn was situated on the right hand side travelling up from the market place, it was on the corner of King Street just passed Jubilee Gardens and backing on to Middleton Parish Church.

1887 the house was purchased by the brewers, J.W. Lees & Co.

1929 the licence was referred for compensation.

1965 the house was demolished. JWL. 1848-1855 S; 1858 PO; 1881-1924 K.

**Oldham, 6 Trent Street** 1854 a licence was granted. 1869 the landlord was notified 'that his house had a low rateable value and it was full of notoriously bad characters, thieves and prostitutes', a year later a licence for music was granted, in 1888 the inn was bought by Oldham Brewery.

1892 a wine licence was obtained; at this time there were three drinking rooms, four bedrooms and an enclosed yard. 1965 the inn was closed due to redevelopment; by 1994 a health centre was built on the site. RMg. 1822/3 P.

**Openshaw, 12-14 Wood Street** By 1898 a beer retailer was trading at No. 12.

1922-1971 with the addition of Nos. 13 and 14 it had become the White Horse Hotel. Later the whole area was demolished and described as a derelict wasteland. WA., K. and PO.

**Pendlebury, Oak Street** Situated opposite Swinton Electroplating. The house was nicknamed 'The Oak' and was the smallest pub in the area.

1836 first listed in the Rate Book.

From 1839 this tiny beerhouse seems to have 'escaped the attention of the surveyors' as it does not always appear in the licencing records.

1872 listed as 'The Spinners Arms' but by 1888 the name had changed to 'The White Horse'. 'In the 1930s the pub was still a tiny building, with a room on the right, a vault on the left and a smaller snug at the rear . . . with a bricked-up doorway, indicating that the original beerhouse may have been considerably smaller, holding perhaps as few as a dozen people. In January 1937 it lost its licence due to slum clearance.' RH.

**Prestwich, 466 Bury New Road** 1848 S.

# Rochdale

It is not clear to which of the five White Horse inns in Rochdale the following trade lists refer: 1822-1834 P; 1848-1855 S; 1858 PO; 1881-1924K.

*Blackwater Street* 1848 S.

*160 Broad Lane* The 1891 census lists the landlord, James Ashworth, as a farmer.

*15 or 17 Lord Street* 1851 listed as trading with the landlord Soloman Ashworth being described as a 'wholesale dealer in French teasels[1]'.

1871 listed as trading in Slater's Directory. 1861 S; 1908-1916 JC; 1935K.

*Rooley Moor Road* The 1891 census lists Charles Nuttall, the landlord, as a springmaker.

*Spotlands (a township of Rochdale) Meanwood Brow* 1881 listed as White Horse in the census. 1858 PO.

**Stalybridge, 28 Back Grosvenor Street** Situated on the corner with Trinity Street near the canal bridge.

1842 the premises were opened by Joseph Waite, a painter by trade.

1848 a full license was granted.

1852 the house was renamed 'The Q Inn' – this name was transferred to premises in Market Street and by 1901 it had become the Queen Hotel.

By 1935 the building was empty, the whole area was demolished and the site became a supermarket. RMa.

**Stockport, Hempshaw Lane** Described as 'a converted cottage with a distinctive coat of paint'; it is uncertain if this was either the White Horse or White House; pre-1875 it appears there were five beerhouses in this street.

1875 the White Horse is mentioned in the Brewster Sessions.

1910 and 1923 the house was listed as at 382 and 384 Hempshaw Lane. PH. 1939K.

**Swinton, 384 Worsley Road** c.1610 the Squire of Haughton owned this isolated farmhouse built with hand made bricks over great arched cellars. It is reputed that there were tunnels linking the alehouse to Wardley Hall.

1650 listed in the licensing records; in the late 17th century it was used as a meeting place for the Select Vestry of the Township of Worsley.

1691 Roger Taylor is listed as landlord.

1706 the innkeeper submitted an account to the Overseers of the Poor for 2s 5d which referred to 'seven weeks houseroome to Hugh Lanshaw'.

During the 140 years between 1754 and 1894 there were only three landlords – except for two years when a temporary landlady took over.

From 1834 to 1881 the inn was listed under Singeley and Worsley.

It is Swinton's oldest public house; in 1996 the hotel became a Grade II listed building.

2006 described as a traditional public house with beamed ceilings and a white-washed exterior. M. and RH.

**Tyldesley, Elliott Street** A beerhouse standing between shops dating from 1790. Standing empty for several years, it was used as a place to enrol new members of the light artillery at the outbreak of war in 1939.

By 2000 a DIY and hardware shop had been built on the site.

**Walshaw, 18 Hall Street** The date with the initials 'SJA 1846' can be seen over the door of the hotel. 1858 PO – 1851 Tottington – 1924 K.

**Westhoughton, 259 Bolton Road** 'The White Horse Tavern'. Originally a coaching inn surrounded by farmland which was to become a mining area. The stables and byres were converted into 'cellars' when the A6 road outside was raised by some five feet. 1858 PO; 1898-1924 K.

**Notes**

1. *Dipsacus fullonum* or Fullers' Teasel – this was used for raising the nap on the baize and flannel manufactured locally. JB.

# HAMPSHIRE

## Isle of Wight

*Rules of a tavern for Lemuel[1] Cox's White Horse Inn, Froxfield:*

*4 pence a night for bed*
*6 pence with potluck[2]*
*2 pence for horse keeping*

*No more than five to sleep in one bed*
*No boots to be worn in bed*
*No razor grinders or tinkers taken in*
*No dogs allowed in kitchen*
*Organ grinders to sleep in the wash house*

**Alresford** 1599, 16th September, notice was given that a court case would be heard 'at the signe of The Whit Horse upon Monday the First of October next for the execution of the said comission . . . originating from the Quenes Ma'tes commission out of her Highnes Court of Chancerie . . . '

1794 noted in the churchwarden's accounts for New Alresford; both before and after this date the property is listed merely as 'a house'.

WCM Vol. II Estates No. 13322 and HRO. 1830-1844 P.

**Alton, 94 High Street** A 15th century listed building. The inn sign, erected in 1967 is a portrait of the race horse Mont Blanc II, a pure white thoroughbred, which was painted by G.E. Mackenney at the Epsom stables of the trainer, Mr. Walter Nightingale. The owner Mr. Charles Clore gave his permission for his colours to be shown in the bottom right hand corner of the sign.

2005, 10th February – the inn reopened after refurbishment. 1823-1844 P; 1855-1915 K; 1939 K.

**Ampfield, Winchester Road** Parts of this listed building date back to the 16th century; there are two inglenook fireplaces which have smoking chambers and priest holes.

During the Second World War the White Horse was designated to be a mortuary, but it never had to be used for this purpose.

The Rev. W. Awdry, (1911-1997), author of the Thomas the Tank Engine books published in 1946 was born in Ampfield.

The garden is bordered by the Golf Course on one side and the Cricket Club on the other. 1805 WCO; 1855 Romsey – 1939 K; 1974 C&L.

**Ashton, Beeches Hill (a mile north of Bishops Waltham)** It is said that at one time the whole building was divided into five cottages. In 1812 the present public house was known as the New Cottage.

1827 it was named 'White Horse' by William Hutt who had acquired the adjoining Old Cottage.

1852 Henry Paice, the village shopkeeper bought the inn.

A map on the wall of the bar dated 1871 marks the White Horse.

Although there is one entrance to the pub there are three drinking areas, one of which is the Stable Bar.

1981 this house won first prize in the Whitbread Wessex competition for the best decorated pub on the day of the Royal Wedding; mentioned in the Egon Ronay Pub Food Guide. WhB July 1987. 1844 P; 1855 K; 1875 PO; 1895 K; 1974 C&L.

**Basingstoke** *See Worting.*

**Bridgemary, Nobes Avenue** Built in 1956 and owned by the Brickwood group of companies, it served a large housing estate on the outskirts of Gosport.
HoW 1960; 1974 C&L.

**Droxford, High Street** A 15th century building – a coaching inn, the stables of which have been converted into a bar. A former landlady claims to have seen the friendly ghost 'the Lady in Grey' upstairs by an old linen box, she may have been killed by her son when he pushed her downstairs.

Lord Nelson (1758-1805) is said to have called here when visiting a relative of Lady Hamilton who lived nearby at a house called Fir Hill.

On 3rd June 1944 General Dwight Eisenhower, Winston Churchill, General Smuts and McKenzie King[3] held a secret meeting in Churchill's private train while parked in a wooded siding at Droxford Station[4] finalising the 'Operation Overlord', D Day plans for the Allied Invasion of north-west France, due to take place on 5th June but which was delayed by bad weather for twenty four hours. HRH Prince Andrew stopped here one evening in 1981.

2006 the Droxford Cricket Club met regularly in the courtyard at the rear of the inn.

Popular with walkers, this hotel stood on the seventy mile Wayfarers Walk.

By 2008 it had closed to become a private house.

JSt, JD and Mrs. E. Forbes-Robertson – local historian. 1844 P; 1855 K; 1875 PO; 1895-1939 K.

## Fareham

**Boar Hunt Road** 1939 listed in Kelly's Directory with Albert G. Cudmore as licensee, he was also listed under the White Horse at North Wallington[5].

**High Street** 1793 listed as trading with Richard Coaker as licensee, John Trimbee took over in 1823.

By 1828 Ann Trimbee was recorded as landlady.

The house was demolished during the 19th century by Stephen Barney, a local benefactor, to build 'Lysses' which had become an hotel by the 21st century.

**119 West Street** The buildings in this street were renumbered during the 19th centuries; in 1857 John Nicholson was licensee of the White Horse at No. 44. The Nicholson family continued to run the house until 1927, by which time the address was No. 119.

Between 1931 and 1970 the pub continued to trade.

1914 presumably the house had its own brewery as the Brewery History Society recorded it as 'White Horse Brewery'. It was demolished.

AHa; Landlord of 'The White Horse', North Wallington; Westbury Manor Museum; Malcolm Lowe – local historian and Mrs. Baxendale – local historian. 1974 C&L.

**Froxfield Green** This house was said to be the haunt of highwaymen.

1855 and 1867 listed in the Post Office directory as the 'Troopers Inn' PO.

The original entrance from the road was shut although customers continued to be served at the same bar but from its opposite side. The smaller parlour had originally been the smithy. Landlord.

**Godshill, Isle of Wight** 1855 K; 1875 PO.

**Liphook** 1855 K; 1875 PO; 1895 K.

**Marchwood**, Hythe Road / Main Road A Grade II listed 18th century building of two storeys, built of painted brick with an old tile roof, hipped at one end; there is a two-storey bay window and a modern extension to the right.

Trading since 1765, it was at one time a coaching inn.

It is said that the Admiral of the Fleet would stop here before travelling to Bucklers Hard[6] on the Beaulieu River, some twenty miles away, where he would be rowed out to his ship. It was on this tidal river that the Men of War were built from timber of the nearby New Forest. The stumps of the cradles for three ships commanded by Nelson can be seen in the mud at low tide – they were Agamemnon, Swiftsure and Euryalus.

By 2007 alterations to the windows of the inn had taken place and the address had been changed to Main Road. Landlord and RD. 1855 K; 1875 Southampton PO; 1895 Southampton – 1939 K; 1974 C&L.

**Milford-on-Sea, 16 Keyhaven Road** An 18th century building with a brick façade, which for years was the last building at the eastern end of the village. By 2006 the house had been painted white with black window surrounds and shutters.

1875 Lymington PO; 1895 Lymington – 1939 K.

**Netley Marsh, Ringwood Road** The inn is situated on the corner with Woodlands Road.

1847 listed as trading in the Post Office Directory with John Heard as landlord.

JB and SRL. 1855 Eling K; 1875 Eling PO; 1895-1939 Woodlands K; 1974 C&L.

**Newport, Isle of Wight** 1923/4-1830 P; 1875 PO; 1895-1915 K.

**Otterbourne, Main Road** 'White Horse Tavern'. 1801, 11th May – the *Salisbury Journal* published a notice about a horse which had been stolen from the White Horse at Otterbourne near Winchester.

1855 K; 1875 PO; 1895-1939 K.

**Portsea, St. George's Square** 1823-1844 P.

**Portsmouth, High Street** 1823-1844 P; 1855 K.

**White Horse Street** 1823 P.

**Priors Dean / Dean Priors** The highest pub in Hampshire. A Jacobean building fronting on to the old Portsmouth to Alton Road, which was a coaching inn with blacksmith and meat trader.

It became isolated when the road was straightened; the sign at the cross roads was taken down and never replaced, it is known as 'The Pub with No Name'. There are several White Horse Inns in this area.

c. 1960 the smithy was converted into a bar; cars are parked on the site of the old road.

2006 voted by the Hampshire Life Magazine as the Best Country Pub in the county.

2008 the two bars have log fires, old tables, clocks and agrarian implements.

KSGH and Landlord. 1915-1939 K.

**Romsey, Cornmarket / Market Place** This town was known for the manufacture of shalloons[7], sacking and paper, also tanning, malting and corn milling.

The inn has a Georgian façade concealing the original features of this coaching inn.

Traditionally it was a resting place for visitors to the famous 10th century Benedictine abbey, the stone tunnel-shaped cellars are thought to date from this era; they were renovated when the Tudor building was erected, but as there is a stream running between the inn and the abbey, it is thought unlikely that there was ever a tunnel running between them which has been claimed by landlords over the years.

From the *Salisbury Journal* in 1775:

> 24th July – a sale took place at the White Horse Romsey.
> 28th August – there had been a theft from Mrs. Dixon at White Horse.
> 20th November – the White Hors (sic.) was advertised for sale as a reputable and good accustomed house as any on the western road, late in occupation of Mrs. Mary Dixon: and household goods.

In 1776:

> 24th June, White Horse Romsey For Sale. In the market place. 35 beds, 6 rooms for entertaining company, stabling for near 50 horses and room for four carriages.
>
> 22nd July, a report having been propagated that the White Horse Romsey is almost shut up, and that there are not proper accommodations for the entertainment of Ladies and Gentlemen passing through this town, Elizabeth Dixon (sister to the late Mary Dixon) cannot acquit herself to the public without informing them such reports are groundless, and that the said house is and will be kept open by her, in the same manner as it was by her late sister, where all those who please to honour her with their company may depend on her utmost endeavours to give satisfaction, and the grateful acknowledgements of their very humble servant.

1780 John Dickman had received no wages for five years while working as a post boy for Charles Sibley of the White Horse, twenty years later he claimed a settlement.

1784, 17th October – the *Salisbury Journal* published the following notice:

> The White Horse, Romsey. Ann Sibly widow thanks customers to herself and her late husband Charles. Hopes for their continued custom – a principle surely still felt some 200 years later by the present hotelier.

1787 Isaac Newman was hired by Mrs. Sibley of the White Horse Inn for £1 10s. per year, he later used this information to claim relief on the parish.

In 1800 from the *Sherborne Mercury*:

> 7th July – an advertisement reads '[This inn is] equal to any on the road for the accommodation of noblemen's and other families . . . commodious stabling for upwards of an hundred horses. Bath, Bristol, Portsmouth, Gosport, Southampton, and Salisbury coaches morning and evening. Hearse and mourning-coaches.'

1810 historian Dr. John Latham recorded the White Horse as the principle inn out of the twenty trading in the town. Customers could hire post chaises and horses.

1811 the Paving Commission were persuading house owners to rebuild their frontages if they jutted over the footway – £3 13s. was paid for surveying fees at the White Horse; it is thought that 19th century alterations may have taken place as a result.

1827 complaints were made about the street lamp outside the inn not being alight either at 10 pm nor at 11 pm in October; in 1841 the licensee was paid £1 per annum for lighting the lamp at the same times as the public lamps; in 1851 the licensee was paid 10 shillings every six months for the light to be discontinued.

1879 Vincent Newman, the licensee, died – two years later his five year old daughter died; within six weeks Mrs. Newman provided a banquet in the Orangery of Broadlands for the Master Builders of Portsmouth.

By 2003 a series of Tudor wall paintings had been discovered under layers of paint and paper – one being a black Tudor rose with a curved geometric design and another of Tudor dragons which was part of a larger painting. Features of the inn include oak beamed rooms, original painted plasterwork and traces of an Elizabethan mummers' gallery. Drinks and meals are served in the original stable yard where there was once a cock pit. A ghost haunts the cellars and has also been seen outside the Elizabethan lounge. JB and RO.

## Southampton

1668 'Halfe peny token issued by George Freeman at Ye Whit in Southampton – *a horse ambling*'. GCW No.191.

***90 Bedford Place*** Known to have been trading from 1861 to c.1934 when it closed to be redeveloped as a petrol station.

***King Street*** 1851 listed in the local street directory.

***8 Nelson Street*** 1878 a beerhouse and shop shown on the 'Drink Map of Southampton'; 1890 became the 'Nelson Tavern'; later became a private residence. 1960s demolished when the area was redeveloped.

***17 Winton Street*** 1846 the house was shown on the city map, by c.1880 it had become a private residence.

**Southsea, 51 Southsea Terrace** Originally a 'Court House' where Royal retainers stayed on visits to Portsmouth. Here the Prince of Wales (1841-1910) – the future King Edward VII – would leave his mistress, the actress Lillie Langtry, before taking ship to the Isle of Wight to visit his mother Queen Victoria at Osborne House. During refurbishment of this house an old chest was discovered in a secret alcove behind a loft wall; it was found to contain Edwardian clothing, including lingerie of high quality – presumed to have belonged to Lillie Langtry; also glass plate photographs of this most beautiful woman known as 'The Jersey Lily'. Born Emilie Charlotte le Breton (1853-1929), the daughter of the Dean of Jersey, she was twenty one when she married Edward Langtry.

1988 the house was renamed 'Langtrys'.

2006 the beer garden was one of the largest in Portsmouth; a three-player circular pool or snooker table with three pockets and a circular cushion in the middle was a feature of the house. WN June 1989 and L&M April 1996. 1875 PO; 1974 C&L.

**Thruxton, Mullins Pond** A thatched building – the date 1451 is carved into the beam over the fireplace.

1939 Kelly's Trade Directory lists Mrs. Flora Mullins as landlady, whose family gave their name to Mullins Pond.

1985 the inn closed, to re-open four years later.

Landlord. 1823-1830 Andover P; 1855 East Cholderton – 1915 East Cholderton K.

**Wallington, 44 North Wallington**
Situated on the corner with Drift Road.

At one time there were four public houses in the village, the White Horse was thought to be the earliest building.

1792 listed as trading with Robert Cooker as licensee.

1939 the licensee, Albert Cudmore was also listed in Kelly's Directory under the White Horse in Boar Hunt Road[8], Fareham.

c.1998 the inn sign was changed and depicted the pure white thoroughbred race horse Mont Blanc II; he also appears on the sign of the White Horse at Alton. Previously the sign has shown a placid grey horse in a meadow and at sometime the Invicta prancing horse.

The pub was still trading in 2008.

AHa and Westbury Manor Museum. 1823-1830 P; 1895 K.

**West Meon** 1875 PO; 1895-1915 K.

**Whitwell, Isle of Wight** Originally 'White Well', a mediaeval place of pilgrimage. The building dates from c.1454 and still has some of the original walls, it is one of the oldest buildings on the island. It was formerly listed as Chiddles Tenement.

1547 Thomas Jacob, a publican, was recorded as brewing ale in Whitwell.

It was owned by the Worsley family of Appuldercombe from the 16th – mid 19th century.

1780 Sir Richard Worsley leased the house with an orchard and garden amounting to three stiches[9] of land to a brewer, Richard Cooke – who it is thought may have turned the property into an inn, when perhaps the stabling was added.

1804 recorded as trading under White Horse Alehouse.

At one time the village was a centre for the smuggling trade.

1845 the inn was sold by the trustees of Baroness de Graffenreid Villars, who were Ernestine Marie d'Houdetat, Baroness Fleming of Paris, Effingham Lindsay of Ermatingen, Switzerland and Major-General Charles Henry Oakey, Counsel to H.M. Embassy in Paris.

Before the new railway was opened in 1897, Irish railway workers frequented the inn from early morning.

The Morris family were recorded as licensees from c.1850 until c.1914.

c.1970 the stables were converted into the lower bar and further alterations were carried out in 1982.

1986, 26th September – the house, which was thatched, was badly burned. It was rebuilt but with a tiled roof, to become known locally as "The Kicker".

'Inns and Ale Bonchurch to Chale' – Ventnor and District Local History Society.

D. Jacobs – artist of Mini Pub Signs, his ancestor was the brewer, Thomas Jacob of Whitwell. IoW JER/WA/32/57; IoW JER/WA/31/67; IoW JER/WA/31/72. 1895 Ventnor - 1939 K.

## Winchester

*High Street* Built on the site of several mediaeval tenements owned by the Abbot and Convent of Hyde and the Hospital of St. John, Winchester. Amongst these was the Church of St. Peter Whitbread – it is of interest that this name may have been taken from the commemoration of Pentecost, or Whitsunday (White Sunday) when those being baptised wore white; Whitsun-ale was the important Church-ale celebrated with much revelry and 'horse bread' was the staple diet in mediaeval monastic institutions.

1543, 24th January – 'Quitclaim: William Mede of Salisbury, Butcher . . . [owned] a house called the White Horse in High Street [Winchester] held under lease from disolved monastery of Hyde, England'.

The first two proprieters were elected Mayor in 1573 and 1579 respectively. Situated on the corner of High Street and Jewry Street, it was one of the largest alehouses in the city. The inn was probably destroyed during the Civil War. B. Carpenter Turner in her 'Winchester' published in 1980, recorded that 'in the main High Street there were at least twelve inns' in c.1784. By the 20th century the site had become shops and offices. WCM Vol. III Estates Hyde Street No. 1321, BCT and HRO.

*113 High Street* 1823-1844 listed in Pigot's Directory as the White Horse Cellar. The words 'WHITE HORSE INN' are built into the knapped flint and brick above the first floor frontage of this 19th century building. 1875 PO; 1895-1935 K.

*26 / 29 Canon Street* 1880 first listed as 'White Horse' in the Winchester Directory. It was previously listed under the name of Edward Boyes, a beer retailer living at this address.
Flourishing in 1973 but by 1986 it had been rebuilt as a private house.

*St. Cross, 55 St. Cross Road* 1878, 31st December – the house, formerly known as The Horse and Jockey, was sold by the Master, Bretheren and Trustees of St. Cross Hospital to Mr. W. D. Forder. It is not known when it became The White Horse.

1931 the house was sold to the Winchester Brewery Co. Ltd, a subsidiary of Marston's.

1997, December - this listed building was sold to become a private house. HRO and MTE. 1830-1844 P; 1878 No.3 Front Street, St.Cross – 1920 K.

One hundred and twenty six inns are listed in an undated dogeril called 'A Ramble Through the Ancient City'. Ending with the words 'God Save the King', it probably dates from before 1837 or after 1902. The three 'White Horses' included are likely to have been 113 High Street, Canon Street and St. Cross.

**Worting, 390 Worting Road** Listed as trading under White Horse in 1811, also in Pigot's Directories of 1828, 1832 and 1844. The house used to be a stopping place for the hop wagons as they travelled from Farnham to Weyhill Fair. After the pub closed, the building became a bakehouse and then the village shop. There are oak beams in the cellar. Ms. Margaret Porter – LHS Basingstoke. 1830 Basingstoke P.

## Notes

1. This name is from the Hebrew, see Proverbs 31 v.1. Sometimes used as a Christian name in the 17th Century.

2. In the 16th Century one took luck or chance as to what may have been cooked in the pot.

3. Prime Minister of Canada.

4. The station became a private house.

5. c.1980 the M27 motorway was built through Boar Hunt Road, the southern section running through Wallington was renamed North Wallington Road.

6. The shipyward here was established in 1750, it became redundant in 1800 when steel ships were being built.

7. A light twill-weave woollen fabric used chiefly for coat linings. Col.

8. c.1980 the M27 motorway was built through Boar Hunt Road, the southern section running through Wallington was renamed North Wallington Road.

9. A ridge of land, specially a strip of ploughed land between two water furrows. A form of measurement peculiar to the Isle of Wight. VDL & SOE.

# HEREFORD &
# WORCESTER

*'The cattle of Herefordshire are considered superior to any other breeds in the island; they are mostly of a reddish-brown colour, with white faces . . .*

*The manufacture of salt at Droitwich has existed from time immemorial . . . The carpets of Kidderminster have long held a high reputation . . . Worcester has extensive manufactures of gloves and porcelain.'*
JB.

**Bromsgrove** 1822-1842 P; 1851 K.

**Bromyard, 1 Cruxwell Street** 1822 listed as trading by Pigot. 1889, 18th January – a Trust Deed of The Cheltenham Original Brewery Company Limited gives details of this property consisting of a dwelling house and garden known as 'White Horse Inn'. Standing in about two acres and situated at the junction of Cruxwell (or Sheep Street) with Milvern Lane, it included yard buildings, orchard or meadow land. A workshop, barn or toft[1] of buildings were occupied by Messrs. Colley & Son Coachbuilders.

Spring of 1998 – the pub was closed; the building was converted into three flats named White Horse Mews. One of the 2 bedroom flats was for sale at £99,995.

J. Emberton – local historian. 1822-1844 P; 1868 S; 1900-1917 K.

**Great Malvern, 3-3A Worcester Road** 1872 listed as trading in the Post Office Directory.

1940 White Horse Hotel was listed in elly's Directory.

c.1965 an Estate Agent took over the premises.

Landlord of The Unicorn, Bellevue Terrace. 1912 K.

**Hereford, 32 Union Street** The inn was situated on the corner of Union Street, making a very narrow entrance to Gaol Street.

A report by The Royal Commission on Historic Monuments in 1928 dated this half-timbered building as early 17th century, but the large square panelling and curved braces were considered to be of an earlier date.

1777 it is said to have been the home of Thomas Gomond, Mayor of Hereford.

By 1816 it was known as the 'Saint Catherine Wheel', possibly taking on the name of another inn in the city, the 'Catherine Wheel', which had recently been demolished.

1844 listed as The White Horse in Pigot's Directory.

1858 the inn was shown on the city map.

1931 a plan of the three storied 'White Horse Hotel' shows it to have had three bars and a smoke room with eight rooms above; in 1937 the owners surrendered the licence. It was occupied by the military during the Second World War and demolished soon after to widen Gaol Street. RS. 1900-1917 K.

**Kemerton** 1940 K.

**Kidderminster** 1828/9-1842 P.

**Kington, 19 High Street** The inn building dating from the 17ᵗʰ century, is two-storied, timber framed and plastered. It lies behind the shops and is connected to the High Street by an alleyway.

1796 a notice was published reporting that Thomas Jones of the White Horse Inn had been born.

1840 John Price was listed as publican.

c.1865 the inn sign, which hung over the street, depicted a white carthorse with the name of the publican underneath – Charles Price.

By 1992 the premises had become a bookmaker's office and in 2002 a private house – the bracket for the inn sign was still there. 1822/3-1844 P; 1868 S; 1900-1917 K.

**Leominster** 1868 S; 1900 K.

**Pershore, Church Row** 1919 the White Horse Brewery adjoining the hotel was closed. The hotel continued to trade and was still flourishing in July 2008 with bars and an Italian restaurant.

BHS and J. Emberton – local historian. 1828 P; 1851-1940 K.

**Worcester, Church Road** 1822-1842 P; 1851 K.

### Notes

1. A 'toft' is a homestead with attached arable land.

# HERTFORDSHIRE

*'The Venerable Bede (673-735 AD) explains the name of*
*'Hartford' as a ford where harts and stags cross regularly – or*
*where they gather . . .*

*The produce in wheat, barley, and oats is very considerable, and*
*of the best quality . . . The chief trade of the county is corn and*
*malt, which is conveyed to London by means of the navigation*
*of the river Lea.'*
JB.

**Baldock, 1 Station Road** 'Old White Horse'. 1667 a halfpenny token was issued by Timothy Marley of Baldock – *a horse* – GCW No. 12.

The present inn is the tap of a much larger White Horse Inn which was a busy coach halt and the rendezvous of many hunting and sporting personalities. A parson, James Woodforde (1740-1803), dined here on 21st May 1776, quoting from his diary:

'From Dunstable we went to Baldock thro' Hitchin about 20 miles from Dunstable and there we dined at the White Horse kept by one Kendall . . . A great many soldiers, Dragoons at Baldock today. From Baldock we went on to Royston about 10 miles, there we baited our Horses and selves a little time at the Crown . . . '
*[Such a pity he did not use the White Horse!]*

A passenger using an 1829 Edward Mogg edition of 'Paterson's Roads', amused himself by plotting on the flysheet the towns on his route from Holton-cum-Beckering in Lincolnshire to London, and the mileage between the inns which supplied post horses. The White Horse at Baldock is recorded as being 8 miles from the inn at 'Bigleswade' and six miles from the Swan at Stevenage.

c.1864 the main building was closed and became a school; shortly afterwards most of the school was burned down. Part of this original building which survived the fire became a private house, standing on the corner with White horse Street. The pub was still trading in 2008. WBJ and JW. 1826 L; 1845 PO.

**Bishop's Stortford** 1845 PO; 1937 K.

**Bourne End, London Road** A listed building dating from c.1756. The original bar at the front of the building is haunted by 'a nice old man wearing 1920 clothes

– both of us were much aware of him. I personally have not seen the ghost of a dog, but mediums have seen it in this same area'.

1984 the house was purchased by the brewers McMullen & Sons.

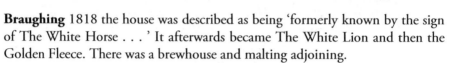

Landlady and McM. 1756 WBJ.

**Braughing** 1818 the house was described as being 'formerly known by the sign of The White Horse . . . ' It afterwards became The White Lion and then the Golden Fleece. There was a brewhouse and malting adjoining.

1877 the pub was bought by a local brewer. WBJ.

**Buckland** 1727 possibly known as the White Horse.

1761 the house was sold to a brewer.

1852 the landlord was recorded as a blacksmith and wheelwright. WBJ.

**Burnham Green, 1 White Horse Lane** A tavern stood here beside the Green prior to the Civil War in 1642. The present inn takes its name from the white stallion belonging to a staunch Royalist, Pennyfather, of Welches Farm in White Horse Lane, who was attacked and beheaded by the Roundheads. When the soldiers tried to steal the horse he "proved as resentful of the Roundheads as his dead master – and his show of spirit cost him his head as well". The ghost of the headless white horse is still said to haunt the village.

According to a local inhabitant, when she was a child in the 1920s, "the local superstition was that to see the white horse without his head – as depicted on the original inn sign – was lucky, but should the phantom figure have a head on, it spelled disaster in some form".

The inn sign was returned after repainting showing a charming white hunter with a head – to the horror of the village!

A local rhyme about beating the parish bounds includes these words:

*"At Burnham Green, just where the recent dead*
*rides a white horse they say without a head"*

1806 was listed as 'White Horse'.

2001, May - irreparably damaged by fire.

2003, 27th February – the premises reopened with a new sign showing a horse with its head. Licensee, WBJ and DWH.

**Bushey, 10 Sparrows Herne** The original building dated from the 16th century.

1694 described as a messuage, though there is little doubt that it was already an inn.

1747 the house was sold, but not until 1756 was it recorded as The White Horse when the property was bought by Henry Goodwin a brewer, it continued in the ownership of the same family until 1897.

1841 the landlord was listed as a victualler, by 1851 he was described as a publican. In this same year on the night of the National Census, there were five lodgers staying at the inn.

1863 it was reported that complaints had been made by the occupants of the adjoining cottage when 'peeping Toms' drilled holes through the wall.

1881 Joseph Parry was listed as a dairyman and publican, his cows probably grazed the 29 acres attached to the inn at this time.

1890 the house was listed as a principal inn.

c.1898 the inn was described as being situated at Clay Hill in the Parish of Bushey; it had yards, stabling, sheds and gardens.

The Bushey Fire Brigade used the outbuildings for their equipment until 1921, although their horses were often out at pasture elsewhere. During this period the adjoining cottage had become a butcher's shop.

1976 the inn was completely refurbished.

By 2006 the pub had closed after a period of decline, the buildings to become private houses. WBJ, BW, and BMT.

**Chorleywood, Rickmansworth Road** 1778 George Wilson paid tithe on land 'opposite the White Horse in Chorley-Hamlet'.
1800-1838 WBJ; 1845 Rickmansworth PO.

**Corey's Mill** 'The Mill'. 1769 Nathan Cooke, a blacksmith, mortgaged his house at Corey's Mill which had lately been known as The Harrow – but had been renamed The White Horse.

1960 it was still trading as White Horse but by 1987 it had become 'The Mill'. The building probably dates from the 17th century, a drawing dated c.1840 shows it as a single storey cottage with a dormer window and two tall chimneys. HCR and WBJ.

**Flamstead End** 1567 possibly known as the 'White Horse'. 1662 described as being 'formerly an inn called Copt Hall, afterwards The Crown and lately The White Horse'.

1701-c.1725 the house was listed under The Ship, but by 1756 it was again listed as The White Horse.

1784 the premises were bought by a brewer. WBJ. 1845 Cheshunt PO.

**Frogmore, 86 Park Street** 1689 recorded as trading.

1722 the house was offered for sale with 45 acres of land - although this could have referred to the 'White Horse' in St. Peter's Street, St. Albans.

c.1780 a land tax was paid and by 1787 it was occupied by a victualler John Seabrook, who stayed here until 1806.

c.1995 the name was changed to 'The Overdraft'. HCR and WBJ.

**Green Tye** 1846 this was possibly a beerhouse but its history is unknown. WBJ and HCR.

**Harpenden** *See Hatching Green.*

**Hatching Green** 1650 John Collins, a brewer, acquired a house at Hatching Green, possibly the 17th Century present-day inn, now much altered. The White Horse Grounds shown on a map of 1623 were situated a mile and a half away near the River Ver, these may have belonged to the White Horse Inn.

By 2007 the rear of the building had been rebuilt, the original building being mainly untouched with its beams and fireplaces. WBJ.

**Hertford, 33 Castle Street** This Grade II listed building was originally the bakehouse for the 10th century Hertford Castle, the inn itself dates back to the 14th century.

The timber framed cottages in this small back street are thought to have been the former outbuildings of the Castle.

By 1838, with the addition of an adjoining cottage, it had become a beerhouse. AAT. 1845 PO.

**Hertingfordbury, Hertingfordbury Road** The original White Horse was a smaller building situated elsewhere in this village; in 1643 it was known as 'The Nag's Head near the Mill'.

1775 it was listed as 'formerly The Nags Head, then The Golden Lion and now The White Horse'.

1808 the licence and inn sign were transferred to Place Farm, a larger half-timbered manor house dating from c.1557 with a fine Jacobean staircase and the original oak beams; a Georgian frontage had been added. The driveway allowed for the arrival of the Reading to Cambridge coaches.

1810 an estate survey mentioned a 'magistrates room' within the building.

A lady recalls her memory of the White Horse during her Edwardian childhood – 'flocks of bicyclists would arrive to refresh themselves, and on summer Sundays a harpist would play outside the door to entertain them, having carried his harp all the way from Hertford'. Hotel brochure.

**High Cross, Great Cambridge Road** Originally two cottages dating back to 1608, this listed building was converted into an inn a century later in c.1700. Landlord. 1756 WBJ; 1845 PO.

## Hitchin

*22 High Street* By 1998 the premises had become Boots Opticians.

*Back Street* The pub was recorded as flourishing in 1867. HCR. 1845 PO; 1898 K; 1902 HCR.

**Kimpton, 22 High Street** 1837 described as a cottage and garden in a row of 18th century cottages owned by a brewer of Hitchin.

By 26th December 1863 it had become two cottages listed as a 'freehold messuage, brew house, grocer's shop, beer house called The White Horse'.

1896, 6th October – the premises were purchased by McMullen & Sons Ltd., brewers.

WBJ and McMullen & Sons Ltd.

**Leverstock Green, Leverstock Green Road** The 1840 Tithe survey did not show the White Horse, there was only a barn on the site in a garden belonging to the cottage which later became the shop and Post Office.

The original mid 19th century beerhouse was listed as trading in 1843, it stood by the village green.

1851 the census recorded James Travell as licensee, he was also a dealer in pigs.

1875 the premises were mentioned in a will.

1901 the census listed Henry Pedley as licensee and carpenter and joiner.

c.1967 when the new village centre was built, the White Horse was demolished and rebuilt by Flowers Breweries on the corner of Green Lane.

1998, September – the name was changed to The Litten Tree. WBJ and WN July 1967.

**London Colney, 182 High Street** 1770 the house was bought by the brewer Thomas Clutterbuck.

1773 the vestry accounts of St. Peter's parish, St. Albans, records 'Dinner at the White Horse on going a processioning £3.4.10.' WBJ. 1845 PO.

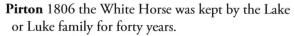

**Pirton** 1806 the White Horse was kept by the Lake or Luke family for forty years.

1840 the property changed hands and was described as 'a small farm with granaries, barns, stabling, cowhouse and piggeries attached'.

1901 the house was rebuilt and set back from its former position. WBJ. 1867-1902 HCR.

**Potters Bar, High Street** 1845 listed as trading under White Horse in the Post Office Directory.

c.1988 the name was changed to 'The Cask and Stillage'. 1886 PO; 1914-1933 K.

**Redbourn, 45 and 49 High Street** The town is situated on the Roman Watling Street, becoming the main route from London to the north – by c.1830 more than eighty coaches passed through.

A 16th century two-storied timber framed coaching inn on the east side of the High Street adjacent to The Bull.

By 1783 a tax form records Mrs. Gould as owner and the Harris family as occupiers of the White Horse Inn until 1826.

1839 Joseph Liley is listed as landlord; three major coaches stopped here daily but this service is thought to have ceased soon after 1840 due to the coming of the railways.

By 1851 the census shows that Joseph Liley, aged 65, had moved and was described as a retired innkeeper. William Lord, aged 27, a wheelwright and victualler was running part of the old inn as a pub with no resident staff or guests. The wheelwright business, thought to have been in the inn yard, would also have declined when the railway came.

1866 the Post Office Directory lists George Lines as landlord but by 1882-1890 Dr. Joseph Ayre had changed the name to 'Brockway' – his middle name.

1892 the house, described as an old fashioned residence, was bought by Dr. Ayre.

1953 described by the Department of the Environment as a house and shop with a 'good early 18th century front in brown brick with red brick dressings and gauged brick lintels. Plain tiled roof. Plastered floor band and wooden eaves band . . . the northernmost window is over a broad segmental-headed coach entrance with 16th century ceiling and wall timbers exposed which can still be seen. Interior has exposed wall and ceiling timbers'. There is an 18th century hipped roof extension to the rear and a 17th to 18th century painted brick and timber frame former stable block, now a shop.

By December 2003 the building was occupied by a firm of solicitors, and is now known as Crown House. Dora B. Wode – member of the Harris family and 'Redbourn's History'.

**Royston, Melbourn Street** 1834 The malting business is extensive here, and the trade in corn considerable. This building adjoined 'The Cave House' which was the entrance to a cavern, 26ft. in diameter and about 40 ft. in height, containing a series of rude and profane carvings. JB. 1823/4-1939 P; 1845 2 listed PO; 1889-1916 K.

**St. Albans, St. Peter's Street.**
1744, 14th May – the *Bath Journal* reported:

'Yesterday a set of brown horses with a coach, belonging to Mr. Williams at the White Horse in Piccadilly ran from thence to St. Albans (which is 20 miles) in two hours and five minutes, for a considerable sum of money, which was much sooner than the time allowed'. BJ.

1962 the house was closed. WBJ.

There appears to be some confusion between this White Horse and that in the nearby village of Frogmore.

**Shenley, 37 London Road** 1756 recorded as having stabling for two horses. The Courts of the manor of Shenley were held here. 1800 bought by a brewer from Hatfield and until 1840 the house was kept by John Hare and his widow.

**Stevenage, 79 Albert Street** 1823 A Labourers' Friendly Society is recorded as meeting here. 1998 and 2008 the pub was still trading. WBJ.

**Tea Green, Stoney Lane** Thought to have been trading as the Red Lion and owned in 1864 by Randall MacDonnell, a blacksmith.

1873 Samuel MacDonnell was listed as licensee.

1877 The White Horse, an alehouse, was sold to J.W. Green, a brewer of Luton.

c.1940 the house was granted a full licence.

1998 still trading. WBJ.

**Tring** 1883 K.

**Wareside** 1787 Thomas Barnard, a victualler moved into this house which was known as Old Hall.

By 1811 when Thomas died, there was a carpenter's shop on the premises and the name had been changed to the White Horse; Thomas bequeathed the house to his wife Mary in trust for their children. Mary later married William Parker who became the licensee.

1860 Thomas Barnard's children sold the public house to a Hertford brewer, Thomas Driver Medcalf. WBJ. 1845 PO.

**Watford** 1845 PO.

**Welwyn, 30 Mill Lane** James Bell in his gazetteer of 1834 described the village as consisting of one principal street with well built houses through which runs the great road from London to York – Welwyn was a day's coach ride from London.

This coaching inn was recorded as trading in 1742, it was originally a cottage standing on its own, in a narrow lane – once known as White Horse Street - just off the old Great North Road. Having stabling and fields for horses, it accommodated the servants and grooms of those staying at the two hotels – The White Hart and the Wellington.

Ghosts have been seen, one of a coachman who was killed in a brawl outside the building.

The present listed building is possibly of Tudor origin; two stables remain of the original stable block at the end of the garden. Landlord and WBJ.

**Weston, Fore Street** 1845 PO; trading 1870-1889. 1933 this house was taken over by Mr Alfred Matthews; the building was in a dilapidated state, with spittoons standing on the sawdusted floor – these were thrown out by Mrs. Matthews much to the disapproval of the locals.

By 1968 the old inn had been replaced by a new building set back from the original site - the inn sign, hanging on a smart new pole in front.

In 1997 it became a private house.

The information was supplied by Janet Shepherd, the daughter of Mr and Mrs Matthews; she was born in the original building.

# KENT

*"By far the most civilised inhabitants are those living in Kent."*
*"Longe sunt humanissimi qui Cantium incolunt"*

Julius Caesar's 'Gallic Wars' 5.14 written in 44 BC.

*"Kent in the Commentaries Cæsar writ,*
*Is term'd the civil'st place of all this isle:*
*Sweet is the country, because full of riches;*
*The people liberal, valiant, active, wealthy . . . "*

So said William Shakespeare's Lord Say to Dick the Butcher in
Smithfield Market *Scene 7 of King Henry VI Act IV, 1595.*

**Bearsted, The Green** This building stands beside the village green. Since 1938
The Bearsted Rifle Club has leased an adjoining building from the pub owners.
1970, 5th May – Whitbread purchased the premises from Fremlins Limited.
1981, May – the pub became a Beefeater Steak House. WP. 1851-1866 PO; 1870-1915 K.

**Bilsington** The public bar is early Georgian, the saloon bar
and pool room on either side being early Victorian.
In the brick walls of the pool room the edges of
wooden planks are still visible in the brickwork;
these would have been placed by the foreman to
show the required height and number of courses
that the bricklayer had to lay in a day – at this period
many workmen would have been illiterate. Licensee. 1859-1866 PO; 1870-1938 K.

**Borstal, 86 Borstal Road** 1850 PO; 1862 Rochester K; 1887-1938 K.

**Boughton-under-Blean** The pilgrims travelling with Geoffrey Chaucer (1343-
1400) tell of seeing the Canon's yeoman's grey horse at 'Boghton under Blee':

*"At Boughton-under-Blean, we saw a hack*
*Come galloping up . . .*
*The hackney horse he rode was dappled grey*
*And sweating hard; it was a sight to see.*
*It must have galloped miles it seemed to me".*

1833, 31st May – The Courtnay Riots took place in nearby Bossenden Wood.
These were instigated by the self-styled 'Sir William Courtenay, Earl of Devon' –

in reality the thirty five year old madman, John Nichols Thom/ Tom (10th November 1799-1838) born at the White Horse at St. Columb Major in Cornwall, the son of Cornishman, William Tom, who at one time also kept The Ship Inn. Having made his way across southern England to Kent, he was nominated Member of Parliament for Canterbury in 1832, but a year later he was convicted of perjury and placed in Barming[1] Heath Lunatic Asylum. Four years later, upon his release he lived in a farmhouse between Canterbury and Faversham and declared himself to be the Messiah, King of Jerusalem, showing stigmata; he attracted disciples comprising of about 100 local illiterate farm workers around Boughton-under-Blean. The culmination came on 31st May in Blean Woods when being served with a warrant by the unarmed village Constable Mears, Courtenay immediately opened fire and shot him dead. Alerted by three terrified survivors of this attack, a detachment of troops from the Forty-fifth Regiment in the charge of Lieutenant Bennett were sent from Canterbury. The Lieutenant called upon Courtenay and his men to surrender but Courtenay raised his pistol and shot him. One volley from the troops massacred thirteen of Courtenay's men, he himself was mortally wounded.

1838 The White Horse Inn was chosen as the location for the several inquests on the thirteen people who were killed in the Courtenay Riots.

1984 this 16th century inn was partially gutted by fire but has since been restored. CT, DNB, MDM, SBG, HLD, CGH. Mrs. Glanville – Recorder of the St. Columb Old Cornwall Society. 1845-1866 Faversham PO; 1870-1938 K.

*Dear Reader, should you wish to know of Mad Tom's early life, please turn to the White Horse Inn at St. Columb Major, Cornwall.*

**Bridge, 53 High Street** Said to be over 400 years old; a staging post for coaches where horses were changed for the London to Dover run. A famous patron, Christopher Marlowe (1564-1593) the Elizabethan dramatist and the son of a Canterbury shoe maker, became involved in a brawl and was thrown out into the street, landing in the gutter. During renovations an open hearth was found behind a Victorian fireplace; old coins were also discovered under the floor boards.

1948 the inn sign was designed by Kathleen M. Claxton for Whitbread at Wateringbury Brewery. WN July 1965. 1851 PO; 1870-1938 K; 1952 and 1964 C&L.

## Canterbury

Over the years there have been at least five White Horse inns or taverns in this city. As the tavern at 36 High Street was described in 1450 as 'very old' – it would seem possible that Geoffrey Chaucer (1343-1400) may have had this house in mind when, in 1387 in his Canterbury Tales he wrote of the peregrinations of Hubert, the Friar thus:

> *"There was a Friar, a wanton one and merry,*
> *A Limiter, a very festive fellow . . .*
> *He knew the taverns well in every town*
> *And every innkeeper and barmaid too*
> *. . . nothing good can come*
> *Of dealings with the slum-and-gutter dwellers,*
> *But only with the rich and victual-sellers"*

The majority of the following details have been taken from a card index and its references compiled by a former City Surveyor, Mr. H.M. Enderby, held by the Canterbury Local Studies Library.

**36 High Street** In the parish of St. Mary Bredman – Westgate Ward. Situated on the corner with Jewry Lane, to become White Horse Lane.

1450 the Mayor, William Bennett, gave his tenements beside Jewry Lane to the city; they were 'the Tigre and the other the White Horse, very ancient both and known to belong unto the City this day'. 1640 Sumner p.181.

1466 'Grant of two houses in the parish of St. Mary Bredman boundaries Kings highway north, Jewry lane west, house of Prior and Convent of Christ Church belonging to their sacristy east, House of Mayor and commonality south, Endorsed now the White Horse'.

1586-1598 corner of Jewry Lane lease from Mayor and commonality. The landlord, Thomas Graddell and his wife, Dorothy the younger sister of Christopher Marlowe (1564-1593), were recorded in a collection of parchment rolls of pleas in the Court of Record as being involved in disorders and fights in The White Horse Tavern; Dorothy herself was involved in one of these, whilst Thomas Graddell had accused the landlord of The Lion, on the other side of the High Street of stealing five of his 'pottes'.

1617 just before Christmas, Thomas Graddell bought two 'turkie cockes' for 3 shillings each from Mr. See of Herne, as he was unable or unwilling to pay, Mr. See sued him for the 6 shillings. Dr. W. Urry – City Archivist's Report 17.12.1958.

1625 the inn is referred to under this date by Dr. W. Urry in his article published by the Kentish Gazette of 9th January 1950.

1664 a 'halfe peny' token issued by Iarvise Willmatt of Canterbury – a horse. GCW No. 83.

1693 listed as a billet for four soldiers.

1792-1796 licences were issued.

1800 'White Horse Lane'. This street name was designated for the first time 'from the public house of that name adjoining'. Bunce's records.

1803 listed in the Kent Directory – later listed under 5 High Street.

1806 Probably known as 'The Painters Arms' to become the 'General Havelock'.

It is surprising that this name was chosen, as General Sir Henry Havelock was known for his strong religious convictions – the drunkenness of other regiments were constantly contrasted with the sobriety of his troops; an army Temperence Medal was named in his honour. He married a Baptist missionary's daughter; he died in 1857 at Lucknow in India. The site of this house in White Horse Lane became the Salvation Army Citadel.

1882-1889 listed as White Horse, it closed the following year.

**15 High Street** in the Parish of All Saints. Formerly 'The Chequers' but from 1841-1848 licensed as the White Horse when it became 'The Bell'. By 1889 it had been reinstated as 'White Horse'. CSD.

**Longport Ward** Possibly situated in Broad Street. 1692 Listed; 1693 listed as a billet for four 'souldiers'.

**Northgate Ward** 1692 listed as trading.

**St. Dunstan's Street** 'White Horse Without Westgate'. 1687-1692 listed as trading. 1693 listed as a billet for four soldiers. MGH.

**Chatham, 80 Chatham Hill** 1845 2 listed PO; 1862 K; 1866 PO; 1870-1938 K.

**Chilham, The Square** c.1422 originally built as a thatched farm dwelling, later to become an alehouse used for festivals held at the adjacent church. This building dating from the 12th century is supposed to be haunted by the ghost of one Elija Frog whose body was found in the building during the 16th century. An inglenook fireplace dated c.1460, with a Lancastrian rose carved at the end of the mantelbeam, was discovered during building work in 1956; from when it was first opened up the landlady reported seeing a grey-haired man in a long black gown standing with his hands behind him in front of the old fireplace – "we have seen him quite distinctly out of the corner of our eyes as we were setting up the buffet". Two male skeletons were also found under the back kitchen floor at a depth of 2ft; it is thought they could be soldiers killed at the battle of Chilham which was fought between Wat Tyler's followers and the local militia, having been pursued after their sacking Canterbury during The Peasants Revolt of June 1381. *Wat Tyler was beheaded in London later that year.*

1790 the premises became a public house.

The inn sign, designed by K.M. Doyle of Wateringbury Brewery in 1949 is based on the George Stubbs's portrait of 'A Grey Hunter with a Groom and a Greyhound at Creswell Crags' painted c.1762-4 which hangs in the Tate Gallery. Licensee, WN – May 1967 and C&L Whitbread Inn-Signia – No. 50 of the 4th series 1953. 1938 K.

**Cranbrook, Carriers Road** Between 1845 and 1847 the house was listed as an hotel, an inn and a tavern. 1862 K; 1866 PO; 1870-1938 K.

### Dartford
The following information was provided by C. Baker of the Dartford Borough Museum.

***Lowfield Street*** 1688 Deed No. 157 records a lease or mortgage of the White Horse between Thomas Jackson and John Round.

1690, April – a probate inventory of Francis Eldredge, an innholder of Dartford referred to a set of ninepins and bowls etc. in the yard; the name of the inn is not mentioned but on 23rd April the Dartford Assessment of the 'Releife of the Poore' records 'Lowfield Widow Eldred (sic.) per White Horse £5 sixpence in the pound 2/6d'.

1707 listed in the Alehouse Keepers Recognizances – the exact site is not known.

***Dartford Road*** Between 1867 and 1905 the White Horse was listed in Kelly's Directory of Kent and Snowden's and Perry's Directories of Dartford.

By 1998 the house had become an Indian takeaway.

### Deal

***Middle Street*** 1679-1757 this was two tenements, the site was then leased as one tenement and outhouse – later known as The White Horse, the rent being 2 shillings yearly.

***1 Upper Queen Street*** 1828 listed as White Horse. 1852 Richard Orrick of 13 Middle Street advertised:

> 'White Horse Inn near the railway terminus. Licensed to hire open and closed carriages. Commodious Gigs, Clarences, Landaus. On the most reasonable terms. Post and Saddle Horses to let. Orders sent to The White Horse or to No. 13 Middle Street will meet with immediate attention'.

1951 the house was bought by a brewer.
1956 listed as trading in the Deal Directory.
Lambeth Palace Library, reference TA 233/1-20. 1859 PO; 1862-1895 K.

**Dover, St. James' Street** Situated on the corner with Hubert Passage leading to the castle; in 1365 the building was the residence of the Warden of St. James' Church part of the priory of St. James; at this date the sea washed to the foot of the Church.

1574 became the residence of the Ale Taster to the Port of Dover. 1630 Came under the control of the Constable of Dover Castle.

An undated ¼d token was issued by Robert Gallant at The White Horse in Dover – *a horse prancing GCW No. 214.* This would apply to the years between 1648 and 1679, the only period when trade tokens were used.

1652 a licence was granted to sell 'Ales and Cyders' on the premises.

1653 due to the licensee having retrieved the nameboard of the American ship, 'The City of Edinburgh' which had been wrecked off the Straights of Dover, the house was registered under this name.

In August 1818 the pub once again was listed as the White Horse and became the meeting place of actors and players of The Dover Theatre.

Inquests were held here, mainly of bodies found in the sea, one being Henry Palmer, a clerk from East India House drowned in 1826.

For ten years from 1890 horse-drawn coaches ran every day – except Sunday – to and from St. Margarets-at-Cliffe, some six miles away, negotiating the steep Castle Hill; the inn opened at 5am for dockers and shift workers.

1952 workmen were carrying out alterations to the house when they discovered a program from the Dover Theatre dated 1809 for Harlequin and Mother Goose.

Licensee. 1845-1859 PO; 1862 K; 1866 PO; 1870-1915 K.

**Tower Hamlets** 1842-1846 DP.

**Buckland** 1847-1852 DP.

**Edenbridge, 64 High Street** A timber framed building which was faced with brick in the 18th century. The date 1574 and two pairs of hand shears are carved into the first floor bressumer[2], this not only records the union of Rafe Shears and a member of the Holmden family whose father gave them the land for building, but also the date a Royal Charter was granted by Queen Elizabeth I stating that the building should be used only as an inn.

It was a staging post for coaches with stabling for about six horses and a blacksmith on the premises. It was here where the daily coach to Westerham started in the early 19th century, the coach would wait in Westerham for the

London coach to arrive and then bring passengers on to Edenbridge. At some time it was known as The White Horse and Market Hotel.

By 2006 the bar and restaurant had been refurbished and the stables had been converted to living accommodation; in this year the Grade 2 listed inn was recorded in the CAMRA Good Beer Guide. 1859 PO; 1862 K; 1866 PO; 1870-1938 K.

**Eythorne, Church Hill** The original building is said to date from the early 18th century. Landlord. 1845 PO and 1887-1899 K were listed under Dover; 1915-1938 K.

**Faversham, 99 West Street** Situated on the corner with North Street. The White Horse was known to have been trading by 1800.

c.1895 a photograph shows the White Horse as a small brick building.

c.1900 the ground-floor frontage was renewed by the brewers, Shepherd Neame; the name boards at roof level were removed some years later.

1910 William Henry Wash was listed as landlord until 1922.

By 1998 the house had become the offices of United Brewery Shepherd Neame, the building now being larger and bearing little resemblance to the original, possibly taking in some of the adjacent building, the rest of which was demolished to widen the road.

2008 the words 'The White Horse' could just be made out beneath the rendering between the windows on the first floor. FLH. 1845-1938 PO.

**Preston Street** 1829 P.

**Finglesham.** 1859 PO and 1862 K were listed under Sandwich. 1939 the signboard was designed by the artist Violet Rutter at Wateringbury, which was reproduced in 1951 by Whitbread Brewery as one of their inn sign miniatures.

c.1983 the building was burnt down. C&L.

**Gravesend, High Street** fl.1796-1801 when this house ceased to trade.

1802, 20th September – a document described 'a piece of land whereon stood the White Horse in the upper part of the High Street . . . '

1845 listed as trading in the High Street; a description dated 1834 gives a picture of the street:

> 'At the termination of High-street a spacious quay has been erected for the convenience of landing goods and passengers, which in consequence of the steam boats which leave London every morning and return in the evening, are very numerous. Every passenger at landing or embarking, pays here one penny of pier-dues. The easy

distance from London, the salubrity of the air, the bathing machines recently erected, the rich gardens, and the fine views of the Thames and Medway afforded by the walks in the neighbourhood, render this a fruitful source of revenue, the visitors through the summer season averaging considerably above 100,000.'

ERG and JB. 1845-1859 2 listed PO; 1899 PO.

**Harbledown.** 'About a mile from the west gate of Canterbury in the wood of Blean'.
1692, 21st November – this White Horse is listed in the earliest complete list of Canterbury's inns and taverns. JB and MGH.

**Harvel** The eye-catching words 'WHITE HORSE' can still be seen on what is now a private house; the words are set on an ornate terra cotta background – framed in brickwork. 1899 Meopham PO; 1915-1938 K.

**Hawkinge** *See Uphill.*

**Hawkhurst** 1870 two White Horse public houses were listed as trading.

**Headcorn, North Street** The original thatched farm dwelling dating from 1560 went through many vicissitudes. In 1682 it was leased to a grazier and wheelwright. In 1714 he was left the property, part of the will reading ' . . . occasioning of his dethe shale inherite his messuage at Hedcorne with its lande and staybles theretoe adjouninge.'
1757 the property was sold to a horse dealer of Headcorn, well known for his fine stock of horses, who by 1772 had become an auctioneer at Ashford.
In 1812 the property was bought by a fly proprietor, whose small one-horse carriages were let out for hire.
1839 the property was thought to have been enlarged when it became a grocer's shop.
1878, 10th August it was registered as a beerhouse to sell ales and ciders, but bore no sign.
By 1924 it was fully licensed as the 'White Horse'. Licensee. 1938 K.

**Herne Bay, 13 Avenue Road** c.1995 the pub closed to become a private house. 1866 PO; 1887-1938 K; 1950 C&L.

**Luddesdown** A small village, six miles from Rochester.
1847 listed as trading with Richard Luggett as a victualler at the White Horse. BDK.

**Maidstone, 46 London Road** 1838 recorded in Courage Eastern MSS. The original barn has been modernised to accommodate the bar and kitchen.

1845 the Post Office listed four White Horse inns under Maidstone, probably referring to those in outlying villages – possibly Bearsted, Otham, or Sandway.
KCC. 1847 BDK; 1866 PO; L&M 27.5.96.

**Minster-in-Thanet, Church Street** According to legend the Black Prince (1330-1376) rode his royal white steed to Minster Abbey in order to attend Mass. His black war horse was entrusted to the owner of the White Horse Inn, Mr. Cobb, who was famed for his knowledge of horse flesh and his skill in brewing fine ales.

Nuns of the Order of St. Benedict have run a working farm at the Abbey since 1937.

1974 ceased trading and became a private house.
MDM and WA. 1851-1859 PO; 1862-1887 K; 1899 PO; 1915-1938 K.

**Otham, White Horse Lane** 1604 first listed as the White Horse Inn. By 1851 it had closed and become a private house. 1853 a new inn was built on a different site but still in White Horse Lane and named The Horse Shoes, which may have drawn attention to the blacksmith, who used to work in the barn which still stands in the garden. Three years later in 1856 it was renamed White Horse. Three times a week a carrier had his regular run from this inn to Maidstone. Landlady. 1851-1866 PO; 1870-1938 K.

**Rainham, 95 High Street** 1732 the White Horse was mentioned in the Vestry Minutes of St. Margaret's Church.

c.1895 Mr. Holloway, the lamplighter, took a party of some fifteen ladies – all dressed in their best finery with feather hats – in his four wheeled wagonette and pair from outside the White Horse Inn to the beauty spot at Sutton Valance, some 12 miles away. This place is described in James Bell's Gazetteer of 1834:

> 'The remains of the strong castle occupy the brow of a hill, and the crumbling rubbish, overgrown with ivy and trees has a most picturesque appearance'.

A photograph dated c.1895 of the wagonette with the ladies is held by Gillingham Library. Maybe the outing was to celebrate Queen Victoria's Diamond Jubilee on 22nd June 1897.

The local Police Station was at one time situated to the rear of the public house.
MALS and RAB. 1829 P; 1845 Chatham – 1866 PO; 1898 K.

**St. Lawrence, 16 High Street** c.1885 a photograph of the High Street shows the original site of the White Horse with the shop of 'Gibbs Saddler' adjacent. A South Eastern Railway notice board by the door advertises 'Cheap Fast Trains Every Weekday To London . . . in 2 hours . . . From Margate, Ramsgate . . . ' *Compared with the time taken by the steam trains, the modern trains are only fifteen minutes faster – some taking well over the two hours.*

c.1960 the inn was rebuilt in modern style on a new site in the High Street.

MLS. 1845-1859 PO; 1862-1870 K – these dates were listed under Ramsgate; 1887-1938 K.

**Sandway** In 1678 the owner of nearby Chilston Park, Richard Douglas, Esq., built this house as a lodge for his bailiff. In 1825 the building was sold to John Collins who in December 1830 opened the premises as a licenced beerhouse – but it bore no sign. *The 1830 Beer Act enabled any house-holder of good character to obtain a licence to sell beer from a dwelling by merely paying two guineas to the department of excise.* 1838 a wine and spirit licence was granted and the house was registered as White Horse. Licensee. 1870 Lenham – 1915 K.

**Seal** 1979, 15th December a photograph in the Sevenoaks Chronicle showed the White Horse Inn since when it has become a private house.

SoL and the Licensee of the 'Kentish Yeoman'.

1829 P; 1859-1866 Sevenoaks PO; 1870-1887 K; 1899 PO.

**Sheerness** There were two listed in the High Street.

**Old White Horse** 1824 listed as trading in Pigot's Directory.

**Lower White Horse** 1829 P; 1845 PO.

**Blue Town** 1840 listed as trading in Pigot's Directory but by 1863 it had ceased to trade.

**Sittingbourne, 84 Charlotte Street** Originally two cottages, but by 1908 it had become an alehouse with a Bottle and Jug, a tiny Public Bar and a Saloon. Each bar had an open fireplace, one of which still remains. During the Second World War the beer was served from large barrels behind the bar. According to Mrs. Barbara Cullen who was the landlady from 1967 to 1989, "in the seventies and eighties Christmas time at the pub became quite famous locally. We used to be packed to the doors with people standing outside drinking, the festivities always ended on Boxing Day with a tug of war nearly the length of Charlotte Street" *some 350 yards long*. An extension incorporating the old sitting room has been built onto the back of the building.

1854-1859 PO; 1862 K; 1866 PO; 1870-1938 K – these dates were listed under Milton Regis.

**Smarden** Commanding a central position in the village at the junction of Water Lane, Cage Lane and the High Street. The iron bracket can still be seen on the wall which carried the sign of this beerhouse. At one time it was a bakery with living quarters at the rear of the building, and was still trading as such in 1960. It became an antique shop – with a private house to the rear.

By 1998 the shop front displayed a collection of pottery.

Smarden Post Mistress. 1840 P; 1851 PO.

**Sundridge, 105 Main Road** A 17th century coaching inn situated in the centre of the village.

c.1905 the inn was enlarged and refurbished.

c.2004 the building was modernised. The inn sign shows the legs of a galloping white horse. Licensee. 1845-1859 Sevenoaks PO; 1870-1938 K.

**Thanington, 83 Wincheap Street** A parish in the lower half hundred of West-Gate, Canterbury. According to the brewers Mackeson & Co. this was a 17th century house.

1870 the street is shown as Wincheap in the Canterbury Street Directory.

1878 listed as 75 Wincheap Street.

1888 shown in the Canterbury Street Directory.

1949 listed as a beerhouse.

1969 the pub was closed, the site being redeveloped as private residences.

JB and PME.

**Tonbridge, High Street** 1896, 11th January the premises were bought by Frederick Leney & Sons Ltd. from Augustus Leney. *See Upper Stoke.*

1939 the ledger shows the house was licensed to sell cask beer, bottled beer, cider, spirits and wines – 'nips and flags' are mentioned – nips are a measure of spirits equal to one sixth of a gill, flags is short for flagon.

1951 HoW (Summer). 1862-1938 K.

**Tovil** The Georgian building on the brow of Farleigh Hill had no hanging sign board but a painting of a grey horse flush to the wall between the first floor windows. In 1882 when Henry Jury was a beer retailer at the White Horse, Tovil had six public houses. The site is overlooked by modern-day Tesco. IH and SDM.

**Tunbridge Wells, Goods Station Road** Renamed 'The Victoria Tavern'. 1938 K.

**Uphill, 7 Canterbury Road** c.1698 the original buildings are shown as an inn with stabling on a local map.

1802 listed as trading.

1891, 31st August – the house was advertised for sale as the owner had become bankrupt; the sale brochure described the property as 'A new-built messuage called The White Horse, with stables at Uphill, near South Hawkinge'. It was known as a cockfighting public house – in the centre of a floor of an upstairs room was arranged an oblong grass-plot on which the birds were to fight. The cocks selected to fight were clipped of their plumage and were armed with long, sharply-pointed steel spurs, fixed upon their legs. The battles were sharp and swift and were usually a horrible and repulsive sight. *Cockfighting was made illegal in 1847.*

Part of the original building was used as a separate dwelling, for storage, and even to lay out the dead when accidents occurred on White Horse Hill. By 1946 this part of the building had been demolished. The 'long room' of the inn was used for village functions before the village hall was built.

2004 the Hilltop Restaurant was opened with seating for sixty people.

FVL and HP. 1845-1866 listed under Folkestone by PO and K.

**Upper Stoke, The Street** Built as a private house and possibly owned by Augustus Leney.

1896, 17th January – the premises were bought by Frederick Leney and Sons Ltd. from Augustus Leney, Esq. 'with land, containing 2r. 29p$^3$.' *See Tonbridge.*

1866 Rochester PO; 1887-1938 K.

**Wierton** Known locally as the White Horse, but since 1920 licensed as The Red House. The original public house at the turn of the century was called White Horse and may at one time have been known as The Drum and Monkey which could have been a recruiting base for the Cornwallis Regiment. This building was burnt down; the present building dates from 1880.

c.1900 Mr. Wallis was the licensee. His daughter, Minnie Wallis remembers hop-picking time when people were always fighting outside the house and the gipsies "who used to trot their ponies along the way to sell them. London people used to come down and there used to be fights over the prices". Her mother recalled that "women would take out their hat pins and shed blood – that was the origin of the White Horse starting to be called the Red House". She also recalled that farm labourers would knock on the door at five o'clock in the morning for a pint of beer to take to work. She describes the house as very old, with one bar "where people had their drinks and took their snuff – to see them taking out their little snuff boxes used to amuse us children". The family kept a horse, cows, pigs, geese, chickens and turkeys. "My brother and other sister had to help serve in the grocer's shop attached to the pub".

1998 the landlord still uses the White Horse logo for his campsite. DT.

**Westerham** 1845 PO.

**Willesborough Lees**, Kennington Road. Shepherds travelling to Ashford Market from the Hastingleigh area of the North Downs would stop at the White Horse in this village. Their sheep were penned within the wall to the rear of the inn where there was a small building for the shepherds; their animals were watered at the lake opposite the inn.

Ashford Market was held in the lower High Street from 1784; the minutes of a public meeting held on 26th February 1856 describe the market as 'scattered from one end of the town to the other . . . stock can scarcely obtain standing. The holding of it in the centre of the town is totally at variance with the sanitary and other regulations now enforced by the legislature'.

Former Licensee and AM. 1870-1938 K.

**Notes**

1. It is thought that the word 'barmy' (insane) could have been derived from Barming Heath Lunatic Asylum in Kent.

2. A beam extending horizontally over a large opening and sustaining the whole superstructure of the wall. SOE.

3. Rod (5½ yards), pole or perch is 5.0292 metres.

# LANCASHIRE

*The Queen is by long custom toasted as 'The Queen, Duke of Lancaster' in the County Palatine and at gatherings of Lancastrians throughout the world.*
The Royal Encyclopaedia 1991.

*'The Lancashire breed of horses were used by Mr Robert Bakewell (1725-1795), as the basis of his improvements . . . they are universally preferred to oxen for the purposes of husbandry. The horses most in request are strong trace horses; stout, compact saddle horses; and a light middle-sized breed for mail-coach and post-horses.*

*The principal manufacture of Lancashire is that of cotton goods in all its branches.'*
JB.

**Bacup, 22 Yorkshire Street** Originally these premises had been cottages; several copper coins dated 1775 were found by workmen carrying out repairs in 1881. At one time the following rhyme was painted on a sign over the door:

*'The White Horse will beat the Buck,*
*And make the Angel fly;*
*The Staggering Man upon its back,*
*Knock out the Dragon's eye'.*

1868 listed as a Beershop.
1869, 11th September – the landlord, Luke Hollis, was bound over to keep the peace after being charged with an assault. The 1871 Census showed there were two lodgers, one a basket maker and the other a winder at the coal pit.

Between 1872 and 1875 the landlord, Abraham Clegg, was fined for selling beer during prohibited hours and twice for permitting drunkenness, as a result he was disqualified from keeping a beerhouse for five years. Both on 3rd August 1878 and a year later, a license for selling beer on the premises was refused.

At the time of the 1881 census the house was unoccupied as repairs were being carried out when the coins were found. Detailed specifications were given of two large ground floor rooms – the Tap Room and Commercial Room, each measuring 15 ft square; above these were two equally large bedrooms together with two smaller bedrooms and an attic. The 33 ft long passage from Yorkshire Street led through to the back yard where there were two entrances. This 6 ft wide yard ran the width of the house and had 'two water closets and a urinal. One of the closets is private'. A report gives the general sanitary conditions as 'good'; facilities for the police supervision as 'good'; the state of repair as 'fair'. The premises became a tied house and a license was granted.

By 1896 the average takings per week were £15 16s 10d, and the trade was described as working class.

1912, 10th June – the licence of the White Horse Inn was objected to as not necessary, and on 21st December the inn was 'closed on account of Compensation' and the contents were sold by auction, including 'about 1,000 cigars – Key West, Bull Dog etc.'

Key West is one of the small islands of Florida Keys which stretch across the Straits of Florida to Cuba. Cigars were introduced into the USA in 1762 after the capture of Havana.

It is more likely that these cigars were cheroots – tobacco from India, rolled with the ends cut square which may have been known as 'Bull Dogs'.

By 1984 the premises had become a fish and chip shop.

The above information is taken from reports in the *Bacup Times* and *Bacup and Rossendale News*. JDs. Vol. 2.

**Bamber Bridge** 1898-1913 K.

**Barton, 913 Garstang Road** This inn is situated beside the main highway running between Preston and Garstang – now the A6 – and a hostelry has been on the site since c.1600, the White Horse being its successor.

By 1900 this hamlet was known as Whitehorse Village, taking its name from the inn.

The car park for the Hotel is on the site of a row of houses which were demolished in c.1940-50.

Mr Wooler – Local Historian, 'Broughton Roundabout' and Landlady. 1855 S; 1881-1924 K. These dates were listed under Myerscough until 1974 when the boundary changes took place.

**Blackburn, North Gate** 1818 The Blackburn Directory lists the licensee of the White Horse as Mary Crook, a victualler.

**Burnley, 17-19 Hammerton Street** 'New' or 'Little' White Horse Inn.
1760 John and James Hargreaves were listed as landlords.
By 2004 the house had been renamed 'Orange House' at 13-19 Hammerton Street.

**Clitheroe, York Street** The old Scandinavian name for this town was thought to have been derived from *klera* – a song thrush.

1660 records show that the parish church wardens held some of their meetings here. 1789 listed as trading. The original White Horse inn was situated in Church Street next door to St. Mary's Church. It was demolished when York Street was built in c.1820 and was rebuilt on the present site, possibly because most of the passing custom used the new road, which had become the main route out of Clitheroe.

At one time it was known locally as 'The Pit'.
Listed as an hotel in trade directories from 1822 to 1924.
These premises used to be a coach house.

OxN and Landlady.

**Colne, 31 Church Street** 1761 is the earliest known record of the White Horse Inn on a map from the Estate Book of Thomas Clayton of Carr Hall. Situated opposite the gates of St. Bartholomew's Church close to the market cross, the town's well and the stone stocks, it was a white washed house with narrow gothic windows; a passage, wide enough for a horse and cart, ran through the middle of the building to the backyard. Mrs Cryer, in her 'Memories of Colne' written in 1910 for the *Colne and Nelson Times* wrote:

> *"How often I have stood as a girl and gazed through that passage*
> *across the sunlit fields and the hills beyond . . . "*

To the rear of the property a narrow strip of land was shown marked 'Burrans' which was a name usually given to sites showing evidence of ancient occupation.

A small field in Colne Lane was called Tenter Croft where finished pieces of cloth were spread out in the form of tents so that they could be naturally dried out. Walker Mill – an ancient fulling[1] mill – on Colne Water near the weir which supplied it with water. By 1834 cotton calicoes[2] and dimities[3] had taken the place of wool.

Mrs Cryer also described wedding parties:

> *"How often have I stood on the road between the toll bar and the*
> *Parsonage and watched them go by.*
> *First came a fiddler, decorated with many-coloured ribbons, and*
> *playing a merry tune. After him came the bridal party . . . when it*
> *was over, and the fiddler struck up his merry tune again as they crossed*
> *the road and entered the White Horse Inn, after one of the Grammar*
> *School boys had said the homily, and received a piece of silver from the*
> *bridegroom and a smile from the blushing bride, while the old bells rang*
> *out a parting peal."*

Her father remembered churchwardens going into the inn with their long poles and forcing the customers out:

> *"Old Lancaster, the sexton, coming out with a stick, pounce upon a crowd of*
> *rebellious boys and lug a couple of them into church, while the others danced*
> *round like wild Indians and shouted "Let 'em go, you old beggar, let 'em go".*

The street was narrow and the inn stood so close to the gutter that pedestrians had to wait each time a cart passed by – on one occasion a child was crushed against the inn wall and was killed. On another occasion the wife of a customer came to take him home and was told he had already left; he was discovered lying at the bottom of an open grave in the churchyard.

1886, October – the building was demolished, having lost its licence.

CRL and Mr C. Wooler. 1822/3-1834 P; 1848-1855 S; 1858 PO; 1881 K.

**Crawshawbooth, 211 Goodshaw Lane** The name is derived from 'Crows wood with dairy-farm' dating from the 14th century. OxN.

Originally built as a farmhouse by woollen merchant in c.1805.

1818 listed as 'White Hart'.

Serving the tiny community of Goodshaw, the inn was occasionally used as a mortuary.

1823-1841 listed as trading.

In the 1851 census the landlord, Thomas Ashworth, was listed as 'innkeeper and farmer of 10 acres' and had a lodger who was a coal miner.

1854 James Hoyle is listed as landlord. The 1871 census shows him to have six children; in 1872 he died and the license passed to his widow, Jane. She remarried in 1878 to Samuel Smith who took the license.

1877, 10th August – 'The house was first lit by gas and the village turned out to see the event'.

1879 listed as White Horse.

1897 an advertisement read 'To be Let, the White Horse Inn, Goodshaw Chapel'; the valuation in 1898 rose from £20 to £25 10s. In 1904 it was again advertised To be Let, and in 1906 offered 'with immediate possession, the fully licensed house, the White Horse Inn, Goodshaw'.

1907, 9th November – Mr J. McDonald Beattie was granted permission to make certain alterations to the house. 'At present customers have to go through the kitchen to get to the back. It was proposed to remove the position of the bar, to where the present bar parlour was situated, and to make a passage through. It was also proposed to widen the entrance and to improve the sanitary arrangements'. These alterations may have included the large banqueting room on the upper floor, which is mentioned in a 1958 report.

1909 The landlord, Amos Foulds, died leaving his widow, Esther, to run the house. On 9th June of the same year an advertisement was placed in the *Rossendale Express*:

'To Let, with early possession, fully licensed house, the White Horse
Inn, Goodshaw Chapel. Apply Masseys Burnley Brewery Ltd.'

1915 the landlord, Herbert Stevenson, was charged in May 1916 'with permitting drunkenness, and a breach of the Liquor Traffic Order. Fined £2 on first charge and £5 on second, with costs of 2 gns. in each case'; in January 1917 he was summoned for supplying liquor during prohibited hours, and for permitting it to be taken from his premises – his case was dismissed.

From 1919 the Hargreaves family ran the inn until 1952; in December 1943 'Hannah Hargreaves, of the White Horse Inn, was fined £8 15s. including costs, for selling drink by agent *[possibly her barmaid]* during non-permitted hours'; and in 1952 Janet Hargreaves was summoned on four charges of supplying intoxicating liquor during non-permitted hours' – her case was dismissed.

1958 a provisional license was granted for the inn with the banqueting room on the upper floor. This large room was later converted into three bedrooms.

1959, 5th January – the inn was closed; it was put up for sale with ten acres, this included a field with a barn opposite the house which was bought by 'the Red Indian'; he eventually died in Canada and the land was claimed by the Council for building.

The inn became a private house – The Old White Horse.

1976 the property was sold to become a Bed and Breakfast. The stone arched cellar and the curved bar have been retained.

The above information refers to 'Goodshaw' and has been taken from reports in the *Bacup Times, Bacup and Rossendale News*, and *Rossendale Express* which have been quoted by J. Davies, in 'Bacup's Hotels, Inns, Taverns and Beerhouses'. Vol. 2.

Mr C. Wooler – local historian.

**Edenfield** c.1760 a fulling mill was built at nearby Dearden Clough, those employed here would have regularly visited the White Horse.

1770 listed as trading under White Horse, sometime later it was renamed the Horse and Jockey.

**Edgworth, 2 - 4 Bury Road** Originally known as Edg'th Moor.

A coaching inn standing on a Roman road, the coach route ran from Bury, through Tottington, 'over the tops' to Blackburn.

The barn and original cobbled stabling which consisted of five stalls with hay racks and rings were still in use at the end of the 20th century – housing two horses.

No. 4, the hotel, has a banqueting room on the top floor.

Landlord. 1858 PO; 1881-1924 K.

**Goodshaw** *See Crawshawbooth.*

**Great Harwood / Cliff** 1776 the Surveyor of the Highway Accounts records:

'To ale at Cliff for people gathering stones in the Calder, 3s. 8d.'.

This ale was bought at the 'Grey Horse', which later became the 'White Horse' and finally 'Old Billy'. John Mercer[4] (1791-1866), the calico printer and chemist, lived here for a while, working on some of his experiments.

1903 the inn lost its license and closed. The building was converted into three dwellings and was eventually demolished.

**Haslingden, 11 Town Gate** The inn, said to date from the reign of Charles II (1630-1685), was three stories high and at one time was known as the 'Stoop[5] House', the name being derived from the two stoops or pillars in front of the doorway; later alterations destroyed many of the building's original characteristics.

1779 William Shaw, a skinner, was listed as landlord.

1788, 22nd April – John Wesley visited the town during the afternoon and preached to a large crowd from the 'horse-steps' – the mounting block in front of the Black Dog which adjoined the White Horse. He wrote of his journey from Padiham 'we went . . . through still more wonderful roads to Haslingden. They were sufficient to lame my horse and to shake my carriage to pieces. N.B. I will never attempt to travel these roads again till they are effectively mended.'

During the 18th century a murderer was captured here. He had murdered his lover, Ellen Strange, after the Haslingden Fair where he had seen her with another man; he became jealous and killed her on the moor near Robin Hood's Well – the scene of the crime is said to be marked by a cairn. He fled to Haslingden so quickly through Helmshire and Flaxmoss that his clogs struck fire at every stride; he was able to prove an alibi but was still hanged on Bull Hill near the scene of the murder. The gibbet was there until c.1810.

1799, 1st January – the Union Society of Women was begun by 39 members 'at the house of Thomas Shaw – the sign of the White Horse in Haslingden' – the admission fee was two shillings and sixpence. They promoted friendship, unity and true Christian charity and raised money for the sick and infirm and other charities. They made hard and fast rules concerning the money raised which included:

> ' . . . moneys should be paid out of the stock box provided the sickness or disability of body did not proceed from debauchery or any other manifest evil course of life.'

and:

> 'If any member happened to bear an illegitimate child no allowance was given, more than one illegitimate child meant expulsion from the Society.'

A number of table fines were imposed including:

Refusing to keep silent being thrice ordered by the stewards – 2d.
Fighting or striking another member – 5s. 0d.
Not providing gloves – 1s. 0d.

The White Horse was listed in trade lists between 1822 and 1898.
1907 the licence was taken away.
c.1930 the building was demolished due to 'clearance schemes'.
*The Haslington Borough News* (10th August 1967) and DH.

**Heath Charnock, 32-36 Chorley Road** Situated in a tiny village south of Chorley near to the Leeds-Liverpool Canal – both Pigot's Directory in 1822 and the Post Office Directory in 1858 listed this house under Chorley.

It is believed that a tunnel once connected the inn to a nearby monastery via the old cellars – still under the original building – whether this was an escape route, or for the monks to secretly buy beer, no one knows.

1866 listed as White Horse Inn with a brewer and victualler as landlord.

1955 Brewery Trust Deed lists four cottages as dwelling houses with this Hotel.

c.1986 two of the cottages were incorporated into the house, the other two being demolished to become a car park. Victorian tiles surround the fireplace and pictures of local scenes and events hang on the walls.

2006 the pub organised a twenty four hour pool marathon in aid of a local cancer charity. They also built a float for the annual Adlington Carnival, the main local fund raising event, unfortunately vandals slashed its tarpaulin cover and daubed paint everywhere. Landlady and MDA. 1881-1924 K.

**Helmshore, Helmshore Road** The name is derived from 'Steep cliff with a cattle-shed' dating from the 16th century.

Once an isolated travellers' inn but now situated on the corner of Free Lane (originally Stake Lane) and Holcombe Road at the southern end of the village.

1947 a Brewery Trust Deed shows the premises as a fully licensed hotel with a plot of land 2 rods 22perches and 3 rods 10 perches.

1953-54 the property, with the plot of land, was valued at £6,950.

1994 the building was described as probably late Victorian and of square-cut, flush pointed sandstone. OxN and AS. 1879 S; 1909-1924 Stakehill K.

**Lancaster, 8 Church Street** The name comes from 'Roman Fort on the River Lune' dating from the 11th century.

1822 Pigot's Directory listed the White Horse as trading.
The following historical records have been provided by the landlord:

The *Lancaster Guardian* of 8th October 1981 reported:

In 1702 this house in Lower Church Street was listed as an alehouse; this building had originally been three separate houses dating from the early 17th century.

1804 payment of rates was first recorded; by 1841 it was listed as having 'spirit vaults, etc.'

'February 1853, Mr George Reginald Kempe, late pupil of the organist of Exeter Cathedral, announces that he has been appointed organist at St. Thomas' Church, and that he is open to receive pupils'. In 1855 he married Agnes, the second daughter of Lieutenant Engleby of the 1st Dragoons, by 1857 they were living at 7 Church Street and he kept the 'White Horse Wine and Spirit Vaults'.

1876 it was listed as White Horse Yard, house and vaults.
1924 listed as trading.
By 1981 the owners were Yates and Jackson of Lancaster, the house was closed for redevelopment; a year later it reopened as 'The Stonewell Tavern'.
1992, March – Brenda Patricia Mulligan had become the landlady and by March 1995 it became 'Paddy Mulligan's Irish Bar', however it later returned to its former name 'The Stonewell Tavern'.
OxN.

**Ormskirk, Burscough Street** 1822/3-1834 P; 1854 MDA.

**Padiham, Burnley Road** c.1850 the building was erected on Plot 508 shown on the 1839 tithe map. The structure is given as:

' . . . punched face 'watershot' stone which is laid at an angle so the horizontal joints face outwards and help to prevent rain entering the building. Troughings[6] are of stone, heavy and moulded at the front but smaller and square at the back; the roof is of blue Welsh slate.'

1854 Three beer houses were listed.
1868 Two beer houses were listed.
1872 One beer house was listed.

1879 First mentioned as White Horse – eleven landlords were listed until 1952.

1931 the site was shown on the 1" Ordnance Survey Map as on the corner of Burnley Road and Sager Street.

1933 listed as trading under White Horse Inn.

1962 The Inn, together with Horne Street, Sager Street and the west side of Riley Street was declared a 'clearance area' and the inn was replaced by the present building. From a local Historian.

**Poulton-le-Fylde** Listed in Pigot's Directory of 1834, but the exact location of the inn is now unknown. According to the retired librarian "the old houses of the Fylde were small, low ceilinged, thatched dwellings, most of which were demolished. The few that I have seen had corrugated iron sheets over the old thatch".

**Preston, 1 Friargate** Situated opposite the Post Office, the White Horse Restaurant was large enough to accommodate funeral parties; it also had several other meeting rooms. The adjoining White Horse public house was listed in Cheapside, presumably because Friargate runs into Cheapside.

1881 the Preston Harriers, an athletics club, was started here.

c.1900 listed as a restaurant.

1822/3-1834 P; 1848-1855 2 listed S; 1858 PO; 1881 2 listed K. *It is possible that these dates may refer to nearby Bamber Bridge and Barton.*

**Rawtenstall, Bank Street** 1740 The White Horse was listed in the area of Tup Bridge and was known as a place of refreshment.

1838 the building was either extended or rebuilt at this time – in order to accommodate the coach passengers for the 'Queen Victoria' which ran between Manchester and Burnley.

1851 the census lists the landlord as an Innkeeper and Blacksmith.

1865, 26th August – the landlord was summoned for having unjust measures and was fined 2s. 6d; 1870 and 1873 two men were fined for refusing to leave the premises.

1877 the inn was sold to a brewer for £4,000.

In the census of 1881 the innkeeper, Ann Walkden a widow aged 32, had three children, had her father-in-law, a retired carrier aged 83, living with them. By 1883 it had become an hotel and was advertising:

> 'Wanted, at once, good General Servant, one used to hotel work preferred. Apply White Horse Hotel'.

1889, 9th March – 'On Thursday afternoon the sudden death occurred of the landlord of the White Horse Hotel who had been ailing for some time. He died in London where he had gone for an operation on his throat. The following year his widow, now the landlady, advertised 'Superior Wines, Spirits, Cigars, etc. Fine Ales. Stabling accommodation'. She had been a 22 year old barmaid at the White Horse before her late husband became the landlord – Mrs Schofield continued as landlady for the next twelve years until her death in January 1893. The following month the new landlord and his wife auctioned four large pieces of mahogany furniture, together with a Tudor bedstead, sewing machine and bedroom furniture. In June 1894 it was reported that the landlord's 34 year old wife had died leaving 'a husband and three children to mourn her loss'.

1896, 28th March – 'Harry Lord age 28 years, a respectably dressed young man describing himself as an engine driver of 14 Byron Street, Manchester – was charged on remand with stealing a horn snuff box valued at 20s'; he was committed to gaol for two months with hard labour.

1898 the old valuation was given as £78, it was revalued at £119. The following year builders were invited to submit tenders for alterations to the premises.

1904, 27th January – 'Margaret Ann Kelly, a middle aged woman of fairly good appearance, who was described as a tramping weaver, was charged with stealing about one gill of rum from a keg in the bar of the White Horse Hotel at about 8.30 am on Friday . . . the value of 1s 2d. Committed to prison for twenty one days hard labour. She was helpless and insensible for several hours'.

1907 Samuel Ashworth, the landlord, provided his own 'White Horse Inn' spirit bottles until his death in July 1911 when he was aged 33.

1915, June – Martha Ann Ashworth was charged with permitting children to be on licensed premises, she was fined £1 or 11 days.

1924, June – Arnold Wiseman was summoned in two instances for permitting gaming on licensed premises, he was fined £1 and 1 guinea costs in each case. In December 1925 the hotel was closed with £2,350 compensation. A month later an auction of the whole contents of the premises was advertised; amongst which there was an eight-day grandfather clock, an excellent satin walnut bedroom suite, a Sykes hydrometer book and glass, an upright pianoforte in walnut case, an antique mule chest and a 'Bex and flat worms' *[as yet unidentified]*.

For the next forty years the premises were used for many purposes, Jack Storey, a local comedian who held a concert party for many years, used the building as a motor accessory shop until it was demolished in 1967 for new road improvements. 1858 PO; 1881-1924 K.

The above information has been taken from reports in the *Bacup Times, Bacup and Rossendale News, Rossendale Free Press, Rossendale Division Gazette* and *Rossendale Express* which have been quoted by JDs. Vol. 2.

**Waterfoot, 136 Edgeside Lane** 1834 listed as trading in Pigot's Directory.

The 1861 census listed the landlord as a beer seller and his lodger as a woollen spinner.

1890, 3rd November – a labourer, John Rothwell, was charged with breaking and entering the White Horse Inn and stealing 24 half sovereigns, a purse and a watch – the police recovered everything with the exception of 4½d.

1898 the old valuation was given as £17, revalued at £21.

1899, 5th August – two labourers were charged with stealing £125 from the landlady 'between 4pm and 7pm on Tuesday last'.

1901, 14th August – the new landlord of the White Horse Inn 'gave his customers a treat when 65 sat down to a splendid knife and fork tea. A social evening was spent afterwards, when songs, games, dancing, etc., were enjoyed, and a pleasant and enjoyable evening was spent'. *The coronation of Edward VII had taken place on 9th August.*

1907, 3rd August – 'On Saturday, customers of the White Horse . . . had a picnic to Morecambe. A splendid motor carriage was chartered from the Burnley Motor Pleasure Company and the drive through such places as Heysham, Lancaster and Cockerham was greatly appreciated'.

1926, 10th July – the house was advertised To Let, due to the death of the landlord the previous year.

1951, 4th August – on two occasions the landlord was fined for supplying intoxicating liquor after hours.

1960 a license was granted for beer only.

1962, 27th October - the landlord was charged twice for aiding and abetting two men to consume liquor during non-permitted hours – fined £2.

1980 Walter Barlow retired as landlord after 18 years.

The above information has been taken from reports in the *Bacup Times, Bacup and Rossendale News, Rossendale Division Gazette* and *Rossendale Express* which have been quoted by JDs. Vol. 2. 1855 S.

**Whalley** 1855 S.

**Whitworth** 1851-1861 S.

**Wigan, 59 Standishgate** Described as a 16th century coaching inn at Worthington. Trade directories list this inn from 1822 to 1924 when it was a hotel.

By 1990 it had become 'Hartley's Emporium', a public house, but was still known as the White Horse. It is the oldest public house in Wigan and part of the originally stabling is now the Company's offices.

1997 a preservation order was put on an original mirror and the mosaic floor of the entrance hall which depicts a white horse.

## Notes

1. To full: to tread or beat (cloth) for the purpose of cleansing and thickening it with soap or fuller's earth. SOE.

2. A white or unbleached cotton fabric with no printed design. Col.

3. Light strong cotton fabrics with woven stripes or squares.

4. A calico-printer, chemist, bobbin-winder and hand-loom weaver. He experimented in dyeing and studied mathematics and chemistry. He discovered dyes suitable for printing calico in orange, yellow and bronze.

5. Archaic or northern British dialect meaning a pillar or post. Col.

6. Gutters.

# LEICESTERSHIRE

*'This is a celebrated hunting county, for which Melton-Mowbray is the grand rendezvous. Mules and asses are much used by the farmers.*

*There are some very large estates, and most of the land is held by freehold. Many persons farm their own land; the rented farms seldom contain more than 200 acres. The greater part of this county is in grass.'*
JB.

**Ashby-de-la-Zouch** 1822-1842 P; 1861 PO.

**Birstall, White Horse Lane / Front Street** Front Street is the oldest named street in the village, forming a 'T' junction with White Horse Lane which runs beside the River Soar; at this point the river was re-routed with a stretch of the Grand Union Canal. Early maps show a considerable amount of land was attached to the inn with access to it from Front Street at the side of the non conformist chapel.

1846 the innkeeper was listed as a victualler and a blacksmith.

By 2004 The Mulberry Tree public house was listed in White Horse Lane. 2008 the name was changed back to the White Horse.

LRO and Landlord. Trade lists: 1828-1941.

**Broughton Astley, 32 Station Road** 1839 the public house was first recorded under White Horse with reference to stables at the rear.

1911 Marstons, the brewers, bought the premises from William Holt Esq.; they were the owners until 1967. MTE. 1888-1941 K.

**Desford, Leicester Lane** 1799, December – the Gentlemen's Magazine recorded a death upon the road within a few yards of the White Horse, Leicester Forest.

There are mortgage and conveyance documents dated 1881, 1885 and 1918.

1922, 22nd August – the house was sold by Samuel William Moore to the Offilers Brewery Ltd. for £1,800. A conveyance describes The White Horse Inn as a tenement or public house with the yard, gardens and orchard, the brew house, barn, stables, sheds, and piggeries belonging together with two closes of land adjoining, the whole being an area of about nine acres two roods and twenty-two perches.

1960 the Offilers Brewery sold the premises to Arthur Clarke.

1967, November – the property was sold to Marston's, the brewers, for £22,500 and described as a messuage tenement or public house known as The White Horse Inn, sometimes known as The Old White Horse Inn, with the yard, garden, garage, sheds and car park fronting to the south-westerly side of Leicester Lane.

By 2006 after refurbishment the name returned to 'White Horse'.

MTE. 1888-1941 K.

**Leire, Main Street** Situated in the centre of the village, this was once a coaching inn. The entrance to the stable yard can still be traced from the original street pavement curb stones.

By 2008 the present building, a free house, had been extended into the yard.

Landlord. 1900-1941 K.

**Leicester, 27 Belgrave Gate / Gallowtree Gate** *Modern maps show Gallowtree Gate leading into Belgrave Gate, and the White Horse Inn was probably situated near the street junction.*

A stone coffin, said to have been King Richard III's (1452-1485), was used as a horse trough at this White Horse Inn. After his death at the Battle of Bosworth, Richard's body was returned to Leicester across a horse's back; it was displayed in the Church of St. Mary of the Annunciation and was buried in Grey Friar's chapel. Some years later Henry VII paid £10 towards the cost of a tomb which was desecrated during the dissolution of the monasteries. It is said that Richard's bones were exhumed and thrown into the River Soare where they were buried in the bank.

It is not now possible to verify that the White Horse Inn was the main coaching inn for London in 1782, but presumably a traveller having walked from Nottingham would have entered Leicester by Belgrave Gate – hence the following description by Mr. C.P. Moritz in his diary of that year:

'Towards evening I arrived in Leicester through a pleasant meadow by way of a footpath . . . I walked down a long street until I came to the inn from which the coaches start. Here I learned that a stage-coach for London would leave that evening, but all the inside seats were booked. I therefore booked a place on the outside of the coach as far as Northampton. The coach started from the courtyard of the inn. The inside passengers got in there, but the roof of the archway leading from the courtyard to the street was too low to permit passengers to be on the top of the coach without danger to their heads, so we 'outsiders' had to clamber up in the street.

My travelling companions on top of the coach were a farmer, a young man quite decently dressed, and a young Negro. Climbing was in itself at the risk of life and when I got on top I made straight for a corner where I could sit and take hold of a little handle on the side of the coach. I sat over the wheel and imagined I saw certain death before my eyes as soon as we set off. All I could do was to take a firmer grip of the handle and keep my balance. The coach rolled along the stony street at great speed and every now and then we were tossed into the air; it was a near wonder that I always landed back on the coach. This sort of thing happened whenever we went through a village or down a hill. Being continually in fear of my life finally became intolerable. I waited until we were going comparatively slowly and then crept from the top of the coach into the luggage-basket behind.

"In the basket you will be shaken to death!" exclaimed the Negro. And I took it for a mere figure of speech!

Going uphill everything was comfortable and I nearly went to sleep between the travelling-boxes and the parcels, but as soon as we started to go downhill, all the heavy luggage began to jump about. Everything came alive! I got so many hard knocks from them at every moment that I thought my end had come. Now I knew that the Negro had uttered no mere figure of speech. My cries for help were of no avail. I had to suffer this buffeting for nearly an hour until we began to go uphill again and, badly bruised and shaken, I crept back on to the roof of the coach and took up my former position.'

*A hundred years later the author's grandfather, John Otter Stephens (1832-1925), son of Rev. Richard Stephens (1785-1871), vicar of Belgrave-cum-Burstal, Leicester, travelled in an open coach by himself aged 11 to Winchester College, to become a scholar in January 1843.*

1797, June. Mr. Taylor who formerly kept the White Horse at Leicester died. The fact that the Landlord was mentioned by name when his death was reported in the Gentleman's Magazine must indicate the importance of this White Horse coaching inn.

By 1999 Nos. 25-27 Belgrave Gate had become the Co-operative Eye Centre.
Richard III Society, CPM and GM. Trade Lists: 1822-1941.

**Loughborough, Bedford Square** Situated next to Packhorse Lane. 1782 Mr. C.P. Moritz wrote in his diary:

*'I was now only four miles from Loughborough, an unseemly small town where I arrived late in the middle of the day and ordered a meal at the last inn on the way to Leicester. They received me at once as a gentleman – rather unexpectedly – and allowed me to sit in the parlour.'*

1783, January – mentioned in GM. Trade lists: 1828-1912. A rent ledger dated 1928 lists this house in Bedford Place – a century later this had become Bedford Street. In August 1994 the inn was renamed 'The Tap and Spile'. *The brand name of a chain of public houses – a spile being a wooden peg used to release carbon dioxide from a cask.*

**Lutterworth** 1828/9-1842 P; 1888 K.

**Market Harborough, 49 Nelson Street** Built in 1900 as the Admiral Nelson.
   1999 the sign was re-named the White Horse.
   2005 was re-instated as the Admiral Nelson. Landlord.

**Oadby, The Parade** 1828 listed as trading in Leister *(sic.)* Road. 1882 in May the 22 ton bell 'Great Paul', destined to hang in St. Paul's Cathedral, passed along the Leicester Road. 'This quaint dilatory procession' will have passed the White Horse Inn on its eleven day journey to London.

1886 the house was shown on the Ordnance Survey map and was still shown there in 1959 with the building extended.

Between 1941 and 1974 it was listed as at 64 Leicester Road.

c.1975 the pub and the old buildings in that part of Leicester Road were knocked down. A new pub was built on the same site and named The White Horse.

1996 listed in the telephone directory as The White Horse, Lloyds, Leicester Road and by 1998 it had become the 'Fraternity and Firkin'.
   TSJ and LCC. 1842 P; 1861 PO; 1888-1941 K.

**Pickwell** An 18th century building which is recorded as trading between 1846 and 1932. 1888-1941 K.

**Quorndon (or Quorn), 2 Leicester Road** 'The Father of Fox-Hunting', Mr. Hugo Meynell (1735-1808) established modern foxhunting upon its soundest basis of all - friendship with the farmers. In 1782 he bought fox hounds from Lord Castle-Haven to set up his own pack known as 'Mr Meynell's' of which he was master for fifty years. Some time after 1840 this became the famous Leicestershire Quorn Hunt; the hunt kennels and stabling are still in the town. LL. 1861 PO; 1888-1941 K.

**Seagrave, 6 Church Street** 1913 the house was mentioned in a will.

1955, March – the pub was offered for sale at £1,800, it was described as a 'Public House with the outbuildings and appurtenances thereto belonging and known as The White Horse'. 1963 a Deed of Grant was agreed concerning the laying of a drain or sewer. MTE.

**Shepshed, Ashby Road** Situated in Charnwood Forest and first recorded in 1842 as a beer shop run by William Henderson in the outbuildings of a framework knitters shop, but known as the White Horse Inn.

1863 the inn had become so busy it took over the knitters' shop.

1903 Sarah Ann Moore sold the premises to Marston's, the brewers for £2,200. MTE. 1861 Sheepshead. 1861 Fenny Hill PO; 1912-1941 K.

**Whitwick, Market Place** Originally a substantial farmhouse thought to date from the eighteenth century. By 13th March 1854 it had been divided into two dwelling houses with 'no rights to minerals or coal'; one of these houses became a shop.

1877, 17th December – a conveyance quoted ' . . . for some time past used as a beerhouse called the White Horse . . . messuage with yard barn and stable with other outbuildings . . . with rights and advantages of the well now sunk between the said premises'. The water from the pump to be reached 'through an access in the wall intended to be built between the two properties'.

1896, 21st April – the inn was auctioned having been advertised as 'White Horse and land totalling 1,134 square yards . . . Containing bar, taproom, lounge, parlour, kitchen, scullery, cellar and seven bedrooms'.

Since 1805 there has been a variety of people involved in the history of this property – a hosier, grocer, ironmaster, shopkeeper, gentleman, land drainer, draper, physician and surgeon, a knight and a dame!

1916-1928 Frank Middleton was listed as landlord – he was a professional footballer and played for Derby County from 1901-1906 and Leicester Fosse from 1906 -1909. Landlord and Professor Jeffrey Knight. 1941 Coalville K.

**Wymeswold, 22 Far Street** The three-storied brick building dates from c.1800. It has a classical symmetrical façade, a central door with a steeply pitched pediment[1] and consoles[2]. The windows have been renewed in the original openings which have stuccoed[3] head with decorative keystones. There are two chimneys, one at each end of the building.

It was a private house until 1874. The following year it was converted into a public house.

A ditty was written by the Landlord and appeared in the Nottingham Evening Post, referring to the many other public houses in the neighbourhood:

> 'My White Horse shall chase the Bull
> And make the Three Crowns fly
> Turn the Shoulder of Mutton upside down
> And make the Fox to cry.
>
> My White Horse shall smash The Gate
> And make the Windmill spin
> Knock the Hammer and Pincers down
> And make the Red Lion Grin.'
>
> BrS.

By 1995 the building was unoccupied and had become neglected. In 1997 it was restored as a private house known as Wolds House – there is still a small white horse painted on the fanlight over the front door. It is a Grade II listed building.

Resident Owner and CBC. 1888-1941 K.

## Notes

1. A low-pitched gable, especially one that is triangular as used in classical architecture. Col.

2. An ornamental bracket, especially one used to support a wall fixture. Col.

3. A weather-resistant mixture of dehydrated lime, powdered marble and glue used in decorative mouldings on buildings. Col.

# LINCOLNSHIRE

## North & North East Lincolnshire

*'The agriculture of Lincolnshire has long been celebrated. There are some large estates, but many farmers work their own ground from four to five hundred acres in extent. The land is chiefly freehold, except in the low districts, and leases are not common.*

*Many extensive rabbit-warrens exist, but are generally on the decline, the rich land of old rabbit-warrens having been found very advantageous for tillage. Geese are also bred to a large extent, chiefly for the sake of their feathers; and the shameful practice of live-plucking prevails very generally.'*

JB.

**Alford, 29 West Street** Originally a coaching inn. 1876 The White Horse was listed as a Commercial and Posting House.

The following comes from an article by Peter Chapman of the Grimsby Evening Telegraph which has been abridged by the authors:

Fifty five years ago I was living in Alford. Every morning a man on stilts tapped on my bedroom window, when I looked out an elephant was having his breakfast drink at a large trough immediately below. It happened every morning as regular as the dawn – aged five I must have imagined it to be perfectly natural; a circus was visiting Alford when war broke out in September 1939 and due to petrol rationing it had to stay put. We were living, my mother and I, at the White Horse 'hotel', a very old thatched public house with stone flagged floors. The speciality of the house was bread and honey in the afternoon and plum duff for supper. The principal 'guests' were the officers of a Royal Artillery unit training on the then new 25-pounder gun, one of which was my father. Lincolnshire with its broad, firm beaches was thought might be a favoured choice for a German invasion. The Mess was a large, long room at the back of the hotel, every night after dinner one of the 'guests would play the piano. By the side of the 'hotel' was a path which led past Soulby's brewery to a large field with a tower windmill. The field was used for practice trench digging by the local Home Guard and soon became a very exciting place for a small chap. Alford in 1943 was blissful, with the absence of all motor traffic except 'the military'; horses were brought back into general service and the railway station was beautifully tended by keen railwaymen gardeners – the platforms had huge churns of milk and piles of wicker baskets containing coo-ing racing pigeons.

One day a band of gypsies passed through, they evidently took a shine to me and me to them – when my mother's guard was down, I went off with them. After much searching and I could not be found, my distraught mother informed my father who turned out as many soldiers as he could muster to find me. As the shades of night began to close in I was discovered sitting with legs a swinging on the tailboard of one of the horse-drawn caravans – by this time beyond Thurlby, some miles away, heading for Mumby.

By 2006 the hotel was Grade II listed building. 1822 P.

**Appleby** 1842 P.

**Baston, 4 Church Street** 1842 listed as trading in Pigot's Directory.
By 1999 it had become The Spinning Wheel.
1842-1937, 1842 P; 1976 PO. These dates were listed under Bourne and Market Deeping.

**Bicker, The High Street** c.1979 the White Horse became a private house.
1850 Donington S; 1876 PO; 1882 WDG; 1913-1937 K.

**Boston, 13 West Street** c.1680 the license was transferred from a former White Horse which stood at the corner of White Horse Lane and High Street. The new White Horse had its own brewery and Tap House in the yard. Until 1766 when the Grand Sluice was opened, country people bringing goods to market were often able to come by flat bottomed boat into the town as far as the White Horse, near the top of West Street. The stabling was used by carriers carts on market days held on Wednesdays and Saturdays. James Bell recorded the following in his Gazetteer:

> 'large quantities of poultry were sold, much of which was sent to the London market . . . cattle were sold, some of which were of an unusual size and quality . . . 11th December for horned cattle only . . . The Mayor was clerk of the market'.

c.1950 the inn was closed and in August 1958 demolished; its unique wall-sign of a large plaster model of a white horse mounted on a plinth was photographed and removed to a garden in Linden Way. The site became The Grandway Shops – later a modern shopping precinct.
*The Standard* – 5th August 1958. Trade lists: 1822-1876; 1892 Commercial Hotel – 1937 K.

**Brigg, 27 Wrawby Street** The inn was originally built as a farmhouse. By 1999 it was a listed building and was no longer trading as an hotel.
Trade lists: 1822 P-1937K.

**Burringham** 1937 K.

**Caistor, South Street** Tradition says that Caistor was rebuilt by the Saxon Hengist, on as much land, granted by Vortigern, as the hide of an ox cut into thongs would compass; and hence, it is said, is derived its ancient name of Thrang Ceastre, or 'Thong Castle'.

Hengist, the 5th century Jutish leader, is said to have brought the iconography of the white horse to the British Isles.

Situated on the corner of Bobs Lane, the arched coach entrance from South Street leads to the original stable yard.

1911 described as one of twenty nine public houses in this town.
Miss K Reader and JB. Trade lists: 1822 P-1907 K.

**Crowland, East Street** Originally a bakery, later to become one of twenty four public houses in this village. This inn would have stood on one of the four streets separated by water courses. There was an extensive trade in wild ducks, which were lured to the ponds by tame ducks and then netted – a thousand at a time.

By 1999 the building was unoccupied. JB. 1828/29 P; 1900-1937 K.

**Dunston Fen** The line drawing of the late 19th century inn is taken from a 1906 photograph showing the original site on the bank of the River Witham.

1926 listed under Southrey Ferry in Kelly's Directory. Two years later the building was demolished and rebuilt further back from the river; the publican at the time had to farm 30 acres, owned 120 horses, managed the chain ferry – and ran the pub.

1965 a caravan park was established as a base for fishermen.

2009 the house was still trading under White Horse and its caravan park for fishermen was thriving. Landlord and KSGH. Trade lists: 1882-1937.

**Folkingham / Falkingham** At one time it was probably a coaching inn – later becoming two private houses. 1900-1937 K.

**Gainsborough, 29 Silver Street** c.1600. The building may have originally been a shop. At the time of the Civil War almost all the inns in Gainsborough had smuggling tunnels; it is said that the one from the White Horse Inn went under the River Trent to the far side. At one time the conglomeration of 'yards' situated near the river were closed at night by strong oak doors as a protection against the

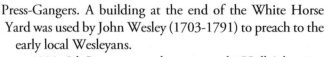

Press-Gangers. A building at the end of the White Horse Yard was used by John Wesley (1703-1791) to preach to the early local Wesleyans.

1838, 5th January – a sale notice in the Hull Advertiser and Exchange Gazette described the White Horse as 'a well-accustomed Inn or Public-House with a frontage of 20 yards and a depth of 100 yards with a number of tenements'. Pre 1900 The White Horse Yard was an important winding passageway running from Victoria Street, through White Horse Yard, here could be found amongst others, a cabinet maker who was also an undertaker; a covered way beside the inn led into Silver Street.

1844 the inn was listed amongst the 45 Hotels Inns and Taverns of the town.

1902, 12th December – The local *Gainsborough & District Gossip Column* reported that 'an excellent and airy site in White Horse Yard' had been offered at a peppercorn rent for the Soup Kitchen which was used to feed the needy in hard times. This new Soup Kitchen was opened on 19th December and the soup was 'manufactured of beef, rabbits, pea flour, peas, onions, celery and various other vegetables'.

1906 a postcard with a photograph of the inn was sent to a London address from a man who was staying at the White Horse, having travelled to Gainsborough to collect a dog. "Dog to hand this morning, all safe. I like the looks of him . . . this house is five hundred years old".

By 1992 the further end of the White Horse Yard had become part of the Bus Station site.

By 1999 it was a whitewashed brick building.

Trade lists: 1822-1937. D, WJ and LL February 1964.

**Grimsby** 1876 Great Grimsby PO; 1892 K.

**Heckington, 61 Church Street** 1776 the site was originally a workhouse which was burnt down in 1813 when an elderly lady inmate took a lighted candle upstairs one night; the ashlar tablet on the front wall records, 'This House of Industry rebuilt 1813'. Following the Poor Law Reform Act of 1834, the building was sold in 1837 to become the White Horse Inn.

1991 closed to become a private house.

*Dick Turpin was hung at York in 1739 for stealing three horses belonging to Thomas Creasey of Heckington.*

Trade lists: 1900-1937.

## Lincoln

***21 Hungate*** Situated on the corner with Motherby Lane.

1697/98 a probate inventory was taken.

By 1998 it had become 'The Tap and Spile'

2006 this traditional pub was known as the home of Circular Chess.

1626 D&W, LAO Ref: INV. 193/416. 1822-1892 two listed LAO; 1960 JM.

The two following sites and dates are recorded as trading by Barrie Cox in his 'English Inn and Tavern Names' published in 1994:

***St. Swithins*** 1626.

***The Close*** 1649.

**Louth, 24 Kenwick Road** 1892 first listed; in 1895 Mrs Nichols paid £1,000 for this house. At one time, due to the circuitous route from the cellar to the customer, it could take 30 minutes to get a drink; the problem was solved by ordering three or four pints at a time.

c.1930 the licensee helped build a light aeroplane with a two-stroke engine which took visitors to Mablethorpe on pleasure flights for 5 shillings.

c.1960 the landlord, a keen gardener – kept 40 pigs, 200 chickens, ducks, an aviary and 2 golden pheasants.

The old stables were eventually sold by the brewery for development.

1926 Newmarket K.

**Mareham-le-Fen** 1937 K.

**Market Deeping, 27 Church Street** 1828 listed as trading, although the building dates from 1837 and the name White Horse has since remained unchanged.

1892 listed as Commercial Hotel and Posting House.

c.1900 a murder took place at a wedding reception in the public bar, the ex-boyfriend of the bride shot the bridegroom through the window.

1926 the house was purchased by John Smith's Tadcaster Brewery Ltd.

By 1999 this listed building was no longer an hotel.

Trade Lists: 1828-1937.

**Marshchapel, Sea Dyke Way** 1876 PO; 1892-1937 K.

**Moulton Seas End** c.1950 a photograph shows the White Horse with an inn sign. Mr T. Royce – The Spalding Gentlemen's Society. 1876 PO; 1882 WDG; 1900-1937 K.

**North Thoresby** 1900 K.

**Old Leake, Fold Hill** Originally a small house. The front doorstep had to be negotiated after many a good drink.

c.1950 the building was demolished.

Local Historian. 1842 P; 1850 S; 1876 PO; 1892-1937 Old Leake K.

**Owston Ferry** 1850 S; 1876 PO; 1892 K.

**Potterhanworth** Once a large coaching inn, it became three private houses – one facing the main road and the other two in Cross Street. 1892 K.

**Scopwick, Brookside** Originally the inn consisted of two rooms - there was no cellar as the stream ran at the rear of the building.

c.1740 this beerhouse, with a smithy attached, had been inherited by two sisters who lived outside the village; the publican who was also a farmer, had to sign an affidavit as to which of the two sisters was the true owner.

1972 it was sold by the brewers, converted to a private house and renamed Brook House. Owner. 1900-1937 K.

**Sleaford, 45 Boston Road** It originally stood on the main road, the centre and back part of the building are original but the front has been refaced; the old frontage is now a car park standing back from the main road; these alterations took place after the last war.

1945 it was purchased by brewers.

Landlady. 1828-1842 P; 1850 S; 1879 PO; 1892-1937 K; 1960 JM.

**Spalding, Churchgate** 'Ye Olde White Horse' is situated on the corner with Church Street overlooking the River Welland. It is said to be the oldest public house in this town.

1553 the house was built from materials salvaged from the demolition of the nearby Benedictine Priory. It was named Berguery House and was occupied by William Willesby whose son, Thomas, founded the Petit or Willesby School.

1732 listed as The George.

1792 John Sills was recorded as the landlord of The White Horse.

1882 White's Directory lists the inn in Cowbit Road – this might have been a short-lived name for Churchgate which leads to Cowbit.

Towards the end of the 19th century the licensee, Thomas Drury, was also an important official of the Spalding Licensed Victuallers Association and a Surveyor of the Highways for the Moulton and Weston district of Spalding. He died in 1903 having been landlord for ten years.

1989 in Pevsner's 'Lincolnshire' the White Horse is described as 'white-painted brick, with a two-storey porch, thatched roof, and prominent chimneystacks. The picturesqueness is enhanced by the irregular lean-to accretions on the main front.' The lean-to was a little shop owned by Mr Bland, a dairyman; later it became a restaurant called 'Isobell's Pantry' before being incorporated into the inn.

1945 the house was bought by the brewers Samuel Smith.

By 2007 the building was Grade 1 listed. MJE, LAO, SaS and T. Royce – The Spalding Gentlemen's Society. 1822 P; 1856-1872 WDG; 19196-1937 K.

**Spilsby, 33 Seasend Road** 1822-1842 P; 1850 S.

**Stamford, 58 East Street** Traded from c. 1837-1909. Became a private house which was demolished in the 1930s. 1850 S; 1876 PO; 1888-1900 K.

**Susworth** The building dates from c.1680 and is situated on the bank of the River Trent. This village inn, the Anchor and Rope, was originally a farmhouse which made and stored ropes.

1716 it was closed down and converted to become the White Horse – possibly as two years previously King George I had arrived from Hanover with his insignia of a white horse.

At one time Sunday Church services, local weddings and criminal trials were conducted here. After one such criminal trial the suspect having been found not guilty celebrated his new found freedom in the pub. After much celebrating he left to make his way home and promptly fell down a nearby well and died. Perhaps it is his spirit which to this day still walks around the house after closing time.

c.1742 it was a 'barge inn' on the bank of the River Trent with a regular service to and from Hull, Gainsborough and Scunthorpe. There was a coal wharf attached to the property which belonged to the inn.

1814-1916 a packet steamer ran regularly between Susworth and Gainsborough for both passengers and goods.

1930 the distribution of coal to local villages ceased, together with the river traffic upon the arrival of the railways.

1947 the river flooded, due to the Eagre bore and the small church was washed away. Congregations were once again invited to hold occasional services in the inn. Between 18th December 1977 and 22nd April 1979 The Reverend Richard Love held six services here; and another service was held outside on 25th June 1978, it poured with rain and the landlord rushed out to invite everyone to continue the service in the inn.

c.1985 this house was renamed 'The Jenny Wren' when the interior was modernised.

The Reverend Richard Love and the Landlord. 1882 Scotter WDG; 1900-1926 K.

**Throckenhall** At one time a Post Office was in part of this building.

1882 Sutton St. Edmund WDG; 1937 K.

**Welbourn, 32 High Street** The White Horse was once an important feature in the life of this village.

The granddaughter of a previous landlord can remember that her grandparents who kept the original alehouse were not allowed to sell spirits; earlier landlords would have brewed their own beer.

1933 at the time of a new tenancy, the inventory included spittoons.

1961 the pub closed to become and was converted into private dwellings, one being Walker's Cottage which retained a local stone fireplace and a working kitchen range.

Mrs J. Birkett and Mr and Mrs T. Walker. 1937 K.

**Whaplode Drove** 1882 listed as trading in W. White's 'History, Gazetteer and Directory'.

1996 the pub closed to become a private house.

**Wrangle** Situated on the boundary with Old Leake – towards the sea.

Once one of twenty public houses in this village. Fairs used to be held behind the building and there was a fish market opposite.

1813 John West VII (1780-1869) became landlord of The White Horse; aged 33, he was a schoolmaster, the Parish Surveyor and would measure land for farmers and labourers in the vicinity when required. He held three acres of Bede Land all the time he was landlord for which he paid £3 5s. 0d. per half year which was raised in 1841 to £4; the rent was paid until 1850 and was collected at the pub. On 12th October 1835 an entry in the rent book read: 'J. West for ale etc. at Rent Day – 5s.0d.' John West always kept a gin bottle full of water which he poured into a glass when a customer invited him to drink so that he never over indulged.

At least one of the local farmers, Mr Render, held the harvest supper for his men at the White Horse, he would probably have invited the local musician John Sherriff, who was deaf, to play his German flute which he always kept in his pocket unscrewed in sections. Sometimes the farm workers would have more than one harvest supper in one night.

John West remained the landlord until he retired when he was seventy in 1850. He is buried in Wrangle churchyard.

By 1999 it had become a private house. Local Historian. 1876 PO; 1882 WDG.

**Wrawby** 1926 K.

# LONDON

*It is important to note that the sites of virtually all the
White Horse Inns of London are from those listed in Bryant
Lilleywhite's original typewritten manuscript of his 'London
Signs', published in 1966 and held by Guildhall Library –
therefore no source has been given.*

*Our research has been greatly helped by the 'A to Z's of London
Past', a series published by Harry Margary in association with
Guildhall Library – which includes John Rocque's map of
1746, together with the 1980 4th Edition of the 'Large London
Map in 9 Sheets' published by Geographers A-Z Map Co. Ltd.*

*When information has been taken from the 1993 Edition of
'The London Encyclopaedia', no reference has been given.*

*We have thus been able to give a picture of the 200 sites in
London where The White Horse Inns, Alehouses or Taverns
have traded over the centuries.*

# LONDON &
# GREATER LONDON

*"I thought of London spread out in the sun,
Its postal districts packed like squares of wheat".*

'The Whitsun Weddings' 1964 by Philip Larkin (1922-1985)

**Addle Hill, White Horse Court, EC4** The White Horse was already trading by 1720.

**Aldersgate Street, EC1** Shown on the 16th century Agas map as running north from the Aldersgate to the junction with Old Street. Part of this main thoroughfare to St. Albans became known as Pickax Street and is shown in the 1746 map.
   1670 listed as a posting house where horses could be hired.
   c.1780 a saddler was listed as trading on the premises.

**Back Side, St. Clements, EC4** An undated ¼d. token issued by Ja. Goles of 'White Horse backside St. Clements I.D.G.' – *a horse*. GCW No.2458. The Church was rebuilt by Christopher Wren in 1683-7. The parishioners were pleased and sent him one third of a hogshead of wine. *Maybe from the White Horse Inn!*

**59 Baldwin's-gardens, EC1** A notorious slum much frequented by criminals in the early 1700s.
   1809-1827 listed as a Tavern.

**Barnaby / Bermondsey Street, SE1** 1746 shown on Rocque's map.
   1826 listed as 85 Bermondsey Street.

**Bartholomew Lane, EC2** 1790-96 'White Horse and Lamb'.
   1794-1801 a tavern used for Masonic meetings. WGT.

**Basen Lane** 1652 a ¼d. token issued by I.G. at 'The Whit Hors in Basen Lane' – *a horse galloping*. GCW No. 148.

**Basinghall Street, EC2** Sir Hugh Myddleton (c.1560-1631), a goldsmith and banker had his office in this street, one of the first British tobacco smokers he would sit in the doorway chatting and sharing a pipe with Sir Walter Raleigh (1552-1618), much to the amazement of passers by.

Two undated tokens – ½d. and ¼d. issued by George Starckey at 'The White Horse in Basinghal' – *a horse prancing.* GCW Nos. 138 and 139.

**530 Bath Road, Longford, UB7 0EE** This 17th century listed building was a popular inn for travellers on the old Bath Road – which was to become the A4.

1831, Saturday 10ᵗʰ March – an article in *The Mirror* reported that a customer who had defaulted on his drink bill offered to 'put some lines together' to advertise the White Horse, it is said he wrote the following:

*My White Horse shall bite the Bear*
*And make the Angel fly!*
*Shall turn the Ship her bottom up,*
*And drink the Three Cups dry!*

An inn sign, painted in 1979 by G. E. Mackenney, shows a saddled grey hackney tied to a sign post pointing to Bath. Trade lists: 1839-1933.

**9 Belmont Road, Uxbridge** 1805 The most important inn amongst several others. This was the year that the nearby Grand Junction Canal was opened.
c.1996 renamed 'Ye Olde Ostler'. WCO. Trade lists: 1826-1933.

**336 Bethnal Green Road, E2 0AG** 1990 listed as a cafe. By 1999 it had become 'White Horse Kebab'.

**Bishopsgate Street Within, EC2** Recorded as trading in c.1670.

**Blackman-street, SE1** A continuation of The Borough. 1667 a ½d. token issued by Edward Salter at 'Ye White Horse'.
1732 listed as trading. MD No.180.

**Bloomsbury Market, WC1** 1739 North end of Lyon Street – later to become Barter Street. The inn was used for Masonic meetings.
1793, November – Mr. Smith, formerly master of the White Horse in Bloomsbury Market died 'of a mortification in his legs, occasioned by hard drinking'. GM.

**The Borough, SE1** This area has always been the main thoroughfare to the south. 'In the 17th century it was full of inns, a continued ale house with not a shop to be seen between'.

1614, August – the Horsham Carriers 'who lieth at the White Horse In Southwark' reported 'A strange and monstrous Serpent or Dragon in Sussex, two miles from Horsham, in a Woode called St. Leonard's Forrest reputed to be nine feet, or rather more, in length. He will cast his venom about four rod from him, as by woeful experience it was proved on the bodies of a man and a woman coming that way, who afterwards were found dead, being poisoned and very much swelled, but not preyed upon'. *This alarming incident will most likely have taken place on the road which is now the A264.*

1660 one of the great annual English fairs – St. Bartholomew – was held in this street. All the nearby streets, courts and inn yards were requisitioned for booths and shows.

1667 a ½d. token was issued by E.M.S. of 'Ye White Horse' – *a horse*. GCW No. 180.

1668 Pepys records 'To Southwark Fair, very dirty, and there saw the puppet show of Whittington which was pretty to see'.

1681 timetables were available at the inn for carriers, wagonners and coaches.

1728 listed as 'the White-horse in Southwark'.

1733 William Hogarth produced his great picture of Southwark Fair, with a dominating central sign showing the wooden horse of Troy depicted as a white horse.

1746-1809 listed as the 'Half Moon Inn' at No.141 Borough High Street.

1746 'White Horse Stables' were shown in Rocque's map as being on the west side of the street. 1826 'White Horse and Half Moon'. Carriers went from here to Kent, Surrey and Sussex. CH and DEL.

**Bow or 'Stratford-le-Bow' (in the County of Middlesex), E3** 1809-1811 listed as a tavern. Trade lists: 1823-1845.

**Bread Street, EC4** 1649 a ¼d. token issued by R.E.R. 'at the White Hores (sic.) in Bread Street' – *a horse galloping*. GCW No. 404. Held by The British Museum.

**Brick Lane, E1** Recorded as trading before 1761. 1790-1796 WGT.

**94 Brixton Hill, SW2 1QN** Trading from the early 18th century, the inn provided accommodation and stabling for travellers on the Brighton road for over a hundred years; one of the oldest pubs in Brixton which held its original name.

c.1921 a photograph shows the inn with a butcher's shop to the left advertising 'Choicest Grass fed Lamb' and The Hill Garage through an archway to the right.

2003, July – a photograph of the inn shows tables outside on the pavement; the garage next door is still trading but the butcher's shop is no longer there.

2005 voted the best night club in Brixton. LA.

**Brixton Rise, SW2** 1881 listed in the census.

There was a swathe of common land extending from the centre of Brixton up

Brixton Hill. Presumably this included Brixton Rise – possibly with Brixton Windmill – which remained open fields until 1850; the Mill continued to work on wind power until 1862 by which time the neighbourhood had become too built up for the windmill to function efficiently. 1874 K.

**261 Brixton Road, SW9 6LH** 'The Whitehors at Bristo . . . Casway' (sic.). The Old White Horse is situated on the corner with Loughborough Road.

1690-1703 listed as a tavern on the Surrey side of Brixton Causway.

c.1760-1789 listed at Rushey Green[1].

1790 the inn is shown on a Route Map, which indicates it had become a coaching inn.

c.1921 a photograph shows the name as Ye Olde White Horse and advertises a 'Grand Billiard Saloon'.

By 2003 known as 'Bar Lorca'; the Grand Billiard Room is used as a Salsa[2] club. The cab rank outside has been there since coaching days. Trade lists: 1809-1874.

**Broad Street, EC2** Listed as a house for the reception of travellers. 1658 a ¼d. token issued by E.A.B. at 'The White Horse in Broad Streete' – *a horse R*. GCW No. 429. This token was found during excavations in the area for the new Royal Exchange after the Great Fire of 1666 when the building was destroyed. It was described by William Boyne as showing a cavalier on horseback, but as a packhorse in the Guildhall Museum Catalogue No. 16. The engraver's mark 'R' shows it was engraved by the chief die maker of Charles I's ambulatory mint.

1677 White Horse Court and White Horse Yard, east out of Broad Street. O&M.

**Broadway, Ilford** 1895 listed as White Horse Hotel with Charles Graham as licensee.

**Brooks Mews, Davies Street, W1** 1768-1820 a tavern or inn used for Masonic meetings.

**Buckingham Gate, Pimlico, SW1** 1802-1806 listed as a tavern.

**236 Cambridge Heath Road, E2 9DA** These premises are described in trade directories as a beer retailer from at least 1870 but it is not until the directory for 1953 that the name White Horse is given. It is quite probable, however, that the name was in use earlier. THL.

**Candelwyckestreete / Cannon Street, EC4** An allusion to the candlemakers who had lived here since the twelfth century.

1589 a messuage called 'le White Horse in Candel-wyckestreet in the parish of Blessed Mary Abchurche'. HAH.

**Carter Lane, EC4** 'Near St. Paul's' – one of the City's main thoroughfares. c.1715 recorded as a tavern.

**Castle-Alley, Cornhill, EC3** 1715 a tavern. 1733 'The White Horse behind the Change' – to the west of the Royal Exchange. c.1760 'The White Horse, Castle alley, Royal Exchange'. Tobacconist.

**21 Castle Street East (north of Oxford Street), W1** 1790-1796 listed as in 'Castle Street'.
   1826 listed as a tavern.

**8 Castle Street (running south to the old Royal Mews), WC2** 1826 listed as a tavern – by 1830 the Royal Mews and most of the adjoining property had been demolished in John Nash's Charing Cross improvement scheme which provided for the building of Trafalgar Square.

**Charlton Road, Charlton** 1881 listed in the census.

**Chick Lane, White Horse Alley (running south into Chick Lane), EC1** Listed as trading by 1720 but in c.1760 the house was listed as a saddler's.

**Chiswell Street, White Horse Yard, EC1** 1746 in Rocque's map two White Horse Yards are shown on the south side of the street. The larger was approached via Powells Alley – the smaller, to the west, leading to the south. In 1750 Whitbread, the brewing firm, moved to the west end of Chiswell Street; here they were visited by George III and Queen Charlotte.

**4 Chitty Street, W1** c.1940. *See North Street.*

**Church Hill, Harefield, UB9 6DX** 1523 the house was sold, it had been owned by John Webb.
   1699 the inn was known to have been trading.
   There were still twenty farms in the Harefield area at the beginning of the 20th century.
   2007, 4th May – the pub re-opened after refurbishment to become a tenanted Punch Tavern.
Landlord. Trade lists: 1839-1933.

**Church Lane, Chelsea, SW3** The building, dating from c.1550, was thought at one time to have been part of the Bishop of London's House; a timber framed building, weather-boarded to the first floor and plastered above. The interior walls were covered with ancient panelling and there were various grotesque ornaments and carving – mainly of human figures in the form of brackets.

Later it became a posting-house; the inn was used for the meetings of the ratepayers and the Parochial Guardian Society as their Vestry Room stood on the east side of what was then Church-lane.

c.1740-1753 listed as 'White Horse Inn behind Chelsea Church'.

An illustration from Old and New London by Walford Vol. 5 Part 1 p.91 – entitled 'Old Chelsea 1750'. Many artists' painted their impressions of this picturesque inn.

1809-1811 listed as in Church-lane.

1819-1824 listed as in Church Street – a tavern or inn on the west side.

1840 it was burned down and rebuilt with a Georgian façade; the hanging sign of the White Horse framed in ironwork was replaced by a large glass lantern; the stable yard had also been renovated. 1872 illustrated by John Timbs in his 'Clubs and Club Life in London'. G, KCL and BG. Trade lists: 1873-1900.

**176 Church Road, Willesden, NW** From 1912 to 1917 this was the terminus for Bus Route No. 8.

In 1998 the public house was closed. Trade lists: 1870-1933. LTM.

**Church Street, St. Paul, Deptford, SE8** Originally known as Depeford – where the deep ford crossed the River Ravensbourne. Deptford was the last stopping place before London for coaches on the Dover Road.

1881 the census listed a White Horse.

**Church Street, West Ham, E15 3HU** 1848 listed as trading.

1998 renamed 'The Angel'. WWH.

**Clare Street, WC2** See St. Clement's Lane.

**18 Cloak Lane, EC4** 1826 listed as a tavern.

**Colchester Street, E1** This street ran parallel to the south side of Whitechapel, the two being connected by the White Horse Inn stable yard.

c.1670 'White Horse Inn south side of Whitechapel near Colchester Street'.

1760 included in LCG; 1809-1811 listed as trading.

It became the 'White Horse and Leaping Bear near Whitechapel Church' and was occupied by a liveryman, Mr. Harrison. AH.

**Coleman Street, EC2** Probably named after the charcoal burners who once lived here.

1666 the house was destroyed in the Fire and rebuilt.

1746 two sites are shown on Rocque's map as White Horse Inns in this street.

1760 included in Dodsley's Complete Guide to the City of London.

1786, March – Mrs. Chapman, the mistress of the White Horse livery stable in Coleman Street, died. LCG and GM.

**Cornhill (in the Parish of St. Michael), EC3** Before 1666 this messuage was known as The White Horse or The Rose and occupied by Henry Nevill.

By 1687, 4th June – 'Ralph and Mary Ayres jointed in a settlement of this messuage' renaming it The Harrow.

Husting Roll 352-16.

**Cornwall Road, Lambeth, SE11** Much of this area has been administered by the Duchy of Cornwall, which was created in 1337.

1826 listed as a tavern.

**Covent Garden, WC2** 1809-1811 listed as a tavern, likely to have been frequented by the new traders in crockery, poultry and bird cages, together with dealers of fruit and vegetables.

**Cow Cross, Whitehorse alley (running from Benjamin Street to Cow Cross), EC1** Listed as trading before 1720.

1826 listed as trading.

**Cripplegate, EC2** Originally a hole left in a wall for sheep to creep through.

1637 'White Horse near the Gate' Inn. Taylor's 'Carriers Cosmography' mentions the 'White Horse Inn where carriers lodged and where the Lincoln Carriers come every Friday'; being a long distance journey these carriers would have lodged within the city walls, those who performed comparatively short journeys started from without the City walls. *These wagons carrying luggage would have left the day before the departure of the coach passengers.*

1647 Nightingale wrote 'Send one of Yor servants to the White Horse without Creeple Gate'.

1677 Recorded as 'White Horse Inn without Cripplegate between the Gate and Fore Street'.

By 1730 the inn had become the main stabling for carriers' horses travelling to and from the north and must have been in a constant state of bustle and activity, which continued for the next hundred years.

The White Horse, or the Great White Horse Yard, as it came to be called, occupied a large area with an entrance through a rough, massive gateway. From this entrance, ponderous, heavy laden wagons, drawn by teams of powerful horses, left daily – sometimes twice daily – for far distant towns. It would seem that the yard – with stabling for seventy horses – ran along nearly the whole length of the north side of the City Wall (which was then standing) from Aldermanbury Postern to the site of the old gate, and had a considerable depth. It had a private entrance into Fore Street.

1702 Carriers left every Monday for Richmond, Durham, Newcastle, Hexham and Alnwick and on Thursday for Bradford and Halifax on Friday the carriers made for Kendal, Otley, Leeds, Pomfret (Pontefract), Tadcaster, and Wakefield. There were White Horse coaching inns in at least ten of the thirteen destinations using this important yard.

1760 the Gate was demolished.

1805 one of three main coaching inns in London for carriers' wagons.

1817-1827 'Cripplegate buildings'. 'At this time . . . the east side being nearly all covered by the White Horse Coaching and Carriers' Inn'.

1826 - c.1832 listed in Pigot's Directory as '10 Cripplegate buildings'.

1832 Daniel Deacon, the occupier and carrier, had an entrance made from the inn yard, through the City Wall into London Wall.

1840 mentioned in a Rate Assessment.

1875 the inn no longer existed, the carrying trade ceasing with the competition of the railways – directories referred only to the 'Cripplegate Tavern' on the east side of the street, but The White Horse Yard was still listed under several carmen's names, who probably used the old inn and occupied premises in the yard.

1878 the yard and buildings were demolished and replaced by the Midland Railway Goods Depot, however Cripplegate Buildings existed until at least 1918.

JJB, NI, LCG, WCO.

**Curtain Road, EC2** 1796, March – Mr. Sanderson died, he was formerly master of the White Horse livery stables in The Curtain-road. GM.

**Drury Lane, WC2** An undated ½d. token issued by Will Neagus of White Horse Yard in Drury Lane – *a pair of scales and a wheatsheaf* – GCW No. 921 – presumably a corn merchant.

In 1646 Oliver Cromwell was living in this fashionable street; in 1667 Samuel Pepys saw Nell Gwynne standing dressed in a smock, sleeves and bodice in the doorway of her lodgings.

1789, 21st September – the death was announced of Mr. Robert Williams, master of the White Horse in Drury-lane. He had belonged to the Hackney Coach-office many years. GM.

**Duke's Street (near Lincolns Inn Fields), WC2** 1714 listed as trading.

c.1820 listed as a tavern in Kings Head Yard, Duke Street.

**East Smithfield, E1** An important thoroughfare running east of the Tower of London. An undated token issued by T.A.G. of The White Horse, East Smithfield – *a horse*. GCW No. 921.

**6 Fan Street, EC1** 1708 'in the Barbican', near Aldersgate Street; a posting house where horses could be hired.

Before 1761 listed under White Horse Yard, Fanns Alley, and by 1826 as a tavern.

**Fenchurch Street, EC3** Trading before 1761.

**88 Fetter Lane, EC4** c.1590 described as a messuage in Fetter Lane later to be known as 'The Great Oxford Horse'.

c.1670 'West from Fetter Lane and South of Barnards Inn'.

1678, 25th April – five Catholic lords held a conference here; Titus Oates (1649-1705) had them committed as traitors to the Tower.

1776 it was here at the White Horse that young John Scott, having just left school, met his brother, William:

"He took me to see the play at Drury-lane, Love played Jobson in the farce; and Miss Pope played Nell. When we came out of the house it rained hard. There were then few hackney-coaches, and we both got into one sedan-chair. Turning out of Fleet Street into Fetter-lane, there was a sort of contest between our chairman and some persons who were coming up Fleet Street, whether they should first pass Fleet-street, or we in our chair first get out of Fleet-Street into Fetter-lane. In the struggle, the sedan-chair was overset, with us in it". JT.

Sir John Scott (1751-1838) was created Lord Eldon in 1821; his brother William (1745-1836) became Lord Stowell in 1821.

c.1782 a painting by the artist, J. Pollard (1755-1838) shows the Cambridge Telegraph (one of the best known coaches) standing in front of the inn; the figure of a large white horse stands above the entrance, advertising 'The White Horse Tavern & Family Hotel'. This watercolour was offered by Christies in their catalogue of 17th June 1966 No. 82; the original print etched by Hanslip Fletcher is entitled 'The White Horse Tavern, Fetter Lane'. This figure of the horse appears to be identical to that at the White Horse Hotel at Sudbury, Suffolk.

1783, 1st August – the death occurred suddenly at the White Horse in Fetter Lane of Mr. John Hich, attorney at law at Rochester in his 73rd year, after a journey from Chatham that afternoon in apparent good health.

1794, February – Mr. Thomas Roberts died aged 18. He was the son of Mr. John Roberts, master of the White Horse in Fetter Lane.

c.1780-90 The Stage Coaches for Oxford & the West Country set out from here, by the early 1800s over thirty coaches left here every day. These included runs to and from Gosport and 'Yarmouth' [*if this referred to Yarmouth on the Isle of Wight, passengers would have taken a ferry from Lymington*].

One of the regular night runs was made by The Chichester & London Patent Mailcoach, which carried four persons inside and two outside and left Chichester every evening at 7 p.m. travelling along the 'new' road via Petworth, Godalming, Guildford, Leatherhead, Epsom, Ewell to arrive at Fetter Lane by 6 a.m. The Arundel coach called here three times a week. These coaches may have used the various inns of this same name when changing the teams of horses. There were White Horse Inns at Epsom, Guildford, Petworth, Chichester and Arundel; also at Cambridge and Reading.

1816 the reminiscences of a coach passenger travelling from Bath:

> "I started at six on a winter's morning outside the Bath 'Regulator', which was due in London at eight o'clock at night. I was the only outside passenger. It came on to snow about an hour after we started . . . The roads were a yard deep in snow before we reached Reading, which was exactly at the time we were due in London. Then with six horses we laboured on, and finally arrived at Fetter Lane at a quarter to three in the morning. Had it not been for the stiff doses of brandied coffee swallowed at every stage, this record would never have been written. As it was, I was so numbed, hands and feet, that I had to be lifted down, or rather hauled out of an avalanche or hummock of snow like a bale of goods. The landlady of the 'White Horse' took me in hand, and I was thawed gradually by the kitchen fire, placed between warm pillows and dosed with a posset of her own compounding".

Two White Horse Inns were listed in Reading; one in 1823 and the other in 1877, both may have been trading at earlier dates.

The 'White Horse' remained in Fetter Lane many years after coaches had disappeared from the roads, and its frontage was considered to be one of the finest left of the old London Inns. But the establishment fell sadly from its high estate, and the ancient building, from whence had set out so many distinguished travellers, finished its career under the name of 'White Horse Chambers' which title did not entirely hide the fact that it was a cheap lodging-house.

1899 the original building was demolished.

HAH, JT, SWA, GM, LGL - Ref. PR 228/FET, and AG. Trade lists: 1895-1966.

**90 Fetter Lane (off Fleet Street), EC4** 1836 a coaching inn listed in Bates Directory from where the High Flyer left for York. W. Chaplin & Co. sent the Royal Mail coach from here to Yarmouth in Norfolk via Colchester and Ipswich, leaving at 7.30pm it took 14 hours. KSGH.

**22 Finsbury Street, EC1** 1826 a tavern. The map of 1746 shows Finsbury Yard from where Finsbury Mews lead into Chiswell Street next to White Horse Yard.

**Fleet Market, EC4** 1707 The White Horse Inn is given as 'by the Ditch Side' and later as 'East out of Fleet Market in Farringdon Ward Without'.

1755, 17th February – The Bath Flying Machines were advertised as calling at the White Horse in Fleet Street.

1760 included in Dodsley's Guide to the City of London. HAH, BJ and LCG.

**Fleet-lane (sic), EC4** 1732 This tavern may have been a Marrying House:

'. . . clandestine marriages were performed, first in the chapel of the Fleet Prison and from the beginning of the 18th century in nearby taverns and houses, several of which bore signs depicting a male and female hand clasped together above the legend 'Marriages Performed Within'. The marriages were mostly conducted by clergymen imprisoned in the Fleet for debt, who were allowed the Liberties of The Fleet. According to James Malcolm, author of Londinium Redivivum, up to 30 couples were married in a day and almost 3,000 marriages were performed in the four months to 12 February 1705. Fleet Marriages were declared void by Lord Hardwicke's Marriage Act of 1753'.

**154 Fleet Road, Hampstead, NW3 2QX** The White Horse stands on a corner on the site of an early 18th century inn. There is a white stone plaque let into the wall which reads:

'White Horse Hotel rebuilt in 1904 by J.P. Davis
Albert Pridmore Architect
C. Gray Hill Contractor'.

1721 listed as trading.

1723-1735 listed under the White Horse at Ye Wells, the name was then changed to The Green Man.

1809 the inn was listed in Pond Street.

1839 listed as in South End Green and by 1899 as 158 Fleet Road.

1904 the hotel was rebuilt to accommodate the new road layout; it has an Italianated stone tiled floor and a wooden partitioned ceiling with central lacquered copper panels. It is a listed building.

2006, 26th May – the White Horse, no longer an hotel, was re-opened after extensive refurbishment which included vintage chandeliers and Venetian mirrors. Hampstead Rugby Football Club met here after matches and their training.

CW and H&H.

**Fleet Street, EC4** 1617 this house was referred to in Harleian Miscellany 6850.

1652-1666 recorded as a letter-receiving house, which was to be 'Consumed in the Fire'.

1676 it had been rebuilt as the 'White Horse Inn north out of Fleet Street, west of Shoe Lane' as shown on the Ogilby and Morgan map of this date.

1681 used by Country Carriers from Surrey, and Berkshire – the draymen frequented the nearby cockpit as did Samuel Pepys.

1703 'the great Elephant' was seen here – this street had long been a show place for freaks, entertainers and large animals.

1708-1755 described as 'between Wine Office court and Peterborough court'.

'White Horse and Bell was situated at the lower End of Fleet Street near the bridge'. It was unlikely to have been a tavern as the premises were used by John Walker, a brazier, the inventor of the oil lamp clock. STR and AH.

**Fore Street, EC2** 'Le Hors in the hop' / The Horse in the Hoop[3]. *This could be interpreted as 'The Horse in the Paddock'. Another explanation may be that the inn sign was painted on the ends of a barrel, the staves held together by metal hoops; this was hung horizontally outside the tavern – an early inn sign.* Perhaps the predecessor of the White Horse – mentioned in the Husting Rolls 1478. Shown on the 1677 Ogilvy and Morgan map.

Rocque's 1746 map gives a White Horse Court, north side of Fore Street near Little Moor Fields.

**103 Fore Street, Edmonton, N18 2XF** 1845 first listed as White Horse Tavern, an alehouse in the Post Office Directory with John Boswell as landlord, but is not listed in the 1841 census under this sign. The rear part of the building was of two-storeys, the front being a single storey protruding on to the pavement; there was an angled front doorway, a large brass lantern above with a figure of a rearing horse fixed to the roof over the entrance. At the back there was a covered yard with stables.

For many years it stood between H. Stiebich, a baker and corn merchant and a builders merchant; the track beside the public house was called White Horse Lane, previously it had been known by other names – Church Road, Meetinghouse Lane and Bridport Road. Many years later when the tavern was demolished, old stone tablets were unearthed, bearing the faint sign 'Church Road' along with a rusty cleaver and a foreleg bone two feet in length of some unknown animal.

The lane led to St. James' Church, where on Sundays carriages and pairs would await the return of their owners, while the working class would make their way along it to the White Horse.

1900, 8th March – the Commercial Brewery Company bought the White Horse from John Giles.

1902, 27th March – the tavern was granted a wine and spirit license.

The original building was replaced by a larger, tiled house with Gothic windows of similar shape. It was described as being compact, furnished and equipped on modern lines with a children's room at the rear, this had an entrance onto the street; there was Austrian oak paneling and a mural depicting the 'fun of the fair' – at one time the Edmonton Fairs were famous. Seven horses heads carved in solid oak were placed in one of the bars; outside, two white horses heads stood each side of the chimney stack and there was a large metal front-half of a rearing horse with its hooves overhanging the pavement which was removed some years later during alterations to the building.

1937 Mr. and Mrs. Arthur Loneragan celebrated twenty five years as being licensees of the White Horse.

1953 the children's room became an off-licence.

1960 a glazed covered shelter was built over the rear paved area and an opening made between the saloon and private bars.

1972 the off-licence and ladies toilet were removed; the patio and private bar becoming lounge bars.

By August 1999 it was rumoured that the owner had become bankrupt and the pub ceased to trade, becoming derelict.

By 2002 the White Horse had reopened.

The following verses were taken from 'The Diverting History of John Gilpin' written by William Cowper and published anonymously in the 1782 edition of the Public Advertiser. They describe the start of the eventful ride to Edmonton. Cowper had been told the story by Lady Austen, who had remembered it from her childhood. She was the wife of Admiral Sir Francis Austen – brother of Jane Austen. A celebrated actor, Henderson, made the story very popular by his recitations at Freemasons' Hall.

The original 'John Gilpin' was a linen draper, Mr. Beyer, who was born in 1692 and lived at the Cheapside corner of Paternoster Row, about seven miles away from Edmonton.

John Gilpin's spouse said to her dear,

*"Though wedded we have been*
*These twice ten tedious years, yet we*
*No holiday have seen.*

*Tomorrow is our wedding day,*
*And we will then repair*
*Unto the Bell at Edmonton,*
*All in a chaise and pair.*

*My sister, and my sister's child,*
*Myself, and children three,*
*Will fill the chaise, so you must ride*
*On horseback after we."*

GBo and G. Dalling, Local History Officer. Trade lists: 1859-1933.

**29 Friday Street, EC4** 1556 le Whyte Horse taverne and Inne; a tenement. *The following details of tenants may be found in Appendix from 'Grants in April 1543' 34 Henry VIII No. 27 p.282.*

'(a) . . . kept by Ric. Eddes, vintner of London – at £5 rent.

(b) By the same, 19 Feb. 19 Hen. VIII., to Wm. Colte, skinner, of London, of a tenement called 'le Whyte-horse Inne' and a tenement adjoining it on the north side, in the parish of St. Margaret in Friday Street, ward of Breadstreet, for 21 years, at 8L. rent.

(c) By the same, 16 June 20 Hen. VIII., to John Style of Higham Ferrers, the messuage and lands which John Pope then occupied (except 'le Bury close') from Mich, 1529, for 21 years, and also the profits of the said ; 'Bury Close' if the said master and fellows should have the grant of it, rent 10s. for the said close and 28s. for the rest . . . '

The Master and Fellows were principals of the late Chicheley's College of Higham Ferrers, Northamptonshire which was founded in 1422 by Bishop Henry Chicheley (1364-1443), a yeoman's son born in Higham Ferrers. In 1823 a White Horse Inn was already listed in Higham Ferrers. EB.

1548 dinners of The Vintners Company sometimes held here – the tenant Richard Eddes, was a member of this City Company.

1553 the inn appears in a list of forty London taverns which were permitted licenses.

In the reign of Queen Elizabeth it was of considerable notoriety among bon vivants, players, playwrights, and roisterers. In the 'Jests of George Peele' it is used as the scene of some of his mad pranks.

An undated ¼d. token issued by H.E.P. of 'Y Whit Hors Tavran in Fryday Streete' – *a horse running.* GCW No. 1132.

1657 ¼d. token issued by E.M.M.of Ye White Horse in Fryday Streete – *a horse current.* GCW No. 1133'.

By September 1666 the tavern had been destroyed in the great fire and rebuilt as an inn with a spacious yard - resuming the old sign together with the tavern which had adjoined it. This substantial property would have consisted of the original tenement, messuage and lands. Over the years it had been listed as an Inn, Tavern and Coffee House, whilst the numbers 29, 31 and 32 had been allocated to this property.

1702-1707 used by The Abingdon Carriers on Thursdays; Wagon on Fridays.

c.1760 the gate was demolished and the street widened.

1800 used for Masonic meetings.

1805 One of the main coaching inns of London. *The carrier's wagons would have left with the luggage the day before the passengers departed.*

1809-1811 'White Horse Coffee House'.

1826 listed in Pigot's Directory as 'The White Horse Tap', 31 Friday Street.

1827 Used by Country Carriers of Brighton, Cambridge, Hampton Court, Pool, Portsmouth, Kingston, Reading, Salisbury, Wallingford and Windsor.

1926-28 Dr. Kenneth Rogers gives an account of the inn which by this time had ceased to be a Coffee House. The buildings in Friday Street were demolished due to the Second World War bombing and street alterations.

GCW, ECM, LMA and WCO.

**Frying Pan Alley (west of The Borough) in Southwark, SE1** In the 17th century Borough High Street was the main highway into London from the south and was noted for its inns – 'a continued alehouse with not a shop to be seen between'.

1769-1776 listed as 'White Horse and Frying Pan'.

**Gilt-spur Street, 'Neere (sic) Newgate', EC1** Once known as Knightrider Street, as the route had been used by the knights on their way to the Smithfield tournaments. Gilt spurs were probably made here later.

1653 the premises appeared in 'A Guide to the City of London' as being occupied by a printer and publisher.

1668 described as the 'White Horse without Newgate map-printer and publisher'.

1670 the map showing all rebuilding after the Great Fire was printed and sold here by John Overton.

1709 listed as Book and Printsellers.

1720 and 1731 new maps were printed by Henry Overton, John's nephew, who had taken over the business from his uncle.

1751, 8th July – after the death of his uncle Henry Overton was print-selling here.

1754, 19th August - listed as a map and print-seller.

By 1770 the premises had become a chemist shop. SM and BJ.

**72 Goswell Street, EC1** 1720 listed as already trading. 'White Horse Yard was a large open place for stabling and coach houses and hath in it some dwelling houses'.

1746 Rocque's map shows an entrance from Peartree Street which ran into the east side of Goswell Street.

1826 listed as a tavern. JS.

**Gray's Inn Lane, WC1** 1791, August – *The Gentleman's Magazine* announced the death in June of Mr. Thomas Field, horse-dealer and master of the White Horse Livery Stables in Gray's Inn Lane.

**Great Chapel Street, Westminster, W1** 1722 listed as trading and shown on Rocque's map of 1746.

**Great Eastcheap**, E1 c.1620 described as a small stable with appurtenances in the parish of St. Leonard.

1761 listed as trading.

c.1836 the buildings were demolished to make the approaches to the then new London Bridge; over a century later the bridge was sold and re-erected at Lake Havasu City, Arizona, U.S.A. The present London Bridge dates from 1972. HAH.

**51 Great Saffron Hill, EC1** 1809-1827 listed as trading. Saffron was grown here which was essential to disguise the taste of the city dweller's rancid meat.

**Green Street (off North Audley Street), W1** 'The White Pony'. A trade Card dated 1773 gives John Morris as a Horse-dealer.

**116 Green Street, Enfield Highway, EN3 7JE** c.1638 three self contained timber cottages were built, the central one becoming the public house – selling Gripper's Stout, Porter and Ales.

1788 the start of an improvised racecourse of some note ran from a point near the inn to finish where Brimsdown Station now stands.

1895, 26th September – transfer of the lease to White Horse Public House with 'yard, garden, pigsty, shed and appurtenances . . . messuage tenement or beershop with the yard garden indenture'.

By 1926 it had become one of the few remaining original inns within easy reach of London. 'Here is undoubtedly one of the finest of gardens to be met with attached to a public house near London' where a tea and dance room had lately been added together with a ladies' dressing room. By 1933 the outer cottages had been added to the licensed premises.

By 2007 the pub had ceased to trade. As it is a listed building it is unlikely to be demolished. HoW July1926, July 1933 and January 1938. Trade lists: 1859-1898 listed under Ponders End.

**Greenwich, SE10** 'The White Horse in Greenwich'. An undated ½d. token issued by Hugh Pudefourd at the White Horse. MD No.333.

1782 a visitor from Germany travelling from Dover to London via Dartford wrote:

> "I was astonished at the great signboards hanging on beams across the street, from one house to another, at various spots. These somewhat resembled town gates – which at first I took them to be – but this is not so; they do no more than indicate the entrance to an inn . . . with one vision after another in such quick succession as to put the mind in a whirl until we arrived near to Greenwich". CPM.

*Still to be seen in Stamford, Lincolnshire.* Trade lists: 1823-1845.

**Hackney Grove, E8** 'Old White Horse' 1809-1811 kept by Sam Mead.

**Hammersmith Broadway, W6** An ancient coaching inn and yard with two parallel tiled roofs.

1656 Nicholas Dauncer willed an annual payment of 30s. to the poor and 10s. for an annual sermon; this 'to be paid out of the profits of the house in Hammersmith called by the name of The White Horse'.

1775, 17th April – John Scott had by now surrendered the license, the inn being renamed The George. 1915 this old building with its modernised frontage had been demolished; but the name 'White Horse Yard' was retained, later to become the site of the Congregational Chapel. EC, HFA. Trade lists: 1823-1870.

**Hampton Wick** Trade lists: 1790; 1851 PO.

**92 Hare Street, Bethnal Green, E2** 1810-1827 a tavern used for Masonic meetings.

**Haymarket, SW1** Listed as trading from 1682-1720. By 1746 the site appears to have been absorbed by the Opera House at the southern end of the Haymarket.

1755 a map showed a White Horse Court close by, leading immediately from the east end of Warwick Street into St. James's Park. *These two sites may have had some connection.* JHM.

**100 High Holborn, WC1** In Rocque's 1746 map 'White Horse stabling' is shown on the north side of the street – opposite the old 'real' tennis court.

1826 listed as a tavern with stabling.

**118 High Road, Chadwell Heath, RM6 6NU** An old coaching inn with extensive stabling.

1577 recorded as trading according to 'The Agrarian History of England and Wales⁴'.

1602 first mentioned in the local assizes.

c.1927 a set of postcards show the extensive and beautifully laid out 'White Horse Gardens', a path leads from what is thought to be the original part of the building.

By 2006 the site of the old stables had become a large function room and is recorded as being the only pub in the country to have its own traffic lights for customers using the car park. Landlady. Trade lists: 1845-1937.

**158 High Street, Hadley** Also known as Barnet, Chipping or High Barnet. Originally a private dwelling and plumber's shop, by 1839 it had become a public house as proved by the will of an auctioneer of Monken Hadley, who had left the house to his wife, 'in which he lately dwelt but now occupied by John Pearce, victualler'.

1890 bought by a brewer.

1895 bought by McMullen's Brewery as a beerhouse.

1972 the pub ceased to trade and was rented to a Building Society, who bought it in 2004. HCR, WBJ and McM. Trade lists: 1859-1870.

**125 High Street South, East Ham, E6 6EJ** This hotel stands on the corner with White Horse Road.

1895 listed as trading with William Appleby Warwicker as licensee.

1917, 12th February was the day the No. 15 Bus Route started to use this terminus displaying the words 'White Horse' on their vehicles, which are still to be seen. A pair of white painted iron bollards on the curb, crowned with horse heads, mark the original bus stop. LTM. Trade lists: 1839-1914.

**Holborne Bridg (sic), EC1** Two undated ¼d. tokens issued by W. A. Birch of 'The White Horse Holborn Bridg' – *a man on horseback* – this could be in connection with the carriage of mail; the other – *a horse saddled.*
GCW Nos. 1368 and 1369.

1707 advertised in the Aldenham Carrier.

1752, 13th November – 'the post chaise from Mr. Clarke's White Horse, Holborn Bridge for Bath'.

1753, 16th April – The Bath Caravan leaving the White Horse, Holborn for the White Lion Bath, on Mondays and Thursdays. BJ.

**Holland Street, Blackfriars, EC4** 1826 listed as a tavern.

**Holywell Street, Strand, WC2** An undated ¼d. token issued by R.S. of The White Horse next to Lyons Inn – *a horse caprioling.* GCW No. 1859.

**86 Horseferry Road, SW1P 2EE** 'White Horse and Bower'.

1746 John Rocque's map shows 'Road to the Horse Ferry'; a building is shown on the site of the present inn surrounded by a large area of open land with trees, this was presumably the older alehouse referred to below as 'The Bower'. It is said that the inn provided stabling for the most important of the London ferries – the Horse Ferry at Lambeth, old London Bridge being the only other means for travellers to cross the Thames. The painting 'Lambeth Palace and St. Paul's Cathedral from the Horseferry' by Jan Griffier (c.1652-1718) shows the ferry – a large punt-like boat propelled by a London Waterman – on board is a pair of white horses harnessed to an elegant carriage.

1812 The Gas-Light and Coke Co. purchased the land adjoining the inn where the tea gardens stood, in order to build their works in nearby Great Peter Street; they had been granted a charter to provide light to the City of London, Westminster and the Borough of Southwark.

c.1820-1833 a tavern named The White Horse was listed in the Licensed Victuallers' Records at 1 Romney Terrace, a small street off Horseferry Road.

1835 White Horse and Bower. On the site of the old alehouse and tea gardens once 'graced with tall poplars' and known as The Bower. R, JES and HEP.

**Houndsditch, EC3** A ditch running 'along the site of the moat that bounded the City wall'. It is so called, according to John Stow (c.1525-1605), 'from that in old time, when the same lay open, much filth especially dead dogges were there laid or cast'. Its name may have been derived from the City Kennels which were in the moat and where the hounds were kept for the City hunts'. *By 1590 the Houndsditch had been levelled and built over.* A ¼d. token issued by I.A.P. of 'The White Hors in Hounesditch' – *a horse prancing; below, a small R* – this shows it was the work of Thomas Rawlins, King Charles I's chief die-sinker. GCW No. 1523. Another undated ¼ d. token issued by L.E.H. of 'The Whit Hors in Hounsdich nea Al gate (sic.)' – *a horse galloping.* GCW No. 1512.

**153-155 Hoxton Street, N1 6PJ** Originally given as 9 Pitfield / Petfield Street. 1826 listed as a tavern.

**4 Hungerford Market, WC2** 1690-1698 'Whitt Hors from Hongerford market Estwas in the Strd' [Strand].

1826 listed as a tavern.

1682 the market was built by Sir Edward Hungerford on land which had once been his garden, his house having been burned down in 1669.

1860 the market was demolished to make way for Charing Cross railway station.

**James Street, Hoxton Market, N1** 1670 listed as 'The White Horse in James Market'.

**Kennington Lane, SE11** 'West part towards Vauxhall'. c.1780 listed as trading.

The many visitors to the nearby Spring Garden at Vauxhall will have brought custom to the inn.

1881 listed in the census under Loughborough Road, Kennington.

1790-1796 WGT.

**Kensington, W8** The house was situated on an important route to the west, standing on the corner of Holland Lane on the west side of the park.

c. 1711 Joseph Addison (1672-1719) wrote several of his 'Spectators' here. He was married in 1716 to the widow of the third Earl of Holland; the union was not a happy one, and Thomas Moore records it as a tradition of Holland House that when the domestic atmosphere was not serene Addison would resort to a little ale house near the turnpike for peace and consolation – the house so described was the White Horse inn. Renamed 'The Horse and Groom'.

1738, 25th October – the earliest known reference in a lease (MS 131) relating to land adjacent to 'The Horse and Groom'; the building is shown on estate plans, surveys and maps until 1770. It was sometimes used as a meeting place of the Court Baron of the Manor of Earls Court.

c.1811-1824 The name was restored to The White Horse Tavern; by this time the inn stood in the newly-named High Road, Kensington.

1824 it was rebuilt using the original mahogany fittings and became 'The Holland Arms'. LH , WWH and KCL.

**Kent Street, Southwark, SE1** 'Ye White Horse and Bucket'. An isolated country inn situated on the important route from Dover to London, which had been the scene of splendid cavalcades and processions; for instance in 1597 the Emperor Charles V accompanied King Henry VIII into London in great state.

An undated ½d. octagonal token issued by Will Williams of 'Ye White Horse' - *a horse and bucket.* GCW No. 294.

1746 in Rocque's map 'White Horse Yard' is shown about a quarter of a mile westward from the White Horse Inn.

c.1780 listed as trading.

**Kilburn, NW6** 1721-1735 listed as trading by E.F. Oppe. H&H 1952.

**King Street, SW1** 1746 Rocque's map shows White Horse Passage running between King Street and Great Swallow Street.

1783-1793 Masonic meetings were held here.

**King Street, Old Street Square, EC1** 1790-1796 listed as trading in the Whitbread Grand Trade Ledger.

1809-11 listed as trading.

**King Street, Westminster, SW1** 1660, 12th March – Samuel Pepys (1633-1703) wrote in his diary 'Then to the White-horse in King's-street, where I got Mr Biddle's horse to ride to Huntsmore'. This was to make arrangements for his wife while he was at sea.

1708 listed as 'White horse Inn east side of King Street Westminster'.

By 1720 the house had been destroyed by fire.

**Leaden Hall Street, EC3** A trade card was issued by Joseph Flight, linen-draper of White Horse and Crown opposite the Pea Market. AH.

**1 Lee High Road, Lewisham, SE13** 5LD Originally a domestic residence, occupied in 1814 by William Moore, a Hairdresser.

1830 the house was converted to sell beer, by c.1840 it was rebuilt as a public house – the White Horse; it was listed in the Property Inventory of Lee, but not in the 1841 census.

By 2006 the name had been changed to 'ONE'.

LLH. Trade lists: 1866-1887.

**Limehous Causway (sic), E14** Included in folio 22 of Harl MS 4716. 1746 Rocque's map shows a distillery in this road. Among the Chinese seamen who settled in this area were a few opium workers and inveterate gamblers – amongst them Oscar Wilde's Dorian Gray.

**9 Little Britain (adjacent to Little St. Bartholomew Gate), EC1** 1667 a ½d. token issued by John Papworth of Little Brittain – *a horse saddled and bridled* – but this could have referred to the Black Horse also listed nearby.

For over two hundred years booksellers traded in this street; from 1666-1699 these premises were occupied by Publishers and Printers.

1750-1751 a Whitbread rent ledger shows The White Horse Alehouse paid £25 per annum as rent; it also records the grant of the original lease from the City of London for the stabling in Grub Street.

1790-1796 again recorded as trading.

1809-1827 listed as a tavern at 9 Little Britain. The inn sign showing a Whitbread dray horse was designed by the artist Miss Violet Rutter, while working in the Wateringbury brewery studio in Kent. HoW, Summer 1950. WGT.

**Little Wild Street, WC2** 1715 listed as 'Ye Whit Hors at ye end of Little Wild Street' off Drury Lane.

**80 Liverpool Road (formerly St. John's Street), N1** This main road to the north was originally for packhorses.

1665 listed as 'White Horse at Islington' and under Clerkenwell some five years later.

In the stage coach era several taverns were trading in this street. By which time this section of the street had become Liverpool Road.

c.1975 renamed 'Minogue's Bar'. ICL. Trade lists: 1826-1878.

**Lombard Street, EC3** Described as an ancient and notable tavern with a garden and yard; south side of Lombard Street and east of Sherborn Lane.

1468 the White Horse Tavern, an ancient house, is probably the one mentioned in the Accounts of Sir John Howard: 'White Hors iij pipes'. *Presumably 112 clay tobacco pipes.*

Between c.1495-1505 a fine of 20 shillings was imposed on Milo of the White Horse in Lombard Street for breaking the price of Rhenish wine, which was 10 pence a gallon.

1499 the Carpenters' Company record a payment to 'ye Whytehorse in lumbard strete for redde wyn and claret wyn xs.viijd.' – presumably 10 shillings and 9 pence.

1519 'The Goodman of the White hors (sic) in Lombard Street was presented for selling wine without licence'.

1553 'The Whyte horse, Lumbard street' was included in a list of Forty Taverns in London permitted by Parliament to continue to trade as Taverns.

1561 the landlord was presented for a hogshead of defective Gascony wine.

This popular tavern was frequented by Samuel Pepys (1633-1703) during the time he was employed in the Navy Office; he became Secretary to the Admiralty in 1672. The following entries are from his diaries:

## 1664

25th November – 'From the 'Change, with Mr. Deering and Lueillin [Llewelyn] to the White horse tavern in Lumbard street and there dined with them, he gave me a dish of meat'.

6th December, 'Thence by appointment to the White-horse in Lumbard Streete, and there dined with my Lord Rutherford . . . and others, and very merry'.

## 1666

1st March – ' . . . at noon to dinner with my Lord Brouncker, Sir W. Batten and Sir W. Penn at the White Horse in Lumbard street where, God forgive us! good sport with Captain Cocke's having his mayde (maid) sicke of the plague a day or two ago and sent to the pest house, where she now is, but he will not say anything but that she is well'. (Two of the staff of the White Horse died of the Plague the previous year).

8th March – 'to the White-horse in Lumbard-street, to dine with Captain Cocke upon perticular business of Canvas to buy for the King. And here by chance I saw the mistresse of the house [Frances Browne] I have heard much of, and a very pretty woman she is indeed – and her husband [Abraham] the simplest-looked fellow and old that ever I saw'.

September – the Great Fire engulfed Lombard Street, including the 'White Horse Tavern'. As a result Frances and Abraham Browne moved to the Beare Tavern.

## 1667

24th February – 'This night, going through [London] bridge by water, my waterman told me how the mistress of the Beare tavern at the bridge foot, did lately fling herself into the Thames and drownded herself; which did trouble me the more when they tell me it was she that did live at the White Horse tavern in Lumbard streete; which was a most beautiful woman, as most I have seen'.

After the Great Fire this important site was bought by Sir Robert Vyner upon which he built a large house in 1678, part of the building was rented to the Post Office at £167 18s. 4d. for a half year.

By 1744 a twenty one year lease had been granted - the site is shown on Rocques contemporary map.

'The White Horse Head' – on the North side of Lombard Street was unlikely ever to have been a tavern; confusion has been made in sundry writings between these two contemporary White Horse signs in Lombard Street.

Lombardy merchants from Italy settled here in the 12th century.

c.1548-1584 the premises were occupied by goldsmiths for almost a hundred years and given as 'Le White Horse head in Lumbard Street, Parish of St. Mary Woolnoth'. (Hustings Roll 266-53.)

1660-65 Clement Punge, linen draper, sold 'The White Horse Head'. (Hustings Roll, 339-98).

1688 Alexander Pope, the poet, was born in this street; his father was also a linen draper.

Upon rebuilding after the Great Fire of 1666 the sign changed from The White Horse Head to The White Horse, 70 Lombard Street. By 1687 Lombard Street had become the Banking centre of London. c.1730 the premises were occupied by Goldsmiths and Bankers. BL, GPO.

**London Wall, EC2** An undated ½d. token issued by John Benion of 'Whit Hors Yrd London Wall neer morgat' – *a horse*. GCW No. 1764.

1664, 21st July – from 'The Newes' on page 468: 'Strayed, or stolen from Oundle in Northamptonshire on the 12th Instant a broad white grey Nag 14 hand high, Mare-faced, all his paces. He that shall give notice of him . . . to Mr. John Benion, at the White Horse London Wall, shall have 40s. for his peyns'. *This would appear to be a proven grey travelling stallion.*

1677 described as 'south out of London Wall, opposite New Bethlehem', in 1708 'White Horse Inn on the South side of London, against Bedlam'. Bethlehem Royal Hospital (Bedlam) was a magnificent building completed in c.1670; in 1800, it was declared unsafe and the hospital was moved to Lambeth.

1746 shown on Rocque's map as 'White Horse and Half Moon Stables'.

1826 listed as No. 54 London Wall, tavern or inn.

1771 William Maitland, the historian, records a Horse Yard in London Wall.

1826 an 'Old White Horse' was listed at No. 61 London Wall.

**166 London Road, Kingston upon Thames** c.1860 according to the Landlady the house was trading at this date.

2001 the premises were closed and by 2003 had been demolished. 1938 K.

**105 Long Acre, WC2** 1826 listed as a tavern.

**Ludgate Street, EC4** 1684 listed as a Bookseller, by the time the gate was demolished in 1760 the address had become 'on Ludgate Hill' – the premises of a Printseller.

**24 Market Place, Brentford, TW8 8EQ** Likely to have been a coaching inn as it is situated on the main road from London to the west and there were once coach houses at the rear of the inn.

The house was known to have traded under White Horse from 1809.

1814, 30th November – the house was bought by the brewers Fuller Smith & Turner Ltd.

The artist J.M.W. Turner (1755-1851) is commemorated in a plaque on the wall which reads 'This Public house was once the home of the famous artist J.M.W. Turner'.

Now trades as The Weir Bar & Dining Room. Landlord.

**36 Market Road, Islington, N1** 1852 The City Corporation acquired 75 acres of the Copenhagen Fields in order to build a new Metropolitan Cattle Market and to close Smithfield Market which had become unable to cope with the amount of livestock flooding into London. 30 acres were designated for the new market on the site of Copenhagen House, adjacent to Caledonian Road, the route which had been used by cattle drovers on their way to Smithfield. Three million paving bricks were used as hard standing for cattle, the whole area being fenced around with iron railings carrying decorative animal masks; buildings for abattoirs and skinning sheds were provided. Horses, pigs, asses and goats were sold on Fridays – there was lairage for some 7,000 cattle and 42,000 sheep which were sold on Mondays and Thursdays.

Four imposing taverns, The Lion, The Lamb, The Bull and The Horse were built at the corners of the new market, designed by J.B. Bunning (1802-1863) who was architect to the City of London; they were built by John Walsh who signed an article of agreement on 24th April 1854.

1856 the house had become The White Horse, a description was given in the newspaper, *The Leisure Hour*.

> 'Everything connected with the market had been designed with a view to strength and 'permanency, not without such ornament as becomes the structure of a wealthy community'.

The six ground floor rooms were furnished with elegant pine panelling and glazed tiling; there was a very large stable yard running 130 ft., the full length of the north side of the building. On 10th April – the first lease was granted to Mr. Packman together with an abattoir in the market – an important condition being that no betting, especially horse racing, be allowed on the premises; some bear and bull baiting and dog fighting were still taking place in the vicinity. c.1880-90 the landlord, Mr. Tooth, lost his licence for keeping an unruly house.

By 1989 the ground floor of this fine building had been gutted and the name changed to The Gin Palace – a den of iniquity – to remain empty for two years from 1995. Various apartments had been constructed on the top floors with a photographic studio on the ground floor - the whole building being resold in 2000. MkD, BS and ICL.

**Market Row, Golden Square, W1** 1826 listed as a tavern.

**Maryland-point, Stratford, E15** 1809-1811 listed as a tavern or inn.

**50 Middle Road, Harrow on the Hill, HA2 0HL**
1950 the caption to a drawing reads 'The White Horse Inn, Roxeth' – *then one and a half miles south west of Harrow on the Hill.* By 1999 it had become a Free House situated on the A312.

**Mile End, Old Town, E1** 1595 recorded as trading by Dunkling and Wright in their 'Pub Names of Britain'.

c.1650 recorded as trading under The White Horse.

1728, April – Edward Fenwick, a sugar baker from London, 'farm lett' to a brewer of Bethnal Green, 'a messuage or tenement and building [in Mile End] with the great yard stables cowhouse barn outhouses shedds hovells & appurtenances thereunto belonging heretofore an inn and called the White Horse alias the Mewse and now or late called or known by the sign or name of the Three Colts'. *In 1746 Rocques map shows Three Colts Yard on the opposite side of the road.* The buildings included 'a Great Hall paved with Free Stone, the greatest part of the front glazed with two Iron Casement and Iron Barrs halfway up'.

1703 Gascoyne's map shows substantial buildings to the north of Mile End old Towne, which became Westfields Brew House and eventually Charrington's Brewery to the north of Mile End Road; Derek Morris (Ref. MoD) records that there was a White Horse Tavern on this site. There was a large yard including a pond and a drive leading south to the Watch House in the tree-lined 'Mile End old Towne'.

1746 the house was shown on Rocque's map as a large inn on the corner with White Horse Lane; but the Ordnance Survey map of 1870 shows no building on this site.

This area was home to wealthy merchants and sea captains; the main industries were farming, particularly fattening up animals for the London markets, with ropeworks and breweries.

1808 it is recorded that 'William Green, brewer, of Whitehorse Street, Stepney hereby agrees to supply and keep the following Drays, Slings etc., the property of Messrs. Charrington . . . ' MoD, THL and HCR. 1790-1796 WGT.

**Moor Lane, Cripplegate, EC2** c.1715 listed as a tavern.
c.1820-1900 trading as the 'Old White Horse'.
c.1901 it was renamed 'Whittington', to become 'The Olde Whittington'.

**Mugwall Street, 'Within Cripplegate', EC2** Possibly a corruption of Monkwell Street.
1660 a ¼d. token issued by Jacob Hickman of the White Horse in Mugwall Street – *a horse saddled and bridled.* GCW No. 1951. This die stamp may indicate that it was a posting-house – or for the carriage of mail.

**New Burlington Street, W1** 1757-1778 listed as an Inn or Tavern, which was used for Masonic meetings. As the inn was sited 'on the corner' it was possibly the same as 169 Regent Street which was trading in 1826; the building of Regent Street took place from 1811 until 1830.

**16 Newburgh Street, W1F 7RY** In 1824 an inquest was held at the 'White Horse, Carnaby Market'. After the Market had closed, most of Newburgh Street was rebuilt.
1830 listed as trading. CoW.

**Nicholas Shambles, EC1** 1636 recorded as a tavern, which would have served the slaughterers and butchers who traded here. After the Great Fire of 1666 this street became Newgate Street.

**North Row, St. Pancras, NW1** 1790-1796 listed as trading. WGT.

**North Street, Barking, IG11 8JE** Originally in North Road, but since 1945 the premises have stood on the corner of North Street and London Road.
Landlord. Trade lists: 1914-1937.

**4 North Street, W1** 1809-1811 listed as a tavern.
1826 listed unders North Street, Fitzroy Square.
c.1843 the area was famous for its little greens and taverns, some of which were tippling houses without licences standing amongst the tumble-down cottages; the tavern was rebuilt in what became Chitty Street.
1922 the house was purchased by the brewers, McMullen & Sons.
1940 listed as trading.
1959 the premises were sold. McM.

**Ockendon Road, North Ockendon** 'Old White Horse'. c.1700 a map held by Chelmsford Record Office shows the tavern standing in its own land. Some walls were constructed with ships' timbers – presumably from Tilbury – which can still be seen.

c.1880 a Friendly Society was registered at the White Horse. TSE. Trade lists: 1839-1863.

**Old Bailey, EC4** 'White Horse and Black Boy', it is uncertain as to whether this was ever a tavern as it was listed under J. Collins, a Watchmaker. AH.

**Old Fish Street, EC4** c.1630 trading as a tavern.

**Old Ford, E3** 1851 listed by the Post Office.

**Old Gravel Lane, White Horse Alley (off Upper Ground on the south bank of the Thames), SE1** c.1750 listed as trading – with a cockpit.

The house was situated at the foot of Blackfriars Bridge, the construction of which started in 1760 taking the name from the Black Fryers Stairs on the opposite bank of the river.

1809 listed as 'White Horse Wine Vaults'.

**Orchard Street, W1** Originally the mediaeval orchard of Westminster Abbey; renamed Abbey Orchard Street. 1809-1827 listed as trading.

**Oxford Street, W1** 1746 Shown on Rocque's map as lying between Great Chapel Street and Angel Hill – the south side of Oxford Street.

**64 Palace Road, Bromley** 1938 listed as trading.

**Park Street, Borough, SE1** 1826 listed as trading.

**1-3 Parson's Green, SW6 4UL** A parsonage used to stand on the west side of this large green area, hence its present name. It is thought to have once been cultivated as a Roman vineyard.

1688 John Haines, Victualler, of Ye Olde White Horse was presented at the local Court Baron for setting up posts before his coaching house at Parsons Green and was charged six shillings and eightpence; he was ordered to remove them by the 1st April next – 'under paine of twenty shillings'. The White Horse by virtue of ancient tenure owned a square foot of ground at the northern end of the Green where the inn sign stood, supported in a curious piece of iron scrollwork.

1712, 19th August – *The Spectator* published an article by Joseph Addison (1672-1719), the essayist, gives a picture of the Parson's Green Country Fair with puppet shows, actor's booths and refreshment stalls 'at the top of the Greene near the White Horse, the fish stalls, oyster and whelks being in great demand'. There were competitions which included hot tea drinking, greasy pole climbing, treacle bobbing and the throwing of sticks at tethered live cockerels.

1777 A boy named Fennell, nicknamed The Giant, had the misfortune to incur the displeasure of John Wright, host of the White Horse, who so severely chastised him that the boy died. Wright became known as 'Jack the Giant Killer'.

c.1800 Billy Button, a well-known clown, gave performances on a horse at the Fair. At the end of its routine, the horse would throw the clown and bolt for the stables of the White Horse inn where it would be fed and watered.

1803 Parson's Green Fair was suppressed by the Magistrates.

1835 *The Times* reported that a cricket match took place on the Green played between eleven married and eleven single women, followed by country dancing. The match was made up for the sum of 10/- and a hot supper provided by The White Horse on the Green. One of the pioneer cricket clubs of England – The Fulham Albion – held their meetings here.

1882 the James family took over the freehold; they also owned the City Arms, the Rising Sun in Battersea and the Marlborough in Chelsea.

1894 the tailor's shop building at No. 3 Parson's Green was incorporated into the White Horse, this is perhaps when the rebuilding of the original house took place.

1900 on the days when polo was played at Hurlingham, between seventy and eighty carriages would be drawn up outside the White Horse and around the green.

1906 the inn sign on the square foot of land at the north end of the green was removed.

1913 the James family sold the freehold for £13,000 to Charrington & Co. of Mile End.

In the Lounge Bar there was a stained glass window commissioned from Spaulls of Chelsea, it was of pre-Raphaelite style and depicted a mediaeval traveller mounted on a white horse. It was blown out by a bomb during the Second World War; to be stored for many years in the cellar until it was sold; by 1984 it had become the property of an antique dealer in Bootle, Merseyside. A photograph of the window hangs by the main fireplace.

1979 Bass Charrington took over the management on the surrender of the lease.

1981 the formidable character, Sally Cruickshank became the manageress, with Mark Dorber in charge of the cellar. During her tenure the small figure of a white horse was stolen from the alcove above the first floor bow window of the old red brick building – sometime later to be re-instated. Sally retired in 1995 when Mark Dorber and Rupert Reeves took over as joint licensees.

2006 also known as 'The Sloaney Pony', there is an art gallery restaurant on the first floor and an outside terraced area; it is reported to offer the largest range of bottled beers in Britain. CJF and HFA. Trade lists: 1839-1870.

**Paternoster Row, EC4** *See Warwick Lane.*

**20 Peckham Rye, SE15 4JR** About a mile from the village of Peckham 'is a green, with handsome houses on one side, called Peckham-Rye'. Nearby there was an 'asylum for decayed victuallers, a spacious building surrounded by 6 acres of ground'.

Cattle drovers used the village as a stopping place before going on to the markets of London. Their herds were put out to graze whilst they themselves took refreshment in the inns.

1881 listed in the census under Peckham Fee.

1809-11 listed as trading. 1867 Queen's Road K. JB.

**Petticoat Lane, E1** 1826 listed as a Tavern.

**Piccadilly, W1** Until c.1900 there appear to have been two White Horse Inns trading at the same time – virtually opposite each other north and south of this main thoroughfare. The one standing on the south side of Piccadilly had been trading from at least 1677, that on the north side from about 1740.

Mr. Bryant Lilleywhite notes the following very true fact in his book 'London Signs' published in 1966 – *the authors of this book entirely agree with him*:

> 'The chronological history of the White Horse Inn, and White Horse Cellars, Piccadilly is difficult to unravel and the task is confused by untidy non-contemporary writings'.

**155 and 157 Piccadilly (south side)** Since the middle of the 18th century, this has been the accepted name for the street extending from Piccadilly Circus to Hyde Park Corner - but this has not always been the case.

In 1662 when King Charles II married the Portuguese Catherine of Branganze, this section of the ancient highway to the west became known as Portugal Street – possibly the date the Crown Lease commenced on the property.

1676 White Horse Inn Portugall Street, Pickadilly (sic.). This site, between St. James's Street and Duke Street is marked on 'A Large and Accurate Map of the City of London' by John Ogilby and his wife's grandson, William Morgan. It is unlikely that the Inn itself had frontage on Piccadilly as it is not marked on Blome's map of 1680.

1699 The ratebooks give Mr. J. Brown as an occupant.

In 1720 this property is described as consisting of the Inn, and nine other houses fronting Piccadilly, two houses on the north side of Jermyn Street, and a 'back house', where it is said Thomas Highmore lived and died. He had been Sergeant Painter to both King William III and King George I who had in 1718 granted him a reversionary lease on the house to end in 1740. Later descriptions show the whole plot as running 167 feet eastward to Piccadilly.

1737 and 1742-45 the inn was used for Masonic meetings.

1742 The Crown lease was renewed to Samuel Rush, to whom Highmore had assigned his interest; the Surveyor General reported at this time that part of the inn had been rebuilt, but that the other buildings, some of which were of timber constructions, were old and needed considerable repairs – along with the neighbouring Elephant Inn and Kings Arms Inn which were described as 'providing indifferent trade'.

1746 White Horse Yard, shown on Rocque's map on the south side of Piccadilly, opposite Old Bond Street with an opening into Dorants Hotel; no licences or recognizances for the White Horse Inn have been discovered after this date, but the large yard off Bury Street was still used as a coaching terminus.

1760 Robert Dodsley (1703-1761), the playwright, bookseller and publisher gives White Horse Yard, Berry Street (sic.) as being 'back of south site' in his Complete Guide. He also lists White Horse Yard, Piccadilly and White Horse Street – this last refers to the present-day White Horse Street opposite Green Park.

Robert Dodsley had been a footman to The Hon. Mrs. Lowther, encouraged by fashionable patrons he set up as a bookseller in 1755; he founded the Annual Register in 1758 and suggested compiling a dictionary to his contemporary Dr. Samuel Johnson (1709-1784).

1762 the Crown Lease was renewed, giving James Mackay as tenant of the inn.

1777, 3rd February – The Salisbury Flying Machine calls at the 'Old White Horse Cellar, Piccadilly'. *This alteration to the sign will have been due to the rebuilding which took place some thirty years earlier.*

1784 a renewal of the lease mentions White Horse Yard but not the inn; Mr. Rix's stage coaches from Putney used this yard, also that on the north side of Piccadilly at the White Horse Cellar.

1795-1799 Horwood's map gives 'Sta Y' [Stable Yard].

1799 No.157 Piccadilly was listed as 'White Horse Cellar Coffee House', by 1903 it is given as No. 155.

1814 'The Old White Horse', Piccadilly, where 'in a pleasant coffee-room passengers can wait for any of the stages, and travellers in general are well accommodated with beds etc. The departure of the mail coaches in the evening also attracted many sightseers'. At this time it 'was famous as a rendezvous of the 'bucks' of the period . . . who repaired thither for their morning rum and milk, and lounged about its doors'. *Gentlemen still rendezvous at their clubs in St. James's to this day; White's was founded in 1693, Boodle's in 1762 and Brooks's, in 1764.*

c.1840-50 the Old White Horse Cellar is shown as being adjacent to, and under the same management as the Bath Hotel at the corner of Arlington Street, having become a railway agency and 'house of call' for a few of the coaches that remained.

1891 'The Old White Horse Cellar' faded before the progress of the railways.

1903 the premises at 155 Piccadilly were demolished and by 1906 the Ritz Hotel had been built on the site.

LCG, OCE, LCC Survey XXIX of 1960, DNB, AG, and BDE Vol. III.

**67 and 68 Piccadilly (north side, at the corner of Dover Street)**
c.1740 'The White Horse Cellar was the terminus for mail coaches
travelling to Oxford and the west of England . . . Those alighting,
did so with more than ordinary relief at dangers passed;
and those setting out, a prey to terrors on account of the
highwaymen who infested the district of Knightsbridge
and Kensington'.

1743 John Williams, 'keeper of the White Horse in
Piccadilly . . . received victuallers' licences for the inn'. He so
named this house in honour of the Accession of the House of
Hanover in 1714 to the English throne. This may have accounted
for many Londoners years later being drawn to this part of Piccadilly on 4th June – the
birthday of King George III – when the White Horse Cellar was decorated with gaily
coloured lamps and the coachmen and guards of the Mail Coaches wore new scarlet coats
and the horses new harnesses, the brass trimmings being brightly polished.

1744, 14th May – 'Yesterday a set of brown horses with a coach belonging to
Mr. Williams at the White Horse in Piccadilly, ran from thence to St. Albans
(which is 20 miles) in two hours and five minutes, for a considerable sum of
money, which was much sooner than the time allowed'. BJ.

1747, 10th April – Horace Walpole (1717-1797) describes an incident which
took place at the inn:

> 'Williams, keeper of the White Horse in Piccadilly . . . gave information
> against the Independent Electors of Westminster for treasonable
> practices . . . Williams, being observed at their anniversary dinner, in
> order to make memorandum with a pencil, was severely cuffed, and
> kicked out of the company. The alleged treasonable practices consisted
> in certain offensive toasts. On the King's [George II] health being
> drunk every man held a glass of water in his left hand, and waved a
> glass of wine over it with the right'.

Walpole was writing to his friend Horace Mann (1701-1789), the British
Envoy at Florence to keep him informed of the results of the crushing defeat by
the Duke of Cumberland's troops at Culloden of Prince Charles Stewart – whose
father, the Old Pretender, was living in Rome at this time.

The incident was an example of the practice amongst Jacobites of drinking the
health of 'The King over the Water' – The Young Pretender Charles Stewart. To
quote from Sir Walter Scott's novel 'Redgauntlet':

> 'My father . . . compromised his loyalty as to announce merely 'The
> King', as his first toast after dinner, instead of the emphatic 'King
> George; . . . Our guest made a motion with his glass, so as to pass it over
> the water decanter which stood beside him, and added, 'over the water'.

1748 a year after Mr. Williams had been 'kicked out of the company' – Mr. Abraham Hatchett had become the owner of the New White Horse Cellar in Piccadilly. *The Gentleman's Magazine* of September 1798 mentions Mr Hatchett's death 'at his house at Wimbledon aged 79 . . . who for half a century kept the New White Horse Cellar in Piccadilly'.

1759, 8th August – Horace Walpole wrote from Arlington Street once again to Sir Horace Mann:

> 'I did send to the White-horse Cellar here in Piccadilly, whence all the stage-coaches set out, but there is never a genie booted and spurred, and going to Florence on a Sun-beam'.

1776, 16th December – The Shepton Mallet, Frome, Trowbridge and Devizes machines call at the New White Horse Cellar opposite St. James's.

1789 Mr Rix's stage coaches from Putney call at the Old and New White Horse Cellars Piccadilly 'where passengers and parcels are booked'.

1798, September – *The Gentleman's Magazine* recorded the death of Abraham Hatchett.

c.1809 a childhood reminiscence giving a vivid description of this inn was written in 1830 by Lord William Pitt Lennox (1799-1881):

> 'Few sights were more amusing than 'The White Horse' Cellars in Piccadilly, in the old times of coaching. What a confusion – what a babel of tongues! The tumult, the noise, was worthy the pen of a Boz, or the pencil of Cruikshank. People hurrying hither and thither, some who had come too soon, others too late. There were carriages, hackney coaches, vans, carts and barrows; porters jostling, cads elbowing, coachmen wrangling, passengers grumbling, men pushing, women scolding. Trunks, portmanteaux, hat boxes, bandboxes, strewed the pavement; orange merchants, cigar merchants, umbrella merchants, dog merchants, sponge merchants, proclaiming the superiority of their various wares; pocket knives with ten blades . . . trouser straps four pairs a shilling; bandana handkerchiefs, that had never seen foreign parts . . . London sparrows, as the coach-makers would say 'yellow bodies' were passed off as canaries, though their 'wood notes wild' had never been heard out of the sound of Bow Bells . . . Members of the Society for the Diffusion of Knowledge were hawking literature at the lowest rate imaginable . . . the Prophetic Almanac, neatly bound, one penny; 'a yard and a half o songs' for a halfpenny; . . . occasionally the music of a guard's horn'.

1812, Saturday 24th October - ' . . . about 7 o'clock in the evening . . . It is to be regretted that some police officers are not stationed near the White Horse Cellar about the hour of the departure of the evening stages, as it is well known that it is considered by those nocturnal depredators a favourable opportunity for effecting their schemes of plunder'. Reported in the Sunday edition of *Bell's Weekly Messenger*, referring to gangs of pick-pockets.

c.1820 The historian William Hazlitt (1778-1830) gives a view of the scene from Down Street:

> '. . . but give me, for my private satisfaction, the mail coaches that pour down Piccadilly of an evening, tear up the pavement, and devour the way before them to the Land's-end'.

A comic song from Country Cousins by the actor, Charles Mathews (1776-1835) describes the inn as 'full of most honest and boisterous fun'.

1824, Thursday 4th March. Writing from The Hague, Harriet Granville, the wife of the newly appointed Ambassador to France, describes their hotel:

> 'We are all bent upon getting into our house as soon as possible, as this hotel is almost too bad to bear, as noisy as the 'White Horse' in Piccadilly. Bad dinner, worse beds'.

1822 listed as Hatchett's Hotel and New White Horse Cellar – the Proprietors are given as Hatchett and Company. It was the stabling place of the four-horse coaches which were run every summer as recorded in the following nursery rhyme written in 1846:

> *"Up at Piccadilly oh!*
> *The coachman takes his stand,*
> *And when he meets a pretty girl,*
> *He takes her by the hand;*
> *Whip away for ever oh!*
> *Drive away so clever oh!*
> *He drives her four-in-hand".*
> GC and JOH.

1826-27 The White Horse Cellar – better known as Hatchetts – was used by Woolcott, Carriers from Exeter and Sherborne.

1835 a caricature by George Cruickshank (1792-1878) entitled 'The Piccadilly Nuisance' shows the life around the building, Hatchetts White Horse Cellar, with The Devonport Mail in the foreground. This inn was also illustrated the following year by the artist James Pollard (1797-1859).

1836 records show that The Rocket departed from the White Horse Cellar to Cambridge, also W. Chaplin & Co. sent coaches to Hastings from this Cellar.

In 1837 the interior was described by Charles Dickens in his Pickwick Papers:

'The traveller's room is of course uncomfortable . . . It is the right-hand parlour into which an aspiring kitchen-fireplace appears to have walked, accompanied by a rebellious poker, tongs, and shove. It is divided into boxes, for the solitary confinement of travellers, and is furnished with a clock, a looking-glass, and a live waiter, which latter article is kept in a small kennel for washing glasses, in a corner of the apartment.'

*This same venue is used in Chapter III of Bleak House published in 1852.*

1866 Larwood and Hotten in their 'The History of Signboards' wrote:

'The White Horse Cellar, Piccadilly, now a tame omnibus office was, for more than a century one of the bustling coaching-inns for the west country coaches . . . '

1888, 13th July – The Brighton coach 'Old Times' was driven from The White Horse Cellar to Brighton and back for a wager of £1,000 to £500 against it being accomplished in eight hours. The coach ultimately arrived back in Piccadilly ten minutes under the stipulated time.

A booklet was published by the proprietor of Hatchett's Restaurant entitled 'Old Coaching Days, a Souvenir of The White Horse Cellar, Piccadilly'. The Cellar was shown as being situated beneath the hotel. The illustration on the back cover was entitled 'Hatchetts in 1870'. This was reprinted by Unwin Brothers after 1888.

By 2007 the site was occupied by Korean Air. CA, VS, GC, AG.

**Pickax Street, EC1** Originally the central part of Aldersgate Street. 1677-1755 listed as trading. This was possibly the inn which served the White Horse Yard in Fans Alley as shown on Rocque's 1746 map.

By 1918 the site had become business premises. HAH.

**Popes Head Alley, Cornhill, EC3** 1608-11 licensed as a print shop, the first to be opened by John Sudbury and George Humble. LH 1898.

**9-11 Poplar High Street, E14** 1690 recorded as a tavern and was probably trading even earlier under the name White Horse. Its situation – which later became No. 11 – overlooked Stonebridge pond, a popular watering-place; the space in front of the modern premises may confirm this fact.

*This tavern has two curious stories, both with the same denouement.*

c.1731 Mary East, a young girl, assuming the guise of a man, called herself James How – another girl acted as her 'wife'. With £500 from a court action settlement together with savings, they purchased this tavern, living as man and wife, evading detection for several years.

1752 Mary, obviously well thought of, was elected an overseer of the poor.

c.1765 Mary's 'wife' died; about this time a woman who had blackmailed Mary in her younger days again threatened her; two 'police officers' also threatened her with exposure. She gave the blackmailers a cheque for £100 – one of the men was caught and sentenced to four years penal servitude. At her death in 1780 she was buried in Poplar Chapel.

*The second story came from 'The Gentleman's Magazine'.*

1766, 21st October – 'A remarkable trial came on at Hicks's Hall in which the mistress of the White Horse was plaintiff, and one William Barwick defendant. During the course of the evidence, it appeared that the defendant, had extorted several considerable sums of money from the plaintiff, for concealing her sex; of which he was convicted to the satisfaction of the whole court. His sentence was to stand three times on the pillor, and to suffer four years imprisonment in Newgate'.

*Hicks Hall was built in 1612 in St. John's Street and was replaced by the Old Middlesex Sessions House in 1779.*

1826 Listed as an inn or tavern; by 1870 it had been rebuilt.

1874 the lessee protested to the Metropolitan Board of Works that it would be fatal to the business to remove the sign – a white horse 'modelled with great verve' of lead and filled with sand, which stands on a post in front of the public house; it has no known history but is thought to date from the early to mid 18th century.

1891 the freehold was sold for £3,750 to a lieutenant in the Hampshire Regiment who sold it in 1921 to brewers for £4,350.

1927 rebuilt, incorporating No. 9 at the corner of North Street which became Saltwell Street. This building had been the hamlet and parish watch-house prior to 1831. THL and SL.

**53 Portpool Lane, WC1** c.1700 listed in 'Perpool' Lane. By 1720 the premises were chiefly used for stabling.

1809-1827 listed as a tavern.

**Portugal Street, W1** 1676 White Horse Inn Portugall Street, Pickadilly sic. *See Piccadilly.*

**70 Princes Street, Leicester Square, W1** 1826 listed as trading.

**Queen Street, EC2** c.1761 listed as trading. Possibly situated off Cheapside.

**23 Queen Street, Rotherhithe, SE16** This may have referred to the 'Litle White Horse nere Ratclif Cross'.
   1881 listed in the census under Rotherhithe Street.

**Ratcliff Cross, E1** The most important station for watermen east of the tower – used by Samuel Pepys (1633-1703). An undated ½d. token issued by Will Baker of 'Litle Whit Hors nere Ratclif Cross' – *a horse*.
   1746 shown on Rocque's map as 'White Horse Yard'. JR.

**Rathbone Place, W1** The inn is thought to have stood to the north, towards Tottenham Street.
   1797 used for Masonic meetings.

**50 Ray Street, Clerkenwell, EC1** 1826 listed as trading.

**169 Regent Street, W1** 1757-1778 trading as an Inn or Tavern and used for Masonic meetings. As the inn was sited 'on the corner' it was possibly the same as New Burlington Street which was trading in 1826; the building of Regent Street took place between 1811 and 1830.

**473 Roman Road, Old Ford, E2** Described as "Very Old" by the Landlord, the house was still trading in 1989. Trade lists: 1823-1845.

**16 Rood Lane, EC3** 1826 listed as a tavern.

**Rosemary Lane, E1** Once an infamous street market for old clothes and frippery. 'Ye White Horse opposite ye Victualling Office on Tower Hill'.
   Two ¼d. tokens, one dated 1667 – *a horse;* the other of 1669 – *a horse saddled and bridled.* Both issued by William Evered of Rosemary Lane. GWC Nos. 2401 and 2402. This die stamp may indicate that it was a posting-house – or for the carriage of mail.
   1746 Rocque's map shows White Horse Court off Rosemary Lane.

**Rotherhithe Street, SE16** 1881 listed in the census.

**Royal Exchange, EC3** This replaced the first Royal Exchange built in 1669. c.1765 'The White Horse under the Piazzas of the Royal Exchange Cornhill' where Wm. Hannell was listed as a Printseller, together with other craftsmen such as milliners, apothecaries and goldsmiths. The whole building was burnt down in 1838.

**45 Rupert Street, Soho (formerly No. 37), W1D 7PB** The fields of this area once abounded with hares. The word Soho originates from hare-coursing, when the greyhound's owner would release his dog with the cry of 'See Ho'; in 1720 the area was described as 'a pretty handsome, well built Street'.

The site of the White Horse public house at No. 45 was occupied by a tavern of that name between 1730-1830.

1998, 21st July – *The Evening Standard* reported:

'A gram of cocaine has been cut on the window sill of the White Horse and is being snorted by two druggies. It's an every day problem around the White Horse . . . the lavatory bowls are made of stainless steel and the cistern lids screwed down to prevent addicts from hiding their stash.' STR.

**2 St. Clements Lane, WC2 2HA** Ye Olde White Horse. In 1700 the historian, John Strype (1643-1737) describes the building as the 'White Horse Inn fronting the market, and the market . . . [which is] very considerable and well served with provisions, both flesh and fish; for besides the butchers in the shambles, it is much resorted to by the country butchers and higglers[5]'.

1809-27 listed as Gilbert Street, Clare Market.

**St. James's Market, SW1** 1670 a ½d. token issued by I.M.H. of 'The White Horse in St. James Market Place' – *a horse.* GCW No. 2533.

**St. John's Street, Clerkenwell, EC1** *See Liverpool Road.*

**St. John Street, Spitalfields, EC1** This road was recorded in 1170, it was originally for pack-horses only, to become the main coaching route north from Smithfield.

This tavern may have stood on the corner with Brick Lane.

1789-1792 listed as trading.

**St. Margaret's-Hill, Southwark, SE1** See The Borough 1708-c.1760 known as the 'White Horse Inn on the W. side of St. Margaret's-Hill, near the middle'.

1746 Rocque's map shows White Horse Stables nearby on the west side of The Borough.

**24 St. Martin's Lane, WC2** Originally trading as tea gardens, it became the White Horse Livery Stables – known as 'Hornby's Livery and Private Stables'.

1820, 10th October – from an advertisement in *The Times*:

'a Pony, Gig and Harness to be Sold, together or separately; the pony is sound, fast, and quiet in harness or to ride; the gig is handsomely and tastily built in the Stanhope style, with drop box and low steps – very little used'. JHM.

**St. Pauls Churchyard, WC2** 1738 Geo. Foster is given as 'at the White Horse a map-printer'; he published and sold 'A New Plan of the City of London'.

**Seething Lane, EC3** 1746 Rocque's map shows White Horse Yard on the east side just south of the Navy Office.

**1 Selhurst Road, South Norwood, SE25 5PP** Situated in old Croydon and was probably part of the ancient manor of Whitehorse which in 1367 was owned by Walter Whitehorse (Whithors).

1839 listed as trading – the year the London and Croydon Railway was constructed.

1841 – the census return gives the licensee as a Licensed Victualler; the household consisting of a labourer aged twenty, two servants – Sarah aged thirty, and the other aged fifteen.

1851 – the census return gives the landlord as an innkeeper who was born in Cheshire, his wife came from Suffolk; Sarah, still the house servant, is shown as aged forty three.

c.1870 Mr and Mrs Paxton took over the inn. The following is an abridged description of life in the inn, given by the nine year old Henry Charles Stevenson. He gives a unique picture of a coaching tavern on the outskirts of London in his youth which he sent to the present owner of the White Horse from his home in America. His maternal grandfather, Mr Paxton, 'maintained' the White Horse Tavern together with his wife. At the age of seven Henry attended the Oval Road School:

> "I had to walk to school from Woodside to Croydon through what was known as the London Fields, which was a series of fields divided by hedges over which there steps that as I remember were called stiles, I have wondered since if the London Fields later became Croydon Airport – *alas they became a large railway junction and depot*. My mother used to visit my grandparents fairly often and would take me with her, sometimes I used to go alone which I enjoyed doing as I liked to be with my grandfather – his principal business was the manufacturing of patent leather. He was a very stout man and had rather short arms. He smoked cigars and a long clay pipe which he had trouble lighting and puffed on, at the same time he had me light his pipe for him . . . I always liked to be at the White Horse Tavern as there was always a lot of activity with people coming and going in coaches and on horseback – they let me do little odd jobs for them. There was one place that always fascinated me called the Washhouse where sand and water were being used to clean the pewter mugs in which the ale had been served".

1874 listed as a tavern in Kelly's Directory. CLS. Further information is available from Surrey History Centre; Ref: QS5/10 1785-1935. Trade lists: 1839-1867.

**Shepherds Bush (sic), W12 8LH** 1798-1811 'White Horse Inn Shepherds Bush'.

1881 there was a White Horse listed in Shepherds Bush Green which was known as Lawn Place. These two could have been connected to the White Horse Brewery and they may have constituted one complex of buildings linked around the junction of the roads. HFA. *See Uxbridge Road.*

**64 Shoreditch High Street, E1 6JJ** The popular derivation of the name Shoreditch was so called after Jane Shore who had died in this ditch.

According to the Dictionary of National Biography of 1997 Jane Shore, mistress of Edward IV (1442-1483), died c.1527. 'There is no foundation for the story that Jane Shore gave her name to Shoreditch. That appellation existed long before her time'.

Situated on the corner with Redchurch Street; the first tavern outside the City walls and therefore not subject to City rules. Standing on the main road to the north, surrounded by marshes and farmland.

1462 first recorded as a tavern. *This is the earliest date recorded for a White Horse Tavern we have discovered in London.*

1547 listed as 'le Whytehorse parish of St. Leonard Shordyche'.

1577 William Shakespeare (1564-1616) started his stage career in the Curtain Theatre next to the White Horse in Curtain Meadows and used the tap room here as a setting for some of his plays.

1826 listed as trading.

By 1935 it had been rebuilt.

c.1985 the sign showed a white carthorse in a marshy meadow, at a later date it fell down in a storm and was not replaced.

c.1985 Visited by the author's researchers who found topless go-go dancing at lunch time. GCW and JTK.

**16 & 17 Shorts Gardens, Drury Lane, WC2** 1826 listed as a tavern.

**50 South Audley Street, W1** 1809-1827 listed as trading.

**Southall** Since 1698 horse auctions were held weekly in Southall market.

1809-1811 listed as trading.

**Spicer (or Spicey) Street** 1826 listed as trading.

**Stable Yard, Westminster, SW1** An undated ¼d. token issued by I.I.N. of 'The White Horse in Stable Yard Westmin' – *a horse prancing.* GCW No. 2940.

The 1746 Rocque map shows a stable yard on the west side of Smith Street.

**1 Stafford Row, Pimlico, SW1** 1809-1826 listed as a tavern.

**Strand, WC2** Originally a bridle path running alongside the north side of The Thames. In this street of just under a mile in length, there have been three White Horse inns viz:

> a) c.1670 listed as 'White Horse, near Arundell House'. The Royal Society held their meetings at Arundel House from 1666 until 1674.

> b) An undated token issued by John James at 'without Temple Barr I.K.I.' – *a horse and sun.* GCW No. 3051. By 1702 these premises had become 'White Horse and Sun' and were occupied by a Bookseller.

> c) 1826 listed as 354 Strand – 'White Horse Tavern, north side just east of Exeter Exchange'. Exeter Change had been built c.1676, where Edward Cross had his menagerie from 1773 to 1829; he kept lions, tigers and monkeys – also a hippopotamus which Lord Byron said looked like Lord Liverpool, and a sloth that looked like his valet. The building was demolished in 1829 and the menagerie moved to the Surrey Zoological Gardens.

**52 Theobalds Road, Redlion Square, WC1** Once the route to King James I's house at Theobalds near Waltham Cross, Hertfordshire. 1809-1827 listed as a tavern.

**Tooley Street, SE1** 1492 listed as 'Near the corner of Tooley Street and the Bridge Gate, Southwark'.
The inhabitants of this area were wealthy citizens, together with priors and abbots. 1746 Rocque's map shows White Horse Court leading into Church Yard Alley and so into Tooley Street.

**Turnmill Street, White Horse Alley** c.1670 listed as trading. 1746 shown on Rocque's map.

**Tyburn, W1** 1751, 28th January – a news entry in the *Suffolk Mercury* tells of a thief who 'mounted his horse and rode off through Hyde Park. When he came to the gate facing Grosvenor Square he dismounted and gave a man sixpence to take the horse to the White Horse Inn in Tyburn Road, and deliver it there to the hostler. The man was suspicious but did not succeed in detaining him'.
Rocque's contemporary map of 1746 shows Grosvenor Gate Lodge opposite King Street, leading to Grosvenor Square. The White Horse Inn would have been one of the un-named buildings situated almost a mile eastward along the main Tyburn road from the triangular gallows, which were not dismantled until 1783.

**Union Street, Southwark, SE1** Possibly the site as shown on Rocque's 1746 map as the White Horse Stables off The Borough. Trade lists: 1809-1839.

**Upper Ground Street, White Horse Alley, Southwark, SE1** 1746 shown on Rocque's map as 'White Horse Alley and White Horse Court'.

**31 Uxbridge Road, W12 8LH** 1722 listed in the licensing records, and in the Licensed Victuallers Records of 1829.

c.1830 some forty stage-coaches passed by on the daily run to and from Oxford.

1988 part of this large premises had become Hucklebury's.

2006 the pub has a very small frontage under an imposing archway, which may have been the carriage entrance. The clientele are mainly Irish; there is a huge screen for showing horse racing. A milestone still stands outside the building. Trade lists: 1829-1870. HFA.

**Wandsworth, SW18** 1809-1811 a waterside inn or tavern.

1881 listed in the census under Waterside. Trade lists: 1823-1870.

**Wapping Wall, E1** An undated ¼d. token issued by Samuell F. Wiseman of 'one Wapping Wall' – *a horse's head bridled.* GCW No. 3354 – this token may have been used by the White Horse Inn.

1720 and 1826 listed as trading.

**Warwick Street, W1** 1755 White Horse Court was shown as leading into St. James's Park. *This house may have stood in White Horse Court, south of Cockspur Street.* JHM.

**16 Warwick Lane / Paternoster Row, EC4** Trading from c.1680-c.1820.

**Welbeck Street, W1** 'The White Horse, corner of Welbeck Street, Cavendish Square, was long a detached public-house, where travellers customarily stopped for refreshment, and to examine their firearms before crossing the fields to Lisson Green'. *Welbeck Street was built in 1720. Lisson Green is entered in the local Rate Books for the first time in 1723.* JT.

**Welling (Hill Grove)** c.1700 The small isolated alehouse stood on a rough track at the edge of the steep north-facing escarpment now known as Hill Grove; overlooking Wickham Lane to Bostall Woods. Here the famous highwayman, Dick Turpin (1706-1739) had a hideout in a cave – Fanny, the proprietress of the White Horse, used to put a light in the window to tell him when the coast was clear. Shooters Hill lies a little over two miles away on the Dover Road where Turpin was reported to have staged his hold-ups having hidden amongst the tangled undergrowth – then to ride 'hell-for-leather for the shelter of Bostall Woods'. As a result of Fanny's escapades, this alehouse became known as 'The White Horse, Fanny on the Hill'.

The illustration shows the original small white building on the left; the brick building and probably the central porch being later additions.

1953, 16th October – the premises became rent free until 1957, it was then demolished and a new public house was built in nearby Wickham Street.

c.1980 the house was renamed 'Fanny on the Hill'.

Beasley's Tenants Register. *See Wickham Street.* WA and CL.

**West Smithfield, EC1** Listed by Bryant Lilleywhite in his 'London Signs', but he was unable to identify the site of this White Horse. *We have taken it to be the one in Chick Lane which was trading in 1720.*

**Wheeler Street, Spitalfields, E1** An undated ½ d. token issued by Alexander Byrchet in Wheelers Street – *a horse saddled and bridled.* GCW No. 3378. This may indicate that it was a posting-house – or for the carriage of mail; the token may have been used by this White Horse Inn. 1736-1811 listed as a tavern.

**Whitcombe Street, Charing Cross, WC2** 1794, 19th August – *The Gentleman's Magazine* reported that 'The White Horse public-house (a recruiting house wherein Edward Barrett, a mariner, had been ill-treated) was saved this evening from destruction by the intervention of the military'.

1906 the house was still trading.

**Whitechapel, E1** c.1670 'White Horse Inn south side of Whitechapel near Colchester Street'.

1760 included in LCG; 1809-1811 listed as trading.

'White Horse and Leaping Bear near Whitechapel Church', occupied by Mr. Harrison, a liveryman. This may not have been a tavern.

**Whitecross Street, without Cripplegate, EC1** 1746 Rocque's map shows White Horse Court.

**1-3 White Horse Hill, Chislehurst** Originally a coaching inn. The stables, with the coachman's quarters, above, are still standing but have been converted to offices.

1936 the inn sign, painted by H. Milton Wilson, was loaned by Whitbread & Co. Ltd. for the Inn Sign Exhibition in New Bond Street.

1975 this hotel was reopened after extensive alterations.

c. 1992, renamed 'The Penny Farthing'. At about this time, the murderer, Steve Wright of Ipswich notoriety, worked here.

Landlord and DTel 22ⁿᵈ February 2008. Trade Lists: c.1742-1938.

**48 White Horse Road / Street, Limehouse, E1 0ND** There is a legend that the body of a Kentish Warrior Chief slain by the Danes was brought along the Thames by riverboat; to be carried up the present White Horse Street for internment in the nearby hollow ground beside the little chapel that once stood here – now the site of St. Dunstan's Church. The effigy of a Saxon White Horse, emblem of the ancient Saxons, was erected nearby.

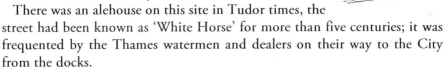

There was an alehouse on this site in Tudor times, the street had been known as 'White Horse' for more than five centuries; it was frequented by the Thames watermen and dealers on their way to the City from the docks.

1595 the inn was described as in 'Whitehorstrete'.

1615 the street is given as White Hart Street, the inn with several acres of land is shown on a map of 'The Worshipful Company of Mercers'.

1668 a heart-shaped ½ d. token issued by Robert I. Beckitt of White [Horse] Street in Stepny (sic.) – *a horse*. GCW – no number.

1682 William Morgan's map shows this thoroughfare had become White Horse Street; a White Horse Lane is also shown.

1760 listed as a coaching inn. It is thought by the landlord that the flagstones in the yard of the building may have dated from this period.

1817 Johnstone's Directory records that the street was 494 yards long with 112 houses.

1938 the street was renamed White Horse Road.

1969, February – the premises were sold to the Greater London Council.

BC, THL and LCG. 1805 WCO.

**11 White Horse Street, W1J 7LL** 1746 shown on Rocque's map as running north from Piccadilly to Shepherds Market; Robert Dodsley (1703-1761) includes this street with others as being 'named from the sign'. *It was he who founded the Annual Register in 1758 and suggested compiling a dictionary to Dr. Johnson.*

c.1970 it had become the Penthouse Club and by 2000 it was the Iceni.

**Wickham Street, Welling** c.1957 built to replace the White Horse alehouse at nearby Hill Grove.

c.1980 renamed 'Fanny on the Hill'. *See Welling.*

**Wood Street, EC2** Formerly the 'Coach and Six Horses'. By 1746 the 'White Horse Inn' appears on Rocque's map on the east side of Wood Street, about midway between Lad Lane and Love Lane.

By 1918 warehouses had been built on this site.

**704 Woolwich Road, SE7 8LQ** 1783-1803 listed as 'White Horse and Star'.

1809 listed as 'White Horse, New Charlton' – an inn or tavern.

1897 Rebuilt. WGT.

**8 & 9 Worple Way, Marshgate, Richmond, TW10 6DF** 1690-98 "Whit Hors" in Richmond. Tavern.

1869 listed as trading but it is thought that this inn was almost certainly established before this date.

1892, February – listed in the Return of Licensed Houses and Beer Houses as being licensed on and off the premises with Jacob Nevard as landlord; it was owned by Mr. Gomms of the Beehive Brewery, Brentford. The brewery was taken over by Fullers of Chiswick in 1908. 1966. SUR, BL.

**Notes**

1. By 2006 named Rush Common which has many mature trees, some having been in the gardens of large houses. The area is signposted as a tree trail.

2. A type of Latin American big-band dance music. Dancers wear high heeled shoes. Col.

3. a) A rigid circular band of metal or wood. b) A large ring through which performers or animals jump.

4. *The Agrarian History of England and Wales* written by A. Everott and J. Thirsk, published in 1967.

5. Those who sell provisions from door to door. Related to 'haggle'. SOE.

# MERSEYSIDE

*In 1938 a Viking longship of Nordic clinker design was uncovered beneath a pub car park on Merseyside, now thought to be a popular Viking settling place.*

*The county of Merseyside was formed in 1974 from areas previously located in Lancashire and Cheshire, taking its name from the River Mersey.*

## Liverpool

1834 P. *It is not known to which inn this date refers.*

**Jordan Street** 1858 PO.

**Parliament Street** 1858 PO.

**Prescot** 1822/3-1834 P; 1848-1855 S; 1858 PO; 1898 K.

**St. Helen's** 1881-1913 K.

**Woolton, 2 Acrefield Road** The village has become a suburb of Liverpool.
The White Horse, a small public house, was the first in Merseyside to have a no-smoking policy. Landlord.

# NORFOLK

*'The name is but slightly altered from the original, Northfolc, a term used by the Saxons to distinguish the inhabitants of the north from those of the south. It was anciently inhabited by a tribe of the Iceni, and subsequently became a part of the East Saxon kingdom under Uffa, about the year 575, who united the counties of Norfolk, Suffolk, and Cambridge, under the title of the kingdom of East Anglia.*

*Norfolk supplies the adjacent counties as well as the metropolis with poultry of all kinds. There are extensive herring and mackerel fisheries. The chief manufactures are those of stuffs[1], silk, cambrics, and calicoes. The town of Worstead is famous for a peculiar sort of woollen goods.'*
JB.

**Ashill** 1795 NoR.

**Ashwellthorpe, 49-55 The Street** c.1744 this inn or alehouse would have been trading, because in June 1794 it was recorded that 'Mr. Christopher Brown of the White Horse in Ashwelthorpe died in his 94th year; he had kept this house upwards of fifty years'.

Originally the building had been three cottages; the front door of one having been bricked up; three people from the same family have reported seeing a ghostly presence of a lady trying to enter through the blocked-up middle door.

From the accounts of William Huggins, Overseer of Workhouse records:

> 1775, 1st May. Expenses at White Horse when the agreement was made with the doctor for curing Chapman's wife 1s. 7½d.

> 30th May. Paid the doctor in part for curing Chapman's wife £2 2s. 0d.

1829, 22nd September – An Agreement of Sale records Robert Colman, the landlord, as a victualler. The Agreement was concerned with the sale of a farm at Wicklewood near Wymondham.

From the Census taken in the following years:

1841 – Thomas Claxton publican aged 39 with his wife Cathy 42 and daughter Mahala aged 4. Also Eliza Maning aged 18, and Rosemund Coleman recorded as a publican, aged 70.

1851 – Rosemund Coleman as 'vitular declined'. Robert Elliot, victualler aged 30, his wife Elizabeth of 37 and the children, Anne Maria 7, Mary Ann 3, and Robert 2. Two lodgers, both widowers aged 62 and 70, and an agricultural labourer, Thomas Ward aged 13.

1861 – Robert Elliot 43 years old, innkeeper, his wife Elizabeth aged 46 and daughter Mary Ann aged 13.

1871 – Robert Elliot innkeeper now aged 53, together with Elizabeth aged 56 had been joined by Mary Ann, their twenty four year old married daughter, and her one year old child, Ann E. Huggins – *possibly connected to William Huggins the workhouse overseer.* There were two lodgers John Shearing aged 66 and Mary Barber a year older.

1864, 18th August – the house was bought by the brewers Youngs, Crawshay and Youngs from St. George's Brewery.

1879 Mrs. Elizabeth Elliott was listed as publican.

1881 – Elizabeth Elliot, innkeeper, aged 68 lived with her son John 29, a carpenter, his wife Mary of 25 and their baby daughter Emily. Three lodgers were railway workers – 22 year old William Slight born in Chatham; William Kemp aged 51 and James Brown aged 45 from Cambridgeshire. *Some persons' ages do not always to concur with the passing of the years!*

1890 John Elliott took over from his mother as licensee, for no more than two years.

1897, 27th November – the freehold was sold by Youngs, Crawshay and Youngs Partnership to their new company for completion in February the following year.

c.1955 Robert and Rose Howes took over as innkeepers.

c.1962 the landlord was listed as . . . Cooper – Rose Howes maiden name was Cooper.

1996, March – the pub became a free house.

GM, NoR. P/C 88/3 Account William Huggins, Mr. and Mrs. Peter Smith. 1789-1799 NoR; 1846 PO; 1858 PO; 1875 NTD; 1900-1937 K.

**Attleborough, White Horse Street, London Road** 1789 listed as trading and is likely to have been a coaching inn.

1830 James Gayford was listed as innkeeper – he was also listed at the 'White Hart' in the same year – this also happened ninety years later when Arthur Jolly was landlord.

1845 William Gayford had taken over as licensee.

1897, 27th November – the Youngs, Crawshay and Youngs Partnership sold the premises to what had now become a company for completion in February the following year.

1900 Arthur Jolly is listed as landlord of the White Horse; however in the gazetteer of Kelly's Directory for this year he is listed under the White Hart, along with Walter Clabburn who is listed in the separate trades section.

During the 20th century an extension was added to this public house which became a restaurant known as White Lodge.

BC. 1789-1799 NoR; 1822-1830 P; 1850 S; 1865 K; 1875 NTD; 1900-1937 K.

**Aylsham, Millgate** 1620 recorded as trading under White Horse with William Kilby as innkeeper.

1723 Thomas and Charles Forster were the landlords.

1734 Mary Forster took over as publican.

1741 William Rannalls is recorded as landlord, but two years later he died and his wife took over. She remained here for almost twenty years.

1794, 11th October – the inn was Lot No. 31 in the sale of St. Martin's Brewery in Norwich.

c.1850 the house was rebuilt when John Nicholls was landlord; he was aged 32 and described as a beerhouse keeper and master carpenter; two years later in 1852 he was described as a butcher, he remained here for 38 years until 1890 when his wife Susanna took over at the age of 68.

1900 Stanley Nicholls took over as licensee.

1906 the house was closed by compensation and the property sold.

KSGH. 1789-1799 NoR; 1830 P; 1875 NTD; 1900 K.

**Blakeney, 4 High Street** In the 13th century King Henry III granted Snitterley a market; by the 14th century this working port had become known as Blakeney. Cargoes were taken down the steep street backwards, their weight being braked by the horses facing up the hill.

The original building, dating from the 15th century, was rebuilt during the 17th century when it became a coaching inn and the market town's first hotel.

1803 the population was 618, by 1828 the port owned fifty vessels with an average of 65 tons burden; the population had reached 929 by 1831.

1830 William Thompson Storey was listed as innkeeper.

1839 Pigot's Directory lists James Thompson Storey as landlord, however Robson's Directory lists W.S. Storey. By 1845 Ann Storey had taken over.

1878, 8th June – the hotel was Lot No. 25 in the sale of Reepham Brewery (Bircham & Sons). It was then let to John Pye by the new owners, Henry Bullard and John Boyce at an annual rent of £18 – this did not appear in the register until 1883. The house had a tap room, bar, bar parlour, large back parlour, pantry and wash house, underground cellar, six bedrooms, two attics and a yard with a cart lodge and stabling for nine horses.

c.1910 Sir Henry Birkin, the former Bentley racing driver, was a frequent visitor; since the 1950s the inn has been frequented by yachtsmen.

1916, 5th June – Herbert Long became the landlord, he remained here for 34 years when Susie Long took over in 1952; she was licensee for 21 years.

1978 a smugglers tunnel was discovered underneath the White Horse car park, it ran from the area of the British Legion car park to the Blakeney Hotel – the site of the former Crown and Anchor.

1980 the courtyard was converted into a family room.

A photograph on the bar wall shows a pre-war lifeboat crew commemorating a rescue – the men's hands having frozen to the oars.

1992, June – the pub became a freehouse; the roof over the family room was removed.

By January 1998 the courtyard and stables had been converted a conservatory and restaurant. RM, JB, KSGH and Mr. and Mrs. Bishop. 1789-1799 NoR; 1846 PO; 1858 PO 1865 K; 1888 *S.O.* K; 1908 *S.O.* – 1937 K.

**Boughton, Church Lane** 1861 Charles Smith, a butcher, was listed as licensee.

By 1917 Frank Shearwood was landlord, he remained here for 28 years until Horace Shearwood took over in 1945.

1929 the owners, Bagges of Kings Lynn sold the premises to Steward and Patteson.

1967, 30th January – as it was no longer thriving, the pub was closed. 1937 K.

**Bradeston** 1789-1799 NoR; 1846 PO; 1858 PO; 1875 NTD; 1908-1922 an hotel K.

**Brancaster Staithe, Main Road** Situated on the marshland coastline by the Norfolk Coastal Path which connects with the Pedlars Way.

1836 Samuel Dowdy was listed as licensee.

1851 John Tebble aged 44, a brickmaker, took over as landlord; his family ran the inn for the next 21 years.

1905, 1st April – the house was purchased by E.C. Quilter, it traded as 'Bidwells'.

The original premises of this hotel became a private dwelling and was demolished in 1964, it stood to the east of the present building which dates from 1934.

1936, 30th May – an article in the *Eastern Evening News* credits the design of the building to Donald C. Chastney, the surveyor for Bullard & Sons Ltd., the builder was R.G. Carter of Drayton and the sign was painted by Miss K. Skelton of Horsford. The house included a fine bar, a good smoke room and a large lounge for parties. There were 3 single bedrooms and 3 double for letting.

1963, 9th December – a supper licence was granted.

c.1990 the hotel was renamed the Lobster Pot.

1997 the new owner reinstated the original name.

By 1999 more hotel bedrooms – with a grass roof, a conservatory for a restaurant and a terrace had been added. GM and Cliff Nye – local historian. 1846-1858 PO; 1865 K; 1875 NTD; 1900-1937 K.

**Bressingham** 1869 listed as trading with William Jolley as landlord, by 1872 he was also listed as a thatcher.

1969 , 2nd October – the house was closed by the brewers Watney Mann Ltd. 1790 NoR; 1937 K.

**Briggate (formerly called Bridge Gate, a hamlet in the parish of Worstead)** Worstead was the centre for manufacture of the woollen fabric 'worstead' during the middle ages. *The author has a jacket and skirt of worstead Jacob tweed manufactured for her by students at the Bradford Technical College in around 1960. It is wonderfully hard-wearing and smart, quite different to everyday tweed suits.*

The house is situated close to the ruins of a large steam and water mill by the North Walsham to Dilham Canal which was constructed in 1823.

1830-1839 the public house was run by the Watson family but owned by the brewer, George Morse, who kept a flock of ornamental sheep at Catton just north of Norwich – which were thought to have been Jacobs.

1831 George Morse merged his brewery with Steward and Patterson – an agreement stipulated that the pubs were to remain 'the separate property of the partners in whom they are now vested'.

1837-1851 George Morse was still recorded as the owner of the inn; he died the following year.

His son Charles held shares in the business but, with the other partners, ceased to own individual pubs.

1856-1883 William Hannant was listed as licensee – he was recorded as a shoemaker in 1861 and a farmer with 7 acres in 1881.

The house continued to trade with a full licence until 28th January 1957 when it was closed to become a private house. In the final year of trading 58⅛ barrels of beer were sold; Reginald Norman Feek, a brewery representative, was the licensee.

The building was later converted into a private house and by 2008 the name was still painted on the door and the lane had become White Horse Lane.

JM and KSGH. 1789-1799 NoR; 1830 North Walsham P; 1846-1858 PO; 1875 NTD; 1900-1937 K.

**Briningham, The Street** 1789 recorded as trading as White Horse.

1836 given as White 'Hart' in Whites Directory. John Maris was the landlord, he was aged 55 and lived here for 33 years; in 1854 he was also listed as a shopkeeper.

1851 John Maris, aged 70, was listed as licensee, by 1854 he was also listed as a shopkeeper.

1869 Henry Barwick, a baker and shopkeeper was listed as licensee.

1884, 12th December – Richard Oliver took over and was the landlord for 30 years.

1897, 28th July – the house was sold by S.H. Brereton, who had leased it to Steward and Patteson; it was Lot No. 4 in the auction at the Hastings Arms Hotel, Melton Constable. The house had a bar parlour, kitchen, cellar, small sitting room, 5 bedrooms, a yard, stables, cart shed, outhouses, offices, a well, gardens, a paddock and a small pightle[2] of land – total 1 acre 2 rods. The premises were bought by the brewers Morgans, who sold them in 1961 to Bullards.

1960 the sales for the year were 60 barrels of beer and 18 spirits.

1966, 27th June - the pub was closed. 1799 NoR; 1846-1858 PO; 1892-1908 *S.O.* K.

**Brisley, Fakenham Road, Wiggs Green** In 1848 the house was owned by Steward and Patteson, brewers.

1861 listed as trading with George Coe as licensee.

1881 Alfred Winearls – or Wineards – took over as landlord, he was also a blacksmith.

By 1895 Elijah Eyres was the owner.

1909, 5th March – the house was referred to compensation and in October it was closed.

## Brundall

*The Street* 1836 listed as trading with Sarah Bailey as landlady.

1845 Sarah Agus was licensee, she was aged 58.

1869 Samuel Fiske took over as landlord, he was also an agricultural labourer.

1874, 13th July – George Dingle became the landlord, he and his family ran the pub for 43 years.

1877, 3rd September – a provisional order was made for the removal of the licence to new premises to be erected near Station Lane. This White Horse pub was renamed the Old Beams.

*Station Lane* 1879 listed as White Horse Railway Hotel and Posting House.

1883 listed as White 'House'.

1888 and 1904 the house was listed as the White Horse Inn.

1916 Kelly's Directory listed Harry Spalding as landlord, but he does not appear in the licence register until 1st January 1917. In 1888, before he came to the White Horse, Mr. Spalding was thought to have been the youngest ever licensee in Norwich. His family ran the hotel for 56 years.

1999 the pub closed and had been demolished by March 2001. 1937 K; 1960 JM.

**Buxton** 1795 NoR.

**Carbrooke, Watton Road** 1836 listed as a beerhouse with George Catton as landlord, he was also a wheelwright and in 1871 was recorded as having 1.5 acres of land.

1881 George's daughter, Miss Elizabeth Catton, took over as licensee.

1908 Henry Jolly became the landlord – in 1900 there was an Arthur Jolly who was landlord of the White Horse in nearby Attleborough, he was also listed in Kelly's Directory as being at the White Hart at the same time.

1938 the house was renamed the Flying Fish and was granted a full licence. 1853 PO; 1937 K.

**Catfield, The Street** 1836 listed as trading with Richard Dye, aged 39, as licensee, he was also a carrier. His family ran the public house for 43 years.

1930, 3rd February – the final licence for the White Horse was issued.

1931 the brewers, Lacons closed the house.

1789-1799 NoR; 1858 PO; 1875 NTD; 1892 *S.O.* K; 1916 K.

**Cawston, New Street** Situated diagonally opposite to the old school.

1794, 11th October – The White Horse was Lot No. 16 in the sale of John Days Brewery.

1850 William Spark, the licensee aged 45, was also a wheelwright.

1861 James Miller the licensee was a dealer.

1865 William Neale, a labourer, was licensee and remained here for the next twenty five years.

1878, Saturday 8th June – the house was Lot No. 9 in the sale of Bircham & sons Reepham Brewery. The description of the property included: a tap and parlour, kitchen, pantry and wash house; also a parlour with cellar underneath, three bedrooms with shelving rooms behind. The large yard which had side gates, included a wood house, a stable with loft above and a large club room with a 9 pin bowling ground beneath. By this time William Neale was paying an annual rent of £11.

On 26th November both the White Horse and the Bell public houses were conveyed to Henry Bullard and John Boyce. Bullards' records state that the Bell was also known previously as the White Horse. *[Not researched by the authors]*

1900 the White Horse was listed as trading.

1906, 10th February – the Norfolk Chronicle reported that the licence had not been renewed at the Aylsham Petty Sessions. It was referred for compensation on 10th June the following year.

KSGH, LHS Cawston. 1789-1799 NoR; 1830 P; 1846 PO; 1858 PO; 1865 K.

**Chedgrave, 5 Norwich Road** Built c.1640 with its own bowling green. The taproom is believed to be the only one in Norfolk which used to be the men's snug.

1826 the public house is shown on Bryant's map.

1836 listed as trading with Philip Smith as landlord, his wife Mary took over in 1839.

1841 James Watson, aged 35, was listed as licensee; he was also described as an overseer, and in 1858, as a market gardener and poor rate collector.

1925 listed as trading.

1937 A.S. Chittock is listed as licensee; in c.1962 a Mr. Chittock was still running the White Horse. KSGH. 1789-1799 NoR; 1846 PO; 1850 S; 1875 NTD.

**Cley-next-to-the-Sea** 1789 NoR; 1830 P.

**Coltishall** 1845 listed as an hotel.

1888 listed as trading in Kelly's Directory.

1900 listed as an hotel.

By 1925 the house was no longer trading under White Horse. KSGH.

**Cranworth, Woodrising Road** 1850 listed as a beerhouse with Robert Cobb as licensee, he was described as a cooper, but the following year the house was recorded as a private dwelling named White Horse House, Robert Cobb being the occupant, he was listed as a farmer of 20 acres.

1854 Robert was still recorded as a farmer and a cooper, but not a licensee.

1871 the census names the house as 'late' White Horse. 1789-1798 NoR.

## Cromer

1665 token issued by Richard Beaney of Cromer – *a horse trotting.* GCW No. 23. This token may have been used by the White Horse Inn in West Street.

*24 West Street* Originally a coaching inn, this listed building is situated on the important main street, which in later years, when it became an hotel led to the Eastern and Midland Railway Station.

1767 the rates list records land 'formerly belonging to the White Horse'.

1789 and 1790 there is no reference to the house in the Alehouse Recognizances.

1792 listed as trading with Phaba Mason as innkeeper.

1884, 24th September – listed as trading with Arthur Crisp as licensee.

1911, 20th March – Frederick Crisp took over the inn until 1927. The Crisp family had run the inn for 43 years.

1892, 28th October – the inn was leased by Emma Watts to Morgan's Brewery for 21 years at an annual rent of £98, the tenants paying for all repairs and insurance.

1893, 15th September – Mary Cooke is recorded as having an interest in the property; it was offered for sale by auction at the Royal Hotel in Norwich on the 6th May. The premises, copyhold to the Manor of Felbrigg, comprised a bar, smoking room, bar parlour, private sitting room, cellar, kitchen, pantry, storeroom, etc., front and back staircases and a WC. On the first floor there was a club room, sitting room, 10 bedrooms and a WC. In the yard there was a stable, 2 loose boxes, slaughter house, pound, closets, etc.

1894, 17th January - William Watts was recorded as having an interest in the property which was sold by the trustees of John Brown to Benjamin Cook on 25th January.

1912, 5th October – the house was sold by the trustees of Benjamin Cook to Arthur Humfrey Mason.

1947, 31st January – Harold Reaney is listed as licensee, he remained in charge until c.1969.

In 1960 the sales recorded by Morgan's, the brewers, were 218 barrels of beer and 94 spirits. Landlord and KSGH. 1792-1799 NoR; 1822-1830 P; 1846 PO; 1858 PO; 1875 NTD; 1900-1937 K.

*It is not known to which of these houses the following dates apply:* 1792-1799 NRO; 1822-1830 P; 1846 PO; 1858 PO; 1875 NTD; 1900-1937 K.

**Garden Street** 1865 listed as trading in Kelly's Directory, it is thought that Thomas Gray was licensee.

**Crostwick, North Walsham Road** 1794, 11th October – the inn was advertised as Lot No. 2 in the sale of Mr. John Day's Brewery.

1797, 13th July – advertised for sale as the property of the late Mr. Day.

1836 William Woodhouse, aged 46, was listed as licensee and shopkeeper; by 1861 Hannah Woodhouse had taken over.

Between 1837 and 1851 the owner is recorded as the brewer, George Morse.

1869 Edward Money, aged 34, was listed as licensee and a carpenter, by 1872 he was also a grocer. He remained the landlord until 1890.

By 1872 the owner was Charles Morse of Aylsham.

1885, 22nd December – the ownership of the premises was transferred to the brewers Steward and Patteson.

The artist Sir Alfred Munnings (1878-1959), acclaimed for his canvas works of horses and landscapes, lodged here in 1900 for 14 shillings a week. He described the house, situated on Crostwick Common, as a thatched and lime washed inn surrounded by farm buildings with its tall sign standing some yards away.

He wrote in his autobiography:

'It was there, liking the place on a ride out of Norwich, that I arranged to paint through the Autumn. I launched this skirmish from my Norwich rooms. With the help of my old friend and open landau driver George Claxton, we came upon the Common, there were donkeys, young and old, a coloured cow or two together with some geese and ponies – and a white horse peacefully standing on a knoll – the wind stirring its tail. What a Common! I had made many such journeys, but the memory of this never-forgotten drive has outlasted all others. The journey ended, the good driver had dinner and beer and left. I did my pictures at Crostwick, undisturbed. An old man, woman or boy fetching a donkey, or roaming children were part of the scene'.

Referring to horses he said:

'Although they have given me much trouble and many sleepless nights, they have been my supporters, friends – my destiny in fact'.

The Landlord, Ted Snelling moved in c.1892 and farmed the surrounding forty odd acres; Sir Alfred described him as a good landlord; he had a round face and was clean shaven, except for a small tuft on his chin.

1915, 16th October – Edward Thaxton took over as licensee, his family ran the inn until 1940. 1790-1799 NoR; 1846-1858 PO; 1875 NTD; 1900-1937 K; 1960 JM.

**Dersingham, Hunstanton Road** 1861 listed as trading with William Smith as licensee, he remained here for 30 years. The house continued to trade until 11th May 1973 when it was closed by the brewers Watney Mann.

1869 K; 1875 NTD; 1900-1937 K; 1960 JM.

**Dickleburgh, Langmere Green** 1826 the White Horse was shown on Bryant's map.

1840, 2nd September – the premises were listed as Lot No. 57 in an auction and bought by Mr. Taylor for £325. The licensee was listed as William Barrett, aged 45, who was landlord here for over 30 years.

1875 John Vyse was recorded as licensee and a carpenter; his wife Mary took over in 1890. 1796-1799 NoR; 1846-1858 PO; 1875 NTD. These dates were listed under Langmere Green.

**Diss, 20 Market Place** 1822 listed as trading in Pigot's Directory with John White as landlord.

1856 George Wright was listed as licensee and a builder. In 1872 he was a builder, but had become a wheelwright and market bailiff.

By 1875 John Nichols had taken over, he was an engineer.

1969 the White Horse was still listed as trading.

1789-1799 NoR; 1830 P; 1846 PO; 1853 Church Hill – *St. Mary's Church stands opposite* – PO; 1875 NTD; 1900-1937 K.

**Downham Market** 1793 NoR.

**East Barsham**

***Fakenham Road*** The White Horse Inn was built during the 17th century midway between Fakenham and Little Walsingham. It is said that King Henry VIII stayed at East Barsham Manor when he visited the well known shrine at Little Walsingham, walking the two miles barefooted to present a necklace of great value to the shrine.

c.1967 a wing was built on to the front elevation of the inn.

2007 a Grade II listed free house.

AAT, KSGH, JB, and inn brochure. 1789-1799 NoR; 1846-1858 PO; 1865 K; 1892 *R.S.O.* – 1900 (Walsingham) –1908 *S.O.* – 1937 K.

**28 Yarmouth Road** 1836 listed as trading with Joseph Holman as licensee.

1858 John Barnes was listed as landlord, James Barnes took over in 1861 and stayed here for at least 30 years.

1878, Saturday, 8th June – the public house was Lot No. 49 in the sale of Bircham & Sons Reepham Brewery. The house was let to John Barnes at £30 per annum. The premises consisted of a tap room, wash house, small bar, a good cellar at the rear with a room over, 3 bedrooms, a large attic, foreground with stable, loft and enclosed cart lodge. Four cottages adjoining the house and a large piece of land on the opposite side of the road were also included. The house was conveyed to Henry Bullard and John Boyce on 26th November.

1979, 29th May – the pub was closed temporarily.

**East Dereham** 1792-1799 NoR.

**East Harling** 1879 K.

**East Rudham** 1836 listed as trading with John Roling as licensee.

1851 William Page, aged 46, was landlord, also a farmer with 15 acres.

1854 trading with Philip Kendall as licensee, and also listed as a butcher.

c.1900 the house was closed. 1789-1799 NoR; 1846-1858 PO; 1875 NTD.

**East Runton, High Street** 1845-1856 Joseph Bird aged 51, a blacksmith, was licensee.

1851 the White Horse was sold by Ambrose Mayes to William Primrose of Trunch Brewery.

1861 Benjamin Thain, also a blacksmith, was landlord, he was followed by James Buddrell who was licensee until 1899.

1895, 30th October – Betsy Neal Primrose and Oldman Carter sold the house to Morgans who owned it until 1961.

1899, 27th November – Robert Burrett/Buddrell is listed as licensee; he was also agent for the Great Eastern Railway Company.

1917, 3rd December – Mrs Martha Lines became the landlady.

1960 Morgans recorded the sale of 160 barrels of beer and 69 spirits.

The pub continued to trade until 1998 when it closed, to re-open in November 2000. KSGH. 1853 PO; 1875 NTD; 1900-1937 K.

**East Ruston** 1789 NoR.

**Edgefield, Cross Way / Norwich Road** 1794, 11th October – the house was advertised as Lot No. 23 in the sale of John Day's Brewery.

1805 the house was referred to in a smuggling case.

1826 the White Horse is shown on Bryant's map.

1836 listed as trading with Francis Woods as licensee.

1841, 14th-17th October – the house was advertised as Lot No. 84 in the Coltishall Brewery sale.

1864 William Broughton was listed in White's Post Office Directory as licensee and a victualler.

1897, 11th October – the premises were conveyed by William Johnson Jennis Bolding to the brewers Steward and Patteson, who on 3rd August 1929 bought eleven other public houses for £17,820.

1959 the sales for the year were 22⅛ barrels of beer, the owners, Watney Mann, recommended immediate closure as the house was in poor condition, the trade was poor, and the toilets and cellar were poor.

1960, 30th May – the licence was removed to the Coachmakers at East Dereham and the White Horse became a private house.

1789-1799 NoR; 1846-1858 PO; 1892 *S.O.* K; 1900-1937 K.

**Ellingham, Yarmouth Road / Bungay Road** 'The Old White Horse'. There was a smithy attached to the inn which flourished until c.1999 when it was taken over by ornamental iron work craftsmen.

1836 listed as trading with David Olley as landlord.

By 1845 James High had taken over the licence, he was also a wheelwright; 45 years later Henry Culley, another wheelwright ran the pub.

1962, 29th May – the brewers Bullards and Steward recorded the sales for the previous year as 82 barrels, no change to the house was recommended.

1969, 10th December – the pub closed to become a private residence.

1858 PO; 1875 NTD; 1900-1937 L.

**Flordon** 1794 NoR.

**Foulsham** 1839 listed as trading with Robert Smith as landlord, the owner was Richard Le Strange, a brewer of Norwich.

1851 Valentine Neale, a farmer aged 52 took over the licence.

By 1879 a wheelwright, John Springall was landlord.

1882, 4th September – an order was made to transfer the White Horse to new premises on the road to the railway station.

1883, 3rd September – a certificate was granted for the new house which was in the Parish of Guestwick; during this year Richard Le Strange sold the premises to Bullard & Son.

1885 John Springall died and his wife Elizabeth took over the licence on 21st December.

1966, 9th March – the pub was closed and sold to become a private house which by 1997 was named Barn Owl House. 1846 PO; 1875 NRD; 1900-1929 Guestwick K.

**Freethorpe** 1790 NoR.

**Garvestone, Church Street** 1836 listed as trading with Ann Jarvis as landlady.

1914, 6th February – the pub, referred for compensation, was closed and sold the following year by Steward and Patteson. 1789-1799 NoR; 1846 PO; 1875 NTD; 1900 K.

**Gayton** 1858 PO.

**Gaywood, 7 Wootton Road** Situated at the north east corner of the entrance to the Salters' Road, used by the salt workers and carriers in the 11th century.

1823 advertised as a well appointed inn.

1830 listed as trading with Matthew Wood as landlord.

1851 Miles Hawes aged 39 became the licensee; Martha Hawes took over in 1854.

By 1861 Edmund Langley was landlord, his wife Mary had taken over by 1865.

1871 James Skerrey was licensee, in 1875 he was described as a beer retailer.

1878, 22nd July – the owner, Elijah Eyre, sold the freehold public house by auction to Morgans the brewers.

1881 White's directory lists Samuel Endledow as licensee; 1883 Kelly's directory lists Sarah 'Engledow' as landlady.

1935 the White Horse was transferred into Kings Lynn Borough.

1960 the sales for the year were recorded as 139 barrels of beer and 55 spirits. HEB, KL&W and Landlord. 1789-1799 NoR; 1830 P; 1846 PO; 1865 K; 1900-1937 K. c.1800.

**Gorleston-on-Sea, 41 Burnt Lane** Originally situated in Beccles Road, the White Horse was thought to have been a coaching inn.

1803, 24th December – the premises were conveyed to either Bells Brewery or Ancestors.

1836 listed as trading with Christopher Marjoram as licensee.

1845 the inn was leased to Steward and Patteson, brewers, who bought it 20 years later on 28th June 1865.

1846 the inn was still listed as trading in Beccles Road.

1850 Elijah Seeley was listed as landlord, his wife Ann took over in 1856 and was landlady for at least 11years.

By 1881 the address had changed to No. 1 Burnt Lane.

1934 the house was described as the last property on the south side of the road. It was damaged by enemy action during the Second World War.

1846-1853 PO; 1865 L; 1890 Southtown WDG; 1900-1937 K; 1960 JM.

**Great Fransham** 1789-1799 NoR.

**Great Hautbois**[3] The name is probably derived from the ridge of high ground, formally wooded and bordering the valley on the east side of the Bure River – lofty trees, as distinguished from shrubs. The manor was granted to Peter Herman who took the name of 'De Alto Bosco' or 'Haytbois'.

c.1235 Sir Peter founded the Hospital of St. Mary here for pilgrims and the poor visiting St. Benet's Abbey.

c.1300 the manor, which became Hautbois Castle, held good messuages and seven cottages.

c.1700 there were considerable number of houses standing by the roadside called the Town of Hautbois Magna.

1728 there is documentation showing the house at this date.

1796, 21st May – the house was Lot No. 3 in the sale of Coltishall Brewery.

1801 the census records the population as 68.

1826 the inn is shown on Bryant's map.

An early Victorian photograph shows the yard with a substantial two storied house in the background covered in creeper.

1831 in 30 years the population had grown to131.

1836 listed as trading with Jeremiah Gaze as licensee.

1864, 6th January – Frederick Press took over as landlord.

1871 Frederick's wife Mary became licensee.

From c.1866 the premises were owned by Bullards.

1912 the bowling green was partly washed away by floods.

1971, 9th February – the hotel, as was closed by the brewers Watney Mann.

2000 the village, which had sometimes been known locally as'Hobbies or Hobboys' was no longer shown on the Ordnance Survey Road Atlas.

JB and SOE. 1789-1799 NoR; 1846-1858 PO; 1865 K; 1875 NTD.

**Great Massingham** 1795 NoR. There was no further listing of a White Horse.

## Great Yarmouth

In 1791 the Royal Mail Service came to Yarmouth, having left London from the White Horse in Fetter Lane at six o'clock – to arrive by 'dinner next day'; which could account for the three public houses using the same sign as that of Fetter Lane. LPT.

**East Street** 1822 P; 1850 S. *Listed under Yarmouth.*

**Gaol Street / 30 Middlegate** 1819 listed as trading with Robert Page as landlord.

1845, 23rd August – the brewers Paget & Co. sold the premises to Steward & Co., later to become Steward and Patteson, who hold documents which record the name being changed to The Red Lion some time after 1854. 1839 P; 1846-1853 PO.

**King Street** At one time known as the Old White Horse Tavern, it was situated on the south east corner with Howard Street.

1819 listed as trading with James Duck as landlord.

1854 Thomas Rolfe was listed as licensee of the White Horse, however by 1865 the name had been changed to The Oxford.

1822-1830 P; 1846 PO; 1850 S; 1858 listed under Yarmouth PO.

**13 Northgate Street, White Horse Plain** Situated near to an old public weighbridge – no longer in use; it is thought to be the oldest hotel in the town. A white stone horse's head featured on the corner of the building.

c.1640 recorded as a coaching inn which in c.1800 was known as The Golden Keys.

1819 Henry Dawson was listed as innkeeper.

1822 the address was given as North End; in 1836 it was Church Plain and in 1846-1854, once again North End.

1865 Mr. D. Snowling was listed as licensee of the White Horse Inn, he and his family ran the inn for 14 years, they were followed by the Hammond family who were innkeepers here for over 50 years.

1934, 15th January – George Hammond was convicted for selling alcoholic drinks out of hours, he was fined £5, or spend one month in detention.

1937 listed at No. 20 Northgate.

2004, April – the hotel was closed and purchased by developers who converted it into three town houses. 1822-1830 P; 1850 S; 1875 NTD; 1900-1916 K.

**Guestwick** *See Foulsham.*

**Hapton** 1826 shown on Bryant's map as the Hapton Hole.

1836 listed as trading with Robert Dix as licensee, in 1845 he was followed by his son John who was aged 28.

1876, 26th May – a document held by the brewers Cann and Clarke record the house as 'late Hapton Hole'.

1890 John Lee Green was listed as licensee, he was also a blacksmith.

1894, 11th May – the brewers Morgans became the owners.

By 1967 the house had closed.

1789-1799 NoR; 1846 PO; 1858PO; 1875 NTD; 1900-1937 K.

**Harleston** This public house had previously been the toll house; in 1654 the owner was recorded as Robert Smyth.

By 1679 the property was owned by Rachel Jacobs, who married John Dove.

Some time during the 19th century the house had become Curl's Shop.

**Hempstead, The Street** 1836 listed as a beerhouse with Thomas Wright as licensee.

1851 George Money, a master shoemaker was listed as landlord, Elizabeth Money took over the licence in 1856.

1865 John Neal, a farmer and overseer, was listed as licensee; he ran the pub until 1893.

1896 John Gurney of Northrepps sold the house to Morgans the brewers.

1910, 25th July – Henry Williamson, the landlord of the white Horse was fined 5 shillings plus 4 shillings costs for permitting drunkenness on the premises.

1920, 21st May – Sidney Riseborough took over as landlord, he remained here for 36 years.

1949, March – the pub was granted a full seven day licence.

1960 the sale records for the year showed 47 barrels of beer and 13 spirits.

1962 26th April – the pub was closed.

1789-1799 NoR; 1846 PO; 1858 PO; 1888-1892 *R.S.O.* K; 1908 *S.O.* 1937 K.

**Hickling, The Green** 1841 Coltishall Brewery advertised the premises for sale as Lot No. 53. William Neal was listed as landlord.

1861 George Abigail, a carpenter was listed as licensee.

1879 listed, probably in error, as the White Hart. Frederick Gibbs took over the house which he ran for 31 years until it was closed.

1910, June – Steward and Patteson closed the pub by compensation and it was sold. 1790-1799 NoR; 1846-1858 PO; 1888 *S.O.* – 1900 K.

**Holme-next-the-Sea, Kirkgate Street** The building dates from the 17th century.

1826 the house was shown on Bryant's map.

1836 listed as trading with Robert Bloomfield as licensee.

1845 listed as a beerhouse.

By 1861 Thomas Bond was the landlord, he and his family ran the house for over 50 years.

1980 listed as trading.

1789-1799 NoR; 1850 Thornham S; 1858 PO; 1875 NoR; 1900-1937 K.

**Horsham Saint Faith / Horsham le Faith, Turnpike Road** 1794 The White Horse was advertised as Lot No. 3 in the sale of St. Martin's Brewery.

1871 listed as trading under White Horse.

From c.1875 Viscount Ranelagh leased the premises to the brewers Steward and Patteson.

1877, 1st December – John Flaxman, the landlord, was fined 10 shillings plus 17 shillings costs for selling alcoholic drinks out of hours.

1883, 27th October – John Rudd, aged 49, moved in with his wife Fanny, aged 39. She took over the licence in c.1903.

1938, 10th May – Henry Ball Watling, the landlord, was given two fines, each of £1 for selling alcohol out of hours.

1956 the sales record for the year show 16 barrels of beer were sold.

1957, 3rd March – the house was closed and the licence removed to the Jolly Farmers at Ormesby. 1789-1790 NoR; 1883 K.

**Ingoldisthorpe, Hunstanton Road** 1820, March – the freehold of the beerhouse was sold by Cornelius Pateman Herbert, a brewer of Wormegay, to George Hogg and Thomas Allen.

1826 the house was shown on Bryant's map.

1845-1869 Isaac Flight was listed as licensee, aged 51 in 1851 while at the White Horse.

The house was renamed the Ship Inn and continued to trade until 11th May 1970 when it was closed by the brewers Watney Mann.

**Kenninghall, Market Place** According to a surveyor, the original building dates from c.1550 and was given a Victorian façade as were the other buildings in the Market Place.

1830 listed as trading in Pigot's Directory with John Garnett as licensee.

1840, 2nd September – the White Horse was Lot No. 37 in the sale of the estate of Robert Sheriffe, it sold for £940.

1881 George Rolfe was listed as licensee, he was also a tailor, draper and grocer.

1898, 1st February – Youngs, Crawshay and Youngs were recorded as the owners.

1962, 29th May – at the first joint committee meeting of Bullards and Steward and Patteson it was decided not to make any changes to the White Horse; the sales were given as 80 barrels of beer and it was stated that the house was capable of further development.

2007, 9th November – the Eastern Daily Press published and article on The White Horse Inn being granted permission to serve alcohol and food, accompanied by music until midnight every night.

KSGH. 1789-1799 NoR; 1846 PO; 1850 S; 1875 NTD; 1900-1937 K.

**Kettlestone** 1836 listed as a beerhouse with John Grimar, or Guymer, as landlord, Ann Guymer had taken over by 1845, she was aged 70.

1895 listed as trading, the brewer Elijah Eyres was the owner.

1861 John Colman was the landlord, his family ran the house for at least 95 years.

1900 the inn was taken over by Morgans Brewery.

1956 the house was closed. 1937 K.

## Kings Lynn

**Grass Market** This was an area between the High Street and Broad Street.

1577 was the earliest reference to the White Horse.

1752 the house was recorded in the St. Margaret's Churchwardens Accounts and was listed under New Conduit Ward.

**King Street / Checker Street** 1548 was the first reference of the White Horse.

1767 the house closed; it was described in the deeds as standing on the east side of Checker Street.

**26 Queen Street** HEB and KL&W.

**Lakenham, 12 Trafalgar Street** 1846 PO; 1859 RD; 1900-1922 – a victualler – J; 1924-1935 K.
These premises were sometimes listed as White House.

**Little Cressingham, Watton Road** This was once a row of old cottages with stables at one end.

1891 Watton Brewery was recorded as having a 'trading interest' in this freehouse.

1845 William Cook was listed as licensee.

1858 the landlord, James Tolman was described as a blacksmith and an innkeeper.

1871 James was also listed as a farmer and engine owner.

By 1879 James' wife Matilda had taken over the inn, she was a farmer and a blacksmith.

1889 Matilda died aged 69 and her son Walter took over the licence.

1895 Walter died aged 34 and his brother John aged 24 became the landlord, he was a farmer and a blacksmith and ran the inn for at least 39 years.

c.1929 the owner was Henry Truman Mills of Hilborough Hall.

By 1935 Reginald Foster of nearby Clermont Hall had become the owner. During this year John Tolman's sister Emily Hoggett took the licence, she was aged 63. By 1939 she had remarried and changed her name to Emily Sutton.

1956, 12th October – Victor Tolman, a relative of Emily, became the landlord.

1958, 14th February – James Hoggett, probably Emily's son, took over the house and remained here to c.1969.

The Tolman family had run the White Horse for at least 111 years.

1962, 9th February – the six day licence restriction was removed.

1964 the owner of the inn, Reginald Foster died and Sir Richard Prince Smith of Driffield in Yorkshire took the ownership.

2004, July – the house closed.

Landlord. 1789-1799 NRO; 1846-1858 PO; 1875 NTD; 1892 *S.O.* – 1937 K.

**Longham, Wendling Road** c.1750 the original house was a clay lump and flint building, with a blacksmith's shop attached; originally there had only been one room.

1836-1845 Thomas Winter, a blacksmith, was listed as licensee.

1851-1856 Richard Winter, also a blacksmith was aged 45 when he became licensee.

1969, June – a report prepared for the brewers Watney Mann recommended immediate closure 'trade is poor – no bathroom and still has earth closets'!

1977 the house was closed and the Killengrey family were obliged to leave having run the pub for 85 years. It reopened as a freehouse after refurbishment.

2000 the old well was excavated and found to be 35 feet deep with lovely clear water.

2008, 14th March – the Eastern Daily Press published a picture of the White Horse with Barry White, the landlord outside – he had put the house, up for sale after twelve years. 1789-1799 NoR; 1846-1858 PO; 1865-1937 K; 1960 JM.

**Morton-on-the-Hill** 1789 the White Horse is recorded as trading by the Norfolk Record Office.

1836 listed as a beerhouse and trading under 'White Hart', in error, with John Blyth, a wheelwright, as landlord.

1845 John was listed as a shopkeeper, wheelwright and victualler.

1851 Frances Blyth, aged 67, was listed as licensee, shopkeeper and farmer of 28 acres.

1861 the landlord, Robert Thomas, was also a carpenter and farmer of 40 acres.

1869 Henry Snelling took over the licence, he was a farmer of 44 acres; his family ran the house until 1882 when another farmer, Robert Arthurton took over.

1900 Robert was still listed as licensee.

1912, 11th October – the brewers Bullards did not renew their lease and the house had ceased to trade by February 1913.

KSGH. 1789-1799 NRO; 1846-1858 PO; 1875 NTD; 1900 K.

**Mundesley, 18 Cromer Road** The Patron was the King, as Duke of Lancaster.

Only listed in 1836 with Robert Summers as licensee – he is also listed under Northrepps as a victualler and carrier. JB.

**Neatishead, The Street** The original hostelry dates from c.1600. Alterations took place during the 19th century. It was known to have been a post house which would have supplied the horses for the postriders and travellers. There was a smithy and wheelwright on the premises and a saddlery on the other side of the road with a notice bearing the words '121 miles to London'. This conjures up the same vivid picture of the village blacksmith given in 1839 by the American poet, Henry Longfellow (1807-1882):

*"Week in, week out, from morn till night,*
*You can hear his bellows blow;*
*You can hear him swing his heavy sledge,*
*With measured beat and slow,*
*Like a sexton ringing the village bell,*
*When the evening sun is low.*
*And children coming home from school*
*Look in at the open door;*
*They love to see the flaming forge,*
*And hear the bellows roar,*
*And catch the burning sparks that fly*
*Like chaff from a threshing-floor."*

1841, 14th to 17th September – Coltishall Brewery, who owned the inn, was put up for sale – the White Horse was Lot No. 46.

1845 listed as trading with Robert Watts as licensee.

1900 George Winston is listed as licensee.

1925 listed as trading with Herbert E. England as licensee.

Standing in the Norfolk Broads, it is 300 yards from the moorings at Lime Kiln Dyke. Two small museums containing items of local interest are on display in the house. There is an interesting old shed outside.

Landlord, KSGH and T&H. 1789-1799 NoR; 1846-1858 PO; 1875 NTD; 1898-1937 Hotel. K.

**New Buckenham, King Street** 1573 this rendered, timber-framed house was already trading as an inn.

c.1600 an addition or rebuild was added to the east end; by 1626 the building had become two dwellings.

By 1750 the larger building had become the White Horse, or Rampant Horse.

1830 John Wells, a carrier called every Tuesday morning on route to Norwich.

1879 Charles Levett was licensee, also a coal dealer and carrier; by 1888 his wife, Elizabeth, had taken over.

Early in the 20th century the inn temporarily ceased to trade and the bowling green became a market garden.

1912 C. Woodrow was licensee and he continued as landlord until 1933.

By 2007 this very old building had become a private house.

Paul Rutledge, Local Historian. 1789-1790 NoR; 1845 W; 1846-1858 PO; 1875 NTD; 1900-1925 K.

## Norwich

In 1760 there were five White Horse alehouses listed in this city. Twenty five years later, when the mail coaches started to run from London to Norwich, almost double this number had started to trade in the seven parishes within the city.

At least three landlords were worstead weavers or were connected in some way with the weaving industry; wool and silk manufacture had been revived by the Flemings during the reign of Queen Elizabeth I, who gave them asylum when they were driven from the Netherlands by the Duke of Alva. Norfolk sheep were considered valuable, their neck wool being considered equal to Spanish fleece, known as 'Merino'.

Coach drivers preferred to use grey horses for the night runs which may have accounted for the popularity of this pub sign. White's 1883 Directory of Norwich gave 584 public houses and 52 beerhouses, hence the traditional jingle of unknown origin:

*'A pub for every day of the year*
*And a church for every Sunday.'*

With the advent of the railway many inns lost their trade and by 1916 only three White Horse inns were trading; by 1936 only two remained and by the year 2000 there were none left in Norwich.

***Back of the Inns*** 1890 WDG.

***Bridge Street, St. Lawrence / Coslany Street*** 1760 recorded as trading with Robert Dack, a worsted weaver, as landlord.

1849, 24th June – William Rix became licensee, he was a dyer.

1854 the address was recorded as Bridge Street, St. Miles, and in 1868 as Coslany Street.

1877 the house was closed, the licence being provisionally removed to be transferred to a new building about to be erected at Garden Road, South Heigham.

1879, 5th August – the licence was officially transferred to the Garden House Tavern. NoR. 1842 Bly; 1846-1858 PO; 1859 RD.

***Castle Ditches, St. Michael at Thorn*** 1822 listed as trading with Simon Goose as licensee. The house was listed under White Horse until 1861.

1822-1830 P; 1842 Bly; 1850 S; 1859 Castle Hill RD.

***Chapel Street*** Listed as trading in 1861 and 1879.

***Coslany Street, St. Michael at Coslany*** *See Bridge Street.*

1822P; 1865 K; 1875 NTD; 1879 JJH.

***1 Crook's Place, St. Stephen's / 190 Essex Street*** 1839 listed as a beerhouse with John Woodcock Selth as licensee.

1904 listed as at Crooks Place, Chapelfield Road.

1925, 19th December – James Gooda, landlord of the White Horse was convicted for selling liquor out of hours, he was fined £2, or 21 days detention.

1963, 1st April – the house was closed and by 26th June had been sold.

1842 Bly; 1850 S; 1859 RD; 1875 NTD; 1900-1922 – victualler - J; 1931-1960 K.

***20 Old Haymarket / The Haymarket, St. Peter Mancroft*** Between 1760 and 1764 there were three landlords recorded, Robert Cattermow a worstead weaver, Joseph Jasy a waterman and Robert Davey a tallow chandler.

From c.1800 to c.1850 the house was known as the Seed Mart.

1802 the address was given as No. 2 Haymarket.

1806 John Johnson, a gardener, was licensee.

1897 the licence was not renewed, the house was closed and demolished.

1763 NoR; 1842 Bly; 1846; 1850 Hay Hill S; 1859 RD; 1875 Old Haymarket NTD; 1883 E.

***King Street / St. Peter Southgate*** Between 1760 and 1764 three licensees are recorded – Francis Reader a barber, John Gay, a labourer and Henry Guyton a labourer.

1836 William Mason, aged 45, was the landlord; his wife Mary had taken over by 1867. Between them they ran the pub for over 40 years.

1932, 21st February – Arthur Wright, landlord of the White Horse was convicted for selling liquor out of hours, he was fined £5 or one month's detention.

1960 the house was still trading. 1858 PO.

***St. Martin at Palace / St. Martin's Palace Plain*** 1760 recorded as trading with John Starr, a victualler, as landlord. NoR.

***84 Magdalen Street, St. Saviour*** c.1845 the freehold was owned by brewers Finch and Steward.

1867 Mary Mason is recorded as the owner and was replaced by Steward, Patteson, Finch & Co. in 1872.

1922 Arthur Wright, the son of Thomas Wright the previous landlord, was recorded by the brewers as being licensee; in fact the Licence Registers do not have him listed until c.1932 when his mother Phoebe gave up the licence.

1955, 29th May – the house was closed, although the brewers Steward and Patteson are recorded as holding the licence until 1964 when the building was demolished.

1822-1830 P; 1842 Bly;1846 PO; 1859 RD; 1879 JJH; 1890 WDG; 1900-1922 victualler J; 1931-1941 K.

***10 St. Andrew's Street / Broad Street, St. John Maddermarket*** 1760 recorded as trading with Edward O'Newton, a worstead weaver, as landlord.

By 1822 Edward Stubbs was listed as licensee, his family ran the house for the next 56 years.

1845 the house was described as being located near the museum.

1850-52 thought to have been listed in error under White 'Hart'.

1872 Grimmer & Co. took over the ownership from Seaman & Co.

1890 listed as The 'Old' White Horse.

1906, 9th February – the licence was provisionally refused and referred to compensation.

By 1909 Lacons, the brewers were the owners.

1910, 29th January – the house was closed. According to local historian G. Kelly, 'Museum Street was of short duration' for many years after the inn had ceased to trade.

This ancient timbered building was demolished in the late 1960s and a bank was built on the site, only to be demolished in the 1980s and in 2002 the site was still derelict.

G. Kelly. 1760-1806 NoR; 1822 P; 1842 Bly; 1846 PO; 1859 RD; 1879 (Museum Street) JJH; 1883 E; 1900-1908 J.

***St. John Timberhill*** 1760-62 listed as trading with Richard Robinson, a baker, as licensee. NoR.

***10 St. Mary's Church Alley, St. Mary's Coslany*** 1760 recorded as trading with Jonathan Barker, a worsted weaver, as landlord.

1845, 25th March – the brewers Tompsons sold the premises to Morgans.

1882, 3rd May – the licensee, William Clayton junior, was convicted of allowing drinking out of hours, he was fined 10 shillings plus 7 shillings costs.

1897, 10th July – Robert Frost, licensee of the White Horse was fined 20 shillings plus 8 shillings costs, or 14 days detention, for keeping open out of hours.

1905, 4th February – the licence was referred to the Compensation Authority and the house closed on 11th January the following year.

1806 NoR; 1822-1830 P; 1842 Bly; 1865 K; 1879 JJH; 1883 E; 1905 J.

***12 Trafalgar Street / New Lakenham Street*** This public house was listed from 1830 as the White 'House'. It was listed in directories as White Horse from 1896 to 1929 although the Licence Registers still recorded it under White 'House'.

**Old Buckenham, The Green** The earlier building, incorporated within the present house, was almost certainly thatched and dates from the reign of King Charles I (1600-1649), retaining most of the original timbers.

1716, 11th October – a brewer made considerable alterations and additions to the building.

1782 listed as trading with James Foulsham as landlord.

1783, 4th January – *The Norfolk Chronicle* published the following advertisement:

'To be Lett, and entered upon immediately, the White Horse . . . an old established public-house, in full trade, now in the occupation of James Foulsham, who has carried on the spirituous liquor trade in the wholesale way, which may be an advantageous branch to the succeeding tenant, with a proper capital, as none but such will be treated with.'

1791 The Enclosure Award for Church Green records 'The White Horse as having one green right'.

1836 George Taylor Holl, aged 36, was the landlord. In 1851 he was also listed as a watchmaker and letter receiver, by 1861 he was a sub postmaster.

By 1865 George's wife Margaret had taken over; she was also listed as a watchmaker in 1872.

1879 George Roger Holl, aged 47, was listed as licensee, he was also a watchmaker. The Holl family continued to run the pub until at least 1929, they had been here for almost a hundred years.

1840 a tithe map shows a porch on the road frontage but this had been demolished by 1904.

1849-1882 various extensions and alterations were made to the building, some of which had gone by 1972. A chimney stack bears the date 1849 and the initials 'G.H.' – George Taylor Holl – the publican, together with 'J. LAN' – John Lancaster, a local butcher who may have shared the premises for a time.

c. 1904 the building was 're-skinned' in brick.

c.1998 the house was closed for major refurbishment; it opened the following year on 28th July and was renamed The Gamekeeper.

GK. 1830 P; 1846 PO; 1850 S; 1875 NTD; 1937 K.

**Overstrand, 34 High Street** An old coaching inn with Victorian and Edwardian additions; it is built of flint with a pantile roof. The stables and hay loft have been converted.

When Victorians came to stay for sea air, peace and quiet at The Pleasaunce with Lord and Lady Battersea[4], the chauffeurs always asked to stay at the White Horse Inn nearby. At Overstrand Station there were two waiting rooms – one for the gentry and one for the servants.

1845 Robert Summers was licensee, he was a fish merchant and also a carrier to Norwich.

1884, 3rd November – John Codling became the licensee, he and his family ran the pub for 66 years.

1897, 25th October – Eybourne Brewery sold the premises to Steward and Patteson.

1900 John Codling was landlord, followed by Thomas Fletcher Codling in 1925. 2000 advertised as a public house with rooms to let.

KSGH and Landlord. 1792-1799 NoR; 1846 P; 1858 PO; 1875 NTD; 1900-1937 K; 1960 JM.

**Rockland St. Andrew** 1792. *Situated within the parish of Rockland All Saints.* NoR.

**Roydon** 1883 WDG.

**Saham Toney, The Street** 1853 listed as trading in the Post Office Directory.

1962, 29th May – the sales were recorded as 68 barrels at the first joint committee meeting between the brewers Bullards and Steward and Patteson; Bullards proposed that the house be closed, but this was not agreed. Seven years later in June 1969 Steward and Patteson recommended immediate closure in their report for Watney Mann and the pub was closed on 6th October. 1865 K; 1892 S.O. – 1937 K.

**Salle, The Street** Pronounced 'Saul', this small, isolated village has a magnificent church built from the wealth of wool.

1836 listed as trading with Joseph Leeds, aged 35, as landlord, he was also a farmer.

1845 listed as trading in White's Directory with Joseph Leeds as licensee.

1870 a stone with this date can be seen on the right hand gable, showing that the house had been rebuilt.

By 1871 James Ashmore, a farmer, had taken over the house. He died in 1888 and his wife Mary became the licensee.

c.1878 the owner was recorded as Major Timothy White of Salle Park, he leased the premises to Morgans Brewery until at least 1929.

1890, 31st November – another farmer, George Derisley became the licensee.

1900 Kelly's Directory lists John Gogle as licensee and in 1925 Frank Laskey.

1987 it is understood that the Lord of the Manor was still renewing the licence annually.

By 2007 the house had been a private residence for some years. The old large-scale Ordnance Survey map displayed outside the church showed the building as a pub. KSGH and SJe.

**Scole** Due to possible misprints the present day Scole Inn is thought to have been The White Hart, confirmed by the impressions of stags to be found in the building, although the trading dates are listed under White Horse.

1796-1798 NoR;1870 POT; 1885 W.

**Sedgeford** 1836 listed under White Horse with John Oughton, a bricklayer, as licensee. By 1881 the name had been changed to the Plough Inn. 1865 K.

**Shipdham, White Horse Street / High Street** 1830 listed as trading with John Morgan as licensee.

In 1881, 1888 and 1891 the address was under High Street.

1894, 11th May – Wymondham Brewery sold the premised to Morgans.

1914, 6th February – the pub was referred to compensation and was closed the following year on 1st November. 1789-1799 NoR; 1830 P; 1846-1858 PO; 1865-1900 K.

**Shropham, The Street** 1836 listed as trading with Jacob Trudgill / Threadgill as landlord.

1869 George Allen, a farmer of 5 acres, was listed as licensee.

1894, 11th May – the premises were included in the sale of Wymondham Brewery to Morgans; the spirit licence was dropped but was granted again sometime later.

1960 the sales for the year were 87 barrels of beer and 49 spirits. Two years later 88 barrels were sold. 1865 K; 1900-1916 K.

**Snettisham, Back Street** 1861 a beerhouse was listed under White Horse with George England as licensee.

**South Lopham, The Street** The building dates from 1658 and was originally three cottages.

1830 listed as trading with S. Harris as licensee, the following year the population was recorded as 729.

1845 listed as trading with John Eaton as licensee.

1854 Joseph Mace, a farmer of 24 acres, became the landlord.

1925 William Pitchers was listed as licensee.

2009 the pub was still listed as trading. KSGH, Landlady and JB.

**Stow Bedon, White Horse Road / Rectory Road** 1836 listed as trading with Robert Osborn as landlord.

1845 John Nurse, an agricultural labourer, became licensee.

1871 Arthur Harvey took over the licence, he was also a dealer.

1875 an agricultural labourer Richard Colby was the landlord.

1889, 25th July – the property was valued at £275.

1890 Robert Bennett, a general labourer, ran the pub.

1961, 1st December – Roy Schofield, aged 37, became the landlord. On 29th May the following year the closure of the house was recommended at the first joint committee meeting of the brewers Bullards and Steward and Patteson. The year's sales were reported as 46 barrels.

1963 it was closed and became a private dwelling.

R. Schofield, tenant. 1875 NTD; 1900-1795 NoR; 1916 K; 1937 NTD.

**Suton, Bait Hill** 1850 listed as trading in Hunt & Company's Directory with John Hubbard as licensee, he had previously been landlord at the White Horse, Damgate Street, Wymondham.

1851 the census gave the address in Suton as Bait Hill.

By 1869 John's wife Hannah had taken over the licence, but by 1875 he was once again listed as licensee. 1846-1858 PO; 1875 NTD; 1900 K.

**Sutton, Church Road** 1789-1792 the Alehouse Recognizances listed John Mower as licensee, Elizabeth Mower was listed from 1793 to 1795, Jonathan Bush took over the following year, he was still here in 1799.

1836 John Julier, a joiner, became the licensee.

1841, 14th to 17th September – The White Horse beerhouse was Lot 54 in the Coltishall Brewery sale, it became a freehouse.

1845 Mary Julier/Juler took over the licence.

1850 Thomas Frosdick, aged 54, was the landlord and also an agricultural labourer.

1892 Kelly's Directory listed the White Horse as a Postal Sub-Office; John Platten, a carpenter, was the licensee. NoR C/Sch 1/16. 1846-1858 PO; 1875 NTD.

**Swaffham** 1793 NoR.

**Tasburgh, Saxlingham Lane / Lower Street / Low Road** The White Horse public house was a wooden building which stood opposite the southern entrance to Rainthorpe Hall.

1845 listed as trading with James Cannell as licensee.

From 1849, 12th December, five steam trains a day stopped at nearby Flordon Station, this linked the area with London, Diss and Norwich, which caused a fall in the heavy stage coach traffic, carriers and wagons which passed through Tasburgh.

1858 Robert Cowell, a cordwainer[5], was licensee.

1869 Henry Rix, a carpenter and farmer of nine acres, was licensee – White Horse Farm is situated on the other side of the road to the public house.

1875 Robert Rix was licensee, he was also a farmer and carpenter.

1891 a carpenter, John Want held the licence.

1914 the lease expired and was not renewed.

1916 The licence was reissued with Arthur Hurry as licensee until a bad fire in c.1925; the roof was replaced with corrugated iron and Mr. Hurry continued to live there until it was renovated by Mrs. Hastings of Rainthorpe Hall; the house was then rented out and was eventually built into a modern larger house.

1993 Bob Lammas, whose mother was born in the White Horse public house, is a survivor of the oldest Tasburgh family. B&M. 1916-1925 K.

**Thetford, 4 Raymond Street** 1822 listed as trading in Pigot's Directory.

1830 Thomas Chilvers was recorded as licensee, by 1836 Elizabeth Chilvers had taken over the licence.

1845 Daniel Davey was the landlord, he was aged 43 and also a cooper.

1889, 25th July – the valuation of the property was given as £900.

1927, 22nd June – the house was referred for compensation and again on 24th December, it was closed on 31st December.

1846 PO; 1850 S; 1870 PO; 1875 NTD; 1885 WDG; 1900 W; 1916 K.

**Thurlton, Loddon Road** Recorded as trading between 1789 and 1799.

1826 the White Horse is shown on Bryant's map.

1836 listed as trading with John Jennis as licensee.

1851 James Banham was the landlord aged 44 and a farmer with 2 acres.

1861 William Smith was listed as the licensee and a farmer; in 1871 he was recorded as a kiddler[6] in 1871 and a carpenter in 1881.

By 1891 another farmer, Arthur Morl, was the landlord.

c.1920 the house was closed. NoR. 1846-1858 PO; 1875 NTD; 1916 K.

**Trowse Newton, The Street** The White Horse Inn was a coaching house standing on the common.

1794, 11th October – the inn was advertised as Lot No. 18 in the sale of St. Martin's Brewery.

1797, 13th July – listed in the sale of the property of the late Mr. Day.

1836 Benjamin Bales was the licensee, he was also a gardener.

1845 the landlord, George Hall, aged 48, was a joiner.

By 1842 a tea garden had been created.

By 1848 the premises were owned by George Morse, the brewer.

1879 Henry Harris held the licence, he was also described as a dealer in 1881 and a pig jobber in 1883.

1931, 17th October – Albert Browne of the White Horse Inn was convicted of permitting gaming. The fine was £10 plus £5 5s. costs.

1942, 8th May – the building was damaged by enemy action. After the War the inn was rebuilt on the opposite side of the road.

2008 the inn was still flourishing with facilities for meetings.

1789-1799 NoR; 1846 PO; 1850 S; 1859 RD; 1875 NTD; 1900-1937 K; 1960 JM.

**Upton, 17 Chapel Road** Recorded as a beerhouse between 1789 and 1798, the building had originally been three cottages which were converted into a coaching inn – a date can be seen in a triangle on the chimney stack.

1836 listed as trading with Thomas Wiseman, a brickmaker, as landlord.

1841, 14th to 17th September – the premises were Lot 34 in the Coltishall Brewery sale. The house was sold to Bullards.

By 1872 the house had been granted a full licence.

1927, 25th May – Samuel Willgress was fined £1 for permitting drunkenness.

1980 The White Horse Taxi Service was run from this public house.

Mr. Ray Norman worked in this inn for almost forty years, twenty eight of which he was landlord and is thought to be the longest serving landlord in Norfolk.

Over the years various alterations to the interior have taken place.

NoR. 1846-1858 PO; 1865-1937 K.

**Walpole St. Peter** 1792. *My grandmother, the Vicar's wife, had the women inebriates to tea once a year. My father and the other children had to hide all the ink pots – in that era writing ink was made with powder and alcohol. My father maintained the hiding places were always discovered.* NoR.

**Weasenham St. Peter, The Green** 1732 deeds record Esau Barrett, a woolcomber of East Dereham as owner and John Kent as the ale house keeper.

1754 listed as the White Horse when Esau Barrett sold the premises to John Carver, a wine merchant and William Collison, a common brewer. The property was described as a public house with stables, barns, outhouses, yards, gardens and orchards.

By 1766 the name had been changed to the Duke's Head; additions to the property description included a building which had formerly been a decayed cottage or tenement, now an outhouse, and 'a new building for a stable erected on the lord's waste near the premises as they abut the highway on the south'.

By 1792 the name had changed to the Fox, the next year to the Fox and Hounds; followed a year later by the Chase and finally back to the Fox and Hounds in 1796.

1949, 11th February – the house was granted a Full Licence, it had previously had a six-day licence.

2007 still trading as the Fox and Hounds.

**West Bilney** 1789-1799 NoR.

**West Dereham, Lynn Road** 1780 the Old White Horse was recorded as an alehouse.

By 1845 William Dent and his wife Elizabeth ran the house, William was a blacksmith, the smithy was an extension to the house; later their daughter Jane Mayers/Mears and her husband took over the licence.

1875 Robert Porter was the landlord, he was also a pork butcher and an assistant overseer; his wife Elizabeth took over in 1904. They ran the house for almost 40 years.

1928 the brewers Bullards became the owners.

1965, 31st may – the house was closed, it became a private house named Bewlah Cottage. Mrs. K. Ballard. 1937 K.

**White Horse Common** Between 1789 and 1799 the White Horse was recorded as trading.

1830 listed as trading with Richard Tinkler as licensee.

1845 the landlord is listed as Edward Sandell.

1848 the inn is shown on Bryant's map, it did not appear on the 1797 Faden's Map of Norfolk.

1850 William Miller / Willer was landlord, aged 30, he was also a cattle dealer.

1865 James Ellis was listed as licensee, by 1890 his wife Elizabeth had taken over. Between them they ran the inn for at least 35 years.

1962, 25ᵗʰ May – the house was closed to become a private house, a horse was still depicted on the flank wall in 2007. NoR and KSGH. 1822-1830 North Walsham P; 1846 PO; 1850 S; 1875 NTD; 1900-1937 K.

## Wymondham

*Damgate Street* 1856 John Hubbard was listed as licensee, previously in 1850 he had been landlord of the White Horse at nearby Suton. 1883-1908 *S.O.* K.

*White Horse Street / Market Street (or Pople Street)* 1738 it was one of several houses in Thomas Randall's estate which he left to his grandsons.

1789-1795 the Alehouse Recognizances record Joseph Haythorpe as licensee; John Cowells took over in 1796 and he was still there in 1799.

1830 listed as trading in Pople Street.

1894, 11ᵗʰ May – the inn was included in the sale of Wymondham Brewery to Morgans Brewery.

1899 John Betts, known as Treacle Jack, drank half a pint of beer here and then jumped off Bait Hill railway bridge into the path of an oncoming train.

1960 Morgans' sales for this year were 126 barrels of beer and 18 spirits.

c.1990 the inn was closed to become a private house.

NoR C/Sch 1/16. 1822-1830 P; 1916-1937 K.

## Notes

1. Woven manufactured materials. SOE.
2. A small field or enclosure. SOE.
3. This name seemingly applies to several unconnected subjects:

   a) A wooden double-reeded instrument of high pitch, having a compass of about 2½ octaves, forming a treble to the bassoon – the modern oboe.

   b) Hautboy / hotboys. A species of strawberry, Fragaria Elatior. The name is probably derived from the fact that the plant bears fruit standing higher than the leaves. According to Mr. Jamers Barnet, writing in 1825 (Account and Description of Strawberries) there are thirteen varieties of 'Hautbois'. In c.1815 the Black Hautbois was received as a new Hautbois from the Royal Gardens of Windsor – from the seed of the Prolific or Conical Hautbois.

   c.1824 William Cobbett describes taking his son Richard to the grounds of Waverley Abbey, where the garden walls built by the monks had totally gone, 'I showed him the spot where the strawberry garden was, and where I, when sent to gather hautboys, used to eat every remarkably fine one, instead of letting it to be eaten by Sir Robert Rich.' WmC Vol. II.

   In 1988 'The Complete Strawberry' by Mr. Stafford Whiteaker maintained that 'it was once the most popular strawberry in England, and the street sellers used to be heard crying out "Hotboys! Hotboys!"

4. Constance Lady Battersea. The daughter of Sir Arthur de Rothschild. She married Lord Battersea in 1877; he died in 1907.
5. A shoemaker or worker in leather. From the Spanish city Cordova where this leather was made from goat-skins or, later from split horse hides. Much used by the wealthy during the Middle Ages. SOE.
6. A barrier constructed of nets and stakes for catching fish in a river or the sea. Col.

# NORTHAMPTONSHIRE

*'The general aspect of this county exhibits a pleasing and interesting variety of vale and upland. The entire surface is peculiarly adapted for agriculture; yet as lately as the year 1818, a large proportion of this land was unenclosed . . . it is not subject to extremes of weather, and is considered one of the healthiest counties in England.*

*The great mail-roads are generally level and wide; the turnpike roads also are kept in good repair, but the cross roads are much neglected.'*
JB.

**Broughton, Church Street** 1936 the house was demolished. 1842 P; 1850 P; 1877 PO; 1898 K.

**Corby** 1830-1850 P; 1898 K.

**Croughton** 1638 owned by John Heynes, a blacksmith.
1880, June – purchased by a brewer with the original brew house for £1,500.
1923 it was sold, eventually to become a private house. B&W. 1850 P; 1877 Brackley PO; 1898-1939 K.

**Daventry, Brook End** "The town derives its name from the British Dwy-avon-tre 'town of the two Avons' denoting its situation between two rivers of that name" JB. *We have been unable to identify Brook End.* 1823-1850 P; 1898 K.

**Great Cransley** 1842 P.

**Higham Ferrers, High Street** The birthplace of a yeoman's son, Archbishop Henry Chichele (1364-1443) who founded both a hospital and a college here – in 1422 – The Bede House for twelve poor men and one woman and in 1424, Chichele College. Some one hundred years later part of the funding for this college came from the rent of a large property in Friday Street, London, which included a 'Whyte Horse Inn'.
1823 listed as trading but the original building of this White Horse Inn in the High Street appears to have been of an earlier date. In 1829 Earl Fitzwilliam was the patron of Irthlingborough – 2¼ miles north-west of Higham Ferrers; Rent

Day Dinners were held at this inn where the rents were collected by his agent.

1851 the landlord was recorded as a painter and plumber.

NRO and JB. 1877 PO; 1898-1914 K.

**Irthlingborough** The Old English derivation of this name appears to be 'the burg of the plowmen . . . an old fort used for the purpose of keeping oxon'.

By 1898 there were White Horse Inns in both Higham Ferrers and Irthlingborough – 2¼ miles away. OxN. 1898-1939 K.

**Kettering, High Street** By 1850 there were two White Horse Inns in the High Street – the Old and the New. There were 22 carriers in the town mostly based at public houses; they only covered the local area.

The 'Old' White Horse was listed in Pigot's Directory of 1823.

1905 the Old White Horse was demolished, the hotel was rebuilt as an impressive building with a dome on the corner with Huxloe Place in front of the *Evening Telegraph* offices, whose journalists frequented the pub.

American servicemen, including the film star Clark Gable (1901-1960), were stationed at two airfields nearby during the Second World War; friction was caused when local girls were given silk stockings which the British servicemen were unable to provide.

1959, 24th November – the hotel closed to be taken over by Montagu Burton the men's tailors, who opened their shop on 23rd February 1962.

During renovation, a hangman's noose was discovered in the attic.

The 'New' White Horse Inn was situated next to the Electric Pavilion, later the Gaumont cinema; the inn was one of the oldest buildings in the town, it had previously been known as The Lord Nelson. The front of the inn advertised good stabling, also accommodation for cyclists; petrol was sold and there was a motor garage and pit. The bar specialised in Scotch and Irish whiskies.

1957, 28th May – the inn closed.

TS, KL, LG and R.W. Kershaw-Dalby – local historian.

**Kingsthorpe, 25 Harborough Road** This village became a suburb of Northampton.
1939 K.

**Lowick, 16 Main Street** The original building dates from c.1530 when it was the Dower Manor House, home of the Countesses of Peterborough until 1671. Drayton House was the home of John, 1st Earl of Peterborough who died in 1642 and the 2nd Earl, Henry, who died in 1697. After the Battle of Naseby in

1645 a white horse – which had been stabled here – was attacked with its rider while returning from the battle; the rider was killed by an arrow but the horse made its way safely back to its stable. A ghost of a white horse has been seen at the inn – hence the name White Horse.

There is supposed to be a mediæval tunnel one mile long leading from the cellar of the inn. Beer was brewed here and the landlord holds original family recipes dating from the 17th century. 1877 Thrapston PO; 1898-1929 K. 1989 renamed The Snooty Fox.

**Northampton, 64 Sheep Street** 1640 recorded as being in the Horsemarket with a malt house attached. 1847-1877 William Edmunds is listed as the owner and brewer. By 1900 the brewing here was no longer recorded and by 1907 the premises had closed.

Pigot in 1823-1850 lists two White Horse Inns and Kelly's Trade Directory of 1854 gives Cotten End, Bridge Street – which runs parallel to the present Cattle Market. Presumably these two White Horse Inns stood at the either end of Sheep Street. B&W.

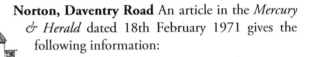

**Norton, Daventry Road** An article in the *Mercury & Herald* dated 18th February 1971 gives the following information:

> In the front of the White Horse is a stone bearing the date 1681, it is presumed this is the date of the three cottages from which the broad-fronted White Horse Inn was constructed. Originally this village had two pubs; on the day the White Horse opened at 12 noon, the Spread Eagle closed at 2 p.m. for ever.

There are reports of ghosts – 'The White Lady' who is said to walk from the lounge bar through the middle of the pub into the public bar was recently seen by one of the bar staff out of the corner of their eye; before turning round to offer a drink, they realised there was no one there. Another ghost, seen by children, is a man dressed in black with a very large hat, presumably a cavalier, who sits on the floor in the corridor that leads to the landlady's rooms. No one seems to know the stories behind these ghosts.

1926 leased to the People's Refreshment House Association as a free house, a large board outside displayed the letters P.R.H.A.

1947 the premises were bought by the brewers Charles Wells.

c.1971 redecoration took place. 1850 Daventry PO; 1877 PO; 1898-1939 K.

**Old, Walgrave Road** The building dates from c.1800. The premises were originally a flour mill; the steam-driven grinding machinery being in a building behind the house. The public house was trading by 1936 when Mr and Mrs Barritt took over the business; their grand daughter who played here as a child, remembers a gentleman who used to come once a year with his two shire horses, take them into the yard and give each of them a pint of beer.

For thirty years from 1957 to 1987 the pub was run by Mr and Mrs Story.

**Paulerspury** There is an ancient cottage named The Old Smithy; on the roadside with the Tow River running nearby.

An extract from a letter written by the owner of The Old Smithy reads:

'I am afraid that I have no knowledge as to whether my cottage was an inn or not but, as you say, the smithy is a hint . . . My cottage was probably built about 1630 as was one across the road which has the date on it. There is evidence that both were constructed with material from a ruined nunnery at Shutlanger. I have consulted an old map of the area (probably 18th century) but it does not unfortunately, mark the inns.' J.M. Oliphant.

In 1805 the village is described as ' . . . on the Tow River in the parish of Paule's Purry, and had its name from the residence of the Danes here . . . Inn, White Horse.' wco.

**Peterborough, Cumbergate** Situated on the north side of the Cathedral – a few buildings of the original monastery still survive particularly the abbots' and priors' lodgings. The inn sign caught in the photograph reproduced in 'Northamptonshire Past & Present' may indicate that a White Horse Inn once stood here, but it has not been possible to substantiate this theory.

AAT and NPP. 1877 PO; 1892-1939 K.

**Silverstone, Stocks Hill** 1934, June – the house was bought by Northampton Brewery Company. B&W. 1850 P; 1898-1914 K.

**Spratton** The White Horse Inn, standing on the Teeton to Brixworth road, measured one rood[1] and one perch[2] and at one time was occupied by Widow Phillips.

1774 a deed of sale records Francis Beynon as the owner.

1821, 12th December – the premises were sold with other land and buildings by Francis Beynon's heir, Francis Beynon Hacket, to the Misses Terry.

1939 listed in Kelly's Directory under Teeton.

It became a private house named Old White Horse Inn. NRO Sale ref. YZ5944.

**Stoke Albany, 1 Harborough Road** There are deeds from the time of Queen Anne (1655-1714), the date 1706 can be seen on a stone plaque in the gable.

1877 PO; 1898-1939 K.

**Tansor** It became a private house called Old White Horse. 1842-1850 Oundle P; 1877 Oundle PO; 1939 K.

**Towcester, 163-165 Watling Street West** From c.1700 this property appears to have consisted of two messuages which at various times were owned by a Baker, two Gents, a Brazier, a Dealer or Chapman[3] – who dealt in the selling of lace and silk stockings – and a Scrivener[4].

1716 all the messuage was sold for £725.

1743 listed as trading under 'The George Inn'.

1784, February 19th and 20th – additional land was purchased 'with the other messuage converted into one messuage or tenament . . . used and occupied as an inn and commonly called or known by the Name or Sign of the White Horse in Towcester'.

1786, February 16th-18th – due to bankruptcy the property was sold 'inter alia All that new erected Messuage or Inn . . . All that piece of ground in Towcester whereon part of the George Inn formerly stood and on which a new built messuage or Tenement Stable and other buildings were then erected called the White Horse Inn with the Garden adjoining the same'.

1797, 16th and 17th June – bought for £100 by George, 3rd Earl of Pomfret (1768-1830) from a mason, John Middleton, and a surgeon, Samuel Deacon. To be sold ten years later, having been refurbished, to William Worth an innholder, together with John Worth and his heirs, as the New White Horse for £1,350.

1840, 28th January – advertised For Sale:

'To be sold by auction in one or more lots. All those valuable freehold premises called the White Horse Inn and Posting House and the licence thereto situate in the town of Towcester. Comprising a substantial built Dwelling House consisting of seven good Parlours, large Market Room, twelve principal bedrooms, six second class bedrooms, water closet and five Attics, Kitchen, scullery, larder, pantry, dairy, beer and wine cellars, brew house, spacious yard, stabling for 50 horses, Coach houses, granary, Men's bedrooms and other convenient out-offices, extensive walled garden, the whole is well supplied with water . . . having a frontage of 80 feet on Sawpit Lane *which became Richmond Road* and 40 feet in Middleton Lane.' NRO Ref. NPL1602.

1842 listed by Pigot as a posting inn and was considered one of the most famous on Watling Street, renowned for its hospitality and the standard of its cuisine.

1855 listed as White Horse Inn and described as a large plot, including the rear yard running back to Richmond Road.

1876, Easter Monday – the opening of Towcester Race Course. This first meeting, a glorified point to point for the four local hunts, did not run as smoothly as expected – all individuals who might have lowered the tone of this promising social event, were excluded; the inn, a considerable property with stabling, would have provided accommodation for racegoers and visitors. The original idea of a race meeting was supposed to have come from Elizabeth, The Empress of Austria[5] (1837-1898) who had rented nearby Easton Neston from Sir Thomas Fermor-Hesketh so that she could hunt with several neighbouring packs of foxhounds. She rode side-saddle and was sewn into her habit every time she did so.

1910 this property was described as two private houses with extensive outbuildings, cottages and stabling to the rear. It was valued as unlicensed, having ceased to be the White Horse Inn.

c.1950 the large three-storied house of six bays wide, was bought by South Northampton District Council; it became a Grade II listed building being built of red brick in Flemish bond with flared headers and a plain-tile roof, the modern double-leaf doors to the left of the centre was the original carriage archway.

By 2001 there was a plan to house a local museum in the buildings at the rear.
NRO Ref. D2552 and NRO Ref. NPL2600; TDL, R&W and Weatherbys Allen. 1877 PO.

**Wellingborough, Pebble Lane** 1800 recorded in the will of Andrew Wilson as 'The White Horse, Market; tenant Josias Denny Roote'.
B&W. 1823-1842 P; 1877 PO; 1898-1940 K.

### Welton, The High Street
1850 Daventry PO; 1877 PO; 1898-1939 K.

**Woodford, 1 Club Lane** The White Horse Inn stands in the centre of the village with views overlooking the River Nene. It has a restaurant to seat 60 people.

A description from the landlady: "Many oak beams – very, very old".

There is an ornate ironwork archway supporting the inn sign depicting a chess piece in the form of a white knight over the entrance to the car park at the rear.

1877 Thrapston PO; 1939 K.

*We have to thank Diana Dalton, local historian, for providing most of the information and illustrations for this county.*

**Notes**

1. The equivalent of a ¼ acre.

2. 30¼ square yards.

3. An itinerant pedlar. Col.

4. A person who writes out deeds, letters, etc. Col.

5. Daughter of the eccentric Duke Maximillian Joseph of Bavaria. She attended the Imperial Riding School in Vienna and had a great love of horsemanship. She was assassinated on 10th December 1898 in Geneva.

# NORTHUMBERLAND

*'The agricultural improvements made in this county within a few years, are scarcely to be equalled in any other part of the kingdom, and are in a great measure owing to the assiduity of the late Mr. Bailey of Chillingham, who invented the swing-plough and many other useful machines.*

*The farmers of Northumberland have long been celebrated for their superior knowledge of breeding young cattle. Oxen are mostly grazed in the eastern part of the county and the vicinity of Whittingham. The long woolled sheep have been much improved by the introduction of the Leicester sheep. Goats are kept on the Cheviot hills. The horses are strong and active, and generally of a middle size.*

*The roads are by no means good. The turnpikes are frequently carried over hills which might have been avoided, and their cross sections are injudiciously formed, and there is a great omission with respect to guideposts.'*
JB.

## Berwick-upon-Tweed

*89 Castlegate* 1876 S; 1897-1938 K.

### Walkergate Lane

**Hexham** c.1604 recorded as trading and considered the most famous old hostelry in Hexham.

1634 three soldiers from Norwich recorded:

"we were as well accommodated with cheap and good fare, sweet lodging, and kind usages, as travellers would desire".

1832 sketched by Thomas Allom in his view of Hexham Market. The inn was situated to the right of the gatehouse. c.1890 the artist, William Bell Scott described in his memoirs his stay in this inn:

' . . . and settled down in a small apartment in the half-timer hostelry called the White Horse. This apartment was over the porch, and the front of it was one continuous narrow casement with a long bunker seat under it, looking out on the quiet market-place and great church . . . This long window, with the casements opened, and this market-place seen without, was my subject, and the landlady's daughter posed to me at full length on the window-seat knitting . . . This hostelry was scarcely ever disturbed by traveller, except on one day, the market-day of the week . . . Towards the end of the 19th century a speculative builder demolished this fine building, replacing it with a row of shops.' FG. 1828/9 P.

**Wooler, The Market Place** James Bell noted in 1834 that this place was:

' . . . remarkably healthy, and it was formerly much resorted to by invalids . . . Goats were formerly much bred here for the sake of the milk, which was used by invalids . . . In 1722, an extensive destruction of property took place by fire.'

Once again in 1863 a 'Great Fire' spread quickly through the thatched buildings of the town in a strong wind. From when the fire was first detected, the nearest fire-engine at Belford – some nine miles away – took seven hours to reach the scene, the second some fourteen miles away at Bamburgh, took fourteen hours. Some of the inns and taverns consumed by the fire had been trading since the 14th century.

Mr. Robert Wheatley, Local Historian wrote on 7th June, 2001:

' . . . We now have our own fire-station manned entirely by volunteers who a week ago were on duty for over 24 hours fighting a 200 acre blaze a mile or so from here on Wooler Common. So we are better prepared.'

According to the History of Wooler, the White horse was not recorded as trading again until after the fire of 1863. 1828 P; 1849-1876 S.

# NOTTINGHAMSHIRE

*'. . . then proceed to 'fair Nottingham', as we used to sing*
*when I was a boy, in celebrating the glorious exploits of*
*Robin Hood and Little John.'*
WmC.

*'. . . this county is not so famous for the breed of any race of*
*animals as for that of pigeons; greater numbers of them being*
*kept here than in any other part of the kingdom.'*
JB.

**Barnby Moor, The Great North Road** Part of the present building is 200 years old.

c.1970 the main business here was a farm with the public house as a sideline; there were several large barns, one demolished to become the car park and one converted into a house, later to become the restaurant named Le Cheval Blanc.

By the year 2000 it had become a Restaurant and Hotel with a garden.

Mrs. Susan Phillips. 1828-1842 Retford P; 1900-1941 K.

**Blyth** 1828/9-1842 P; 1876 PO.

**Misson, High Street** 1876 PO; 1900 K; 1916 Bawtry K; 1988 MA.

### Newark

These two inns listed as trading at identical dates, stood at the opposite ends of the town.

*Barnby Gate* Listed as trading between 1782 and 1958.

1822-1842 P; 1876 PO; 1900-1916 K.

*29/63 Millgate* Listed as trading between 1832 and 1935.

c.1900 William Congreve Hoe became landlord of the White Horse at 63 Millgate. Previously he had worked in his father's pub before becoming a carter. He established the first motorised bus service in the area and later opened two grocery shops.

1935 the pub was closed and was demolished in 1967.

1822-1842 P; 1876 PO; 1900-1916 K.

**New Basford, Mount Street** An article published in Issues 51 and 52 of *The Basford Bystander* gives the following description of the public house:

'Situated near the top of the hill, at the junction with Duke Street, was the White Horse public house, known locally as 'The Oss', which had painted on its front wall a full size painting of this creature rearing slightly, as if shy of the viewer, and about to take off into the open space behind – a public house sign without equal anywhere'.

1965, February – the house was demolished. 1862-1876 PO; 1900-1941 K.

**Nottingham, 14 Barker Gate** In 1782 the European traveller, C.P. Moritz, noted:

'Of all the towns I have seen outside London, Nottingham is the loveliest and neatest . . . A charming footpath leads over the fields to the highway, where a bridge spans the Trent. Not far from this bridge was an inn where I ate my midday meal. I could get nothing but bread and butter so I asked them to make it into toast'.

1822 listed as trading in Pigot's Directory.
1923, March – the house was closed.
1864 WDG; 1876 PO; 1900-1916K.

**Radford, 313 Ilkeston Road** An un-named inn has stood on this site since 1660; the date 1661 was shown on the building.

1777 listed as trading – it was not until the years 1786-1844 that it was given in the deeds of Radford Gasworks as White Horse.

During the 19th century it was a coaching inn with its own brewhouse.

1912 the house was rebuilt with a green tiled exterior. The author, Alan Sillitoe (b.1928) mentions this inn in his novel 'Saturday Night and Sunday Morning'.
WC. 1822-1842 P; 1876 PO; 1900-1941K. 1956-1962 listed in trade directories.

**Ruddington, 60 Church Street** This traditional village pub was built in c.1916 as a 'home alehouse' with little rooms everywhere, these were later altered to two good sized rooms; in 1960 a bay window was added. Since 1980 two out of the three landlords have died in harness.

John Wain, Tenant. 1916-1941 K.

**Wellow** 1876 PO.

**Worksop** 1842 P.

# OXFORDSHIRE

*'Clover, trefoil, and sainfoin, are abundant. The grass-lands are
extensive, especially on the river banks. There are many dairy
farms, and much butter is made for the London market.'*
JB.

*'You will hear more good things on the outside of a stagecoach from
London to Oxford than if you were to pass a twelvemonth with the
undergraduates, or heads of colleges, of that famous university.'*
WHa.

**Abingdon, 189 Ock Street** This Victorian public house has a lovely garden in
front with hanging baskets and window boxes; the interior has recently been
colourfully redecorated. The Assistant Manager and PE. 1877-1891 K; 1899-1939 K.

**Banbury, Sheep's Street** This street ran from Banbury Cross to Calthorp Street
to become the High Street where the White Horse stood near to the famous cross
still remembered in the well known nursery rhyme:

> *'Ride a cock-horse to Banbury Cross,*
> *To see a fine lady upon a white horse;*
> *Rings on her fingers and bells on her toes,*
> *And she shall have music wherever she goes.'*

'The problem of determining the likely age of the rhyme and the identity of the
lady is most difficult . . . Katherine Thomas, the author of *The Real Personages
of Mother Goose* describes a never to be forgotten incident . . . when, standing
beside my mother as she sang 'Ride a cock horse to Banbury Cross' she smilingly
remarked The Old Woman on the white horse was Queen Elizabeth. This
comment [was] made with the certainty of one who repeats a well known fact'.

It has been suggested that the bells on her toes points to the fifteenth century,
when a bell was worn on the long tapering toe of each shoe. The 'goodly Crosse'
at Banbury was destroyed at the turn of the sixteenth century . . . A modern cross
now stands in its place'.

The term 'cock Horse' has been used to describe a proud, high-spirited horse,
and also the additional coach-horse attached when going up a hill . . . To ride a
cock-horse can refer to a small child 'riding' on a grown-up's knee.

1833 a pantomime was performed at Astley's Royal Amphitheatre in London
named 'The Witch and the White Horse, or the Old Woman of Banbury Cross'
by Andrew Ducrow.

1930, 2nd November – the *Sunday Times* published a letter:

It was a customary sight during the latter half of the 18th century for travellers to Banbury and Birmingham to observe a group of children clustered at the foot of Stanmore Hill to witness the be-ribboned and rosetted fifth horse attached to the coach. As the gaily caparisoned jockey flourished his gilt-staff the boys and girls would chant "Ride a cock-horse to Banbury Cross". OPI.

*People travelling from London to Banbury can still use Stanmore Hill, the present day A4140.* 1823-1842 P; 1883 K; 1898 High Street K; 1899 PO; 1939 K.

**Bicester, Churchill Road** 1968 built by Morrell's Brewery Ltd. for a new housing estate.

**Burford, High Street** The White Horse was sold in 1987 and by 1995 it had become a private house. 1823/4 P; 1851-1939 K.

**Duns Tew, Daisy Hill Tew** 'The name implies 'in good health, excellent' – 'the Tew belonging to Dunne'.
  A 17th century coaching inn, thought to have been trading since 1664; it has the original flagstone floors, low beams and an open fire. The stables to the side of the inn have been converted into bedrooms. OxN and Landlord. 1883 K; 1899 PO; 1815-1939 K.

**Forest Hill, Wheatley Road** The White Horse stands at the junction of a Roman road and a Saxon track – it may well have been the site of a hostelry for centuries. This cottage-like building, originally part of the New College estates, dates from 1645 but still maintains its original appearance although alterations have been made to the interior.
  A record of daily village life was given in the diary of Mary Powell, the daughter of the Lord of the Manor of Forest Hill and wife of John Milton (1608-1674) the poet. She referred to Royalist troops passing through the village:

" . . . some were there who would not pay their score at the inn, alleging that they had defended us from the fury of the rebels and deserved to drink *free* . . . "

The inn was substantially rebuilt in the eighteenth century, probably to honour the new Hanovarian Royal Family whose emblem is a white horse.
  Kindly copied by Mrs. Marlene Vellenoweth from a framed history hanging in the pub. 1844 the house was leased to Morrell's Brewery.
  1883 K; 1895 - 1915 Stanton St. John K.

**Headington, 1 London Road** This road was built by the Turnpike Trust in c.1790, together with its toll house; the original public house, built beside the east bank of the Boundary Brook, was thought to date from c.1830; the brook divided Oxford and Headington, by the 21st century it had been piped underground.

1841 the census records that Henry Stanley, victualler, and his wife Sarah were living here.

1847 the Post Office Directory lists Henry as a beer retailer but he is thought to have died by 1850 as the Headington Rent Book lists Elizabeth Stanley as tenant of the White Horse – the first time the sign is mentioned by name. Owned by Henry Purdue, its gross estimated rental was £12 and the rateable value £8 10s.

1851 the census records Elizabeth Stanley, aged 56 living here with her widowed sister Mary Pimm and a 15 year old servant boy.

1861 Daniel White, a victualler aged 31, lived here with his wife Jane and one year old daughter, there were two servants and a lodger; by the 1871 census he had three daughters. He is no longer listed after 1877.

1881 a widow, Mrs. Elizabeth Harris aged 33 ran the pub on her own for eighteen years.

1901 by this date John William Thomas, a 'printer (compositor)', had taken over the pub.

1915 Joseph Skey was the landlord for two years; he was the son of John Skey who had been the central Headington toll-house keeper in 1851. A photograph dated 1916 shows Joseph Skey with his wife and son standing outside the inn.

c.1930 the old inn was demolished and rebuilt.

Joseph Gartside wrote a description of the inn before its demolition:

No. 1 London Road starts at the White Horse. "Owned by Morrells, it had a lovely lawn where the car park is now, and at the back of the lawn was an open pavilion where we could sit and have a drink and watch the children play on the lawn. Behind the pavilion was a lovely garden." 1939 K.

## Henley on Thames

1790-98 the location unknown – the publican is listed as a victualler. UBD.

*Bell Street* White Horse and Star. 1693 a mortgage was taken by Richard Brinklow Gibbs 'on a messuage known as the White Horse'; he had inherited the property from his grandfather in 1679.

1713, March – Richard's will recorded that he had left the property to his son.

1719 a mortgage was taken out, but the following year the property was sold with two stables, a piece of ground and four garret rooms.

1739, 8th June – the Clerk's Book of Burials recorded that 'William Davis at ye Wit Hors and Star' had been buried.

1750 the property was sold to James Brooks.

1772 James Brooks dissolved a partnership with Richard Hayward, but kept the public house as part of his stock in trade.

1789 Daniel Tindall, the landlord who had previously been catering to carriers and wagons, advertised to 'Bargemasters and Cost Bearers'. He intended to keep a good team of horses 'for the purpose of towing barges from Marlow to Henley, and if required from thence to the Kennett's mouth, and that he likewise purposes providing Waggons for the Conveyance of Goods to and from the Boats viz. loading and unloading'.

1826 a brewery inventory lists 'three stables in yard for 26 horses with lofts over, detached wash house and garden'. It is thought there would have been a carriage entrance to the yard. The Land Tax was 20s.

1797, 1st May – The *Reading Mercury & Oxford Gazette* published the following notice:

> 'Whereas, I Noah Ford of Henley on Thames in the County of Oxford, labourer, did on Sunday the 14th instant very much abuse the character of my then Mastr Mr. Thomas Hunt of the White Horse and Star Inn without any foundations whatever, for which he did threaten to proceed at law against me; but upon my publicly acknowledging my fault and asking his pardon, which I hereby do, he hath consented to withdraw all proceedings against me, for which I consider my self under many obligations to him.'

1800, 7th April – the same paper announced a forced sale, by order of the Sheriff, of furniture, 100 gallons of British compounds, 40 of foreign spirits and effects belonging to Thomas Hunt of the White Horse and Star. In June of the same year a Waggon Inn was advertised to let and could be entered immediately. There was stabling for horses 'whose consumption was 11 – 12 quarters of horse corn and 3 tons of hay per week'.

1817, October – the landlord advertised the Wagon Inn with 'good beds and every accommodation for man and horse at reasonable terms. N.B. an ordinary Market Day'.

1830 Pigot's Directory states wagons travelling from Sodbury to London called at the inn twice a day, it is thought that 24 to 32 horses were constantly available here every day.

1847 John Drewett is listed as landlord. He had married the daughter of Thomas Marlow, the previous landlord, and started a wheelwrights and carriage building business in 1849 which eventually became more important to him than the inn – his wife became the landlady.

1887 in the year of Queen Victoria's Golden Jubilee 'The White Horse and Star' closed. A photograph of this date shows the carriage workshop taking up two thirds of the frontage with the sign, 'T.M. Drewett Coachbuilder', the inn had been reduced to a small adjoining house with a single door and probably one room; above the door was a grand projecting clock. Jubilee displays were over the front of the building with a sign, 'Royal Patronage of Her Highness Princess Louise[1]'.

c.1960 the buildings were demolished and a block of flats built on the site.

Ann Cottingham – 'The Hostelries of Henley'. 1823 P; 1883 K.

**100 Northfield End** 'The Old White Horse on the Bridge' is situated at the point where Northfield End Road becomes the Fairmile. The White Horse was attached to the original bridge over the Thames.

1587, 22nd December – a deed states that the Warden granted a newly built tenement upon the bridge to John Lewis at a rent – this is thought to be the White Horse.

1629, 10th April – the Warden leased 'a messuage adjoining the Hermitage at the west end of the bridge' to Roger Shard, a Hosteler for twenty one years.

1631 the innholder, Roger Shard, died and left an inventory showing that he had been fully equipped for catering, brewing ale and taking in lodgers. The house had a hall, 'newe parlour, buttery and a garrett' – as no kitchen is mentioned, it is thought the cooking took place in the hall. Outside there was a backside with one hovel; the cellar was alongside on the riverbank, not under the house as the house projected over the river. At a later date the Henley Rentals lists a cellar as a separate property.

1636 records mention an inn having only a bush for a sign 'at the Bowling Green without the Town in Oxford way' – thought most likely to be the White Horse.

Linen was listed as 'thirteene pairs of sheetes and one odd sheet' valued at £3 7s. 6d. The tablecloths, napkins – some 'flexen' and some of hemp – and towels amounted to 30s. The pewter, valued at 46s. 8d. included platters, dishes, saucers, six 'eared porringers', candlesticks and chamber pots. Brass and metal pots and pans amounted to £2 13s. 4d., while those made from iron were 15s. 4d.

1649, 14th April – the property was leased to Robert Shard, a yeoman, for twenty one years.

1665 the Hearth Tax records the house as having three hearths.

1679 repairs were needed to the house – this is the first time the name White Horse is recorded.

1681, 2nd September – an eleven year lease was granted to William Usher, a bargeman.

1713 the Henley Rentals record 'Widow Usher for an encroachment belonging to the Hermitage'; between 1721 and 1727 they list 'John Usher's house on the bridge' and 1733 Rob Usher had the adjoining cellar and two rooms. Members of the Usher family were connected with the inn until 1775.

1779 was the final rental before the building was demolished to make way for the new bridge.

1781, 7th December – the Bridge Book recorded an agreement for the purchase of the inn authorised by Mrs. Elizabeth Blackman and Mrs. Anne Gear.

1790 the property was held by the brewers, Brakspear, on a normal lease until 1815.

1810 legal documents concerning a wall on the east of the Fairmile recorded that there was a skittle alley at the White Horse – the adjoining farm being Bowling Green Farm.

1815, 21st June – a lease was held for three lives of the following ages: Richard Brakspear, 25 years, William Brakspear, 12 years and Mary Brakspear 15 years. The Land Tax was 10s.

1826 the house is listed in the brewer's inventory as a Lifehold belonging to William Henry Brakspear Esq. – aged 23 – in occupation of Widow Kentfield. It had stabling for five horses and a large garden.

1869 listed as trading.

1874 William Lennox, a tramp, sat in the Tap Room playing tunes on some glasses, he was given some supper by the landlord but later became quarrelsome and was thrown out, he picked up a stone and hurled it through the window. He was summoned and fined 1s. 6d.

c.1882 the property was purchased by the Brakspear company following the death of William Brakspear.

1895 listed as trading.

c.1905 a photograph shows the inn with horses and cart outside.

Between 1930 and 1964 the Rawle family ran the house, except during the war years of 1939-1945.

1938, October – the Old White Horse re-opened having been rebuilt and set back 40ft from the road; the former building had slightly projected into the road.

OA. 1823-1842 P; 1847 PO; 1851 K; 1854 BD.

**Kings Sutton, 2 The Square** Situated on the village green, the building was once three separate cottages, one of which was an alehouse.

c.1763 listed as The Plough and Harrow – for some years it may have been known as The Ship to become The White Horse by 1890.

Roy W. Fisher and Brian McCabe – Landlords.

**North Stoke** 1883 K; 1899 Wallingford PO.

## Oxford

*52 Broad Street* Early in the 19th century the pub was named the Duke of York.

Between 1629 and 1773 the building was owned by the City of Oxford.

The stuccoed timber-framed frontage is thought to date from the 18ᵗʰ century.

1753 recorded as a public house.

1772 the house was sold by the City to Robert Curtis, a cook; the front measurement was recorded as 4 yards 1 foot 9 inches. It had been occupied by a widow, Mrs. Norris, whose husband John had moved there in 1748.

1776, 9th December – an advertisement in the *Salisbury Journal* reads: 'The Oxford and Winchester Diligence leaves from the White Horse Cellar and Coach Office, Oxford'.

1810 the lease of the Duke of York was given to Morrells Brewery for seventy years by the Revd. Edward Turner and his wife.

1841 the census lists James King, a victualler, as licensee.

1851 the census still records James king as licensee; he lived upstairs with his wife Ann and son Thomas, a coach porter – it is thought the building had probably become a cab office.

1861 the census shows Mrs. Ann King as licensee, her husband having died.

1866-1871 John Jones was recorded as licensee, he was also a cab proprietor as well as a beer retailer.

1881 the census lists John Harper as licensed victualler; he lived upstairs with his wife and four young children.

The house is recorded as trading between 1875 and 1952.

1951 The front of the building was rebuilt and a painted wall was uncovered on the first floor.

The small public house standing between Blackwell's two bookshops is near to Trinity College. It was rebuilt in the Victorian era; sporting photographs of undergraduates from various colleges are displayed. The previous swing sign depicted a mounted policeman on the famous white horse which controlled the crowds in 1923 when spectators broke the barriers at the Cup Final between West Ham and Bolton; this sign was removed in c.1983 – the new one features a grey dray horse.

1980 a fire which started in the kitchen badly damaged the building, it was rebuilt and opened again the following year.

The pub has appeared in the Inspector Morse television series.

By 2005 the Grade II listed building was owned by Exeter College.

At one time it was thought this public house was trading from the 15ᵗʰ century and was named The Elephant, this was in fact situated on the north side of Broad Street and believed to have been on the site of No. 37.

Halls Oxford and West Brewery. 1842 P; 1851-1939 K.

**St. Thomas** 1823 P; 1851-1939 K.

**Stonesfield, The Ridings** The premises ceased to trade in mid June 2001. 1895-1939 K.

**Thame, 12 Corn Market** The earliest known references date back to the late 18th century . . . its proximity to the more prosperous Black Horse may have been partly instrumental in the suppression of its licence in 1908.

AHi. 1823-1842 P; 1851-1885 K; 1899 PO.

**Uffington, Broad Street** The Uffington White Horse chalk cutting on the hill can be seen from the inn. At the re-alignment of the county boundaries which took place in 1974, the village of Uffington and the chalk cutting was 'moved' from Berkshire to Oxfordshire.

1995 the premises became a private house. Mor. 1877-1939 K.

**Wantage, Grove Street** By 1907 the house had been renamed 'The Wheatsheaf'. Listed in part as an historic building. 1823-1830 P; 1840 R; 1844 P; 1877-1907 K.

**Woolstone** Reputed to be one of Britain's oldest inns. It is situated close to the ancient Ridgeway which passes through the Vale of the White Horse.

There are deeds dated 1540 referring to two farm cottages; a Grade II listed building with a lancet window.

1857 Thomas Hughes (1822-1896) is reputed to have written part of 'Tom Brown's Schooldays' while staying here.

The inn sign depicts the Uffington White Horse carved on the hill above the village.

1928 Kelly's Directory lists these premises as 'White Horse, People's Refreshment House Association Ltd.'

2007 the building still had a thatched roof and oak beams.

BHS. 1823-1844 Faringdon P; 1877-1939 K.

**Wroxton St. Mary, Stratford Road** The inn stands on Trinity College land and was part of the original endowment of the College by its founder Sir Thomas Pope in 1555.

1697 the earliest extant lease describes the premises as a messuage, cottage or tenement with a backside or yard with a rickyard and a piece of waste ground.

Sometime during the 17th Century it was known as The New Inn; by 1867 it had become The White Horse. TCA. 1842 P; 1895 K.

## Notes

1. (1848-1939) Fourth daughter and sixth child of Queen Victoria and Prince Albert, she was considered to be the most artistic as well as the most beautiful of all the daughters. She married John Campbell, Marquess of Lorne (1845-1914) who became 9th Duke of Argyll in 1900.

# RUTLAND

*Rutland was absorbed into Leicestershire in 1974;*
*it was reinstated as a county in 1997.*

*For over five hundred years it has been the custom for visiting*
*royalty and peers to pay the town a forfeit of a horseshoe. On the*
*walls of the Norman great hall of Oakham Castle this collection*
*of over two hundred horseshoes hang upside-down – it is thought*
*that this will stop the Devil from sitting in the middle.*

*'The management of grazing land is well understood here. The*
*cattle reared are of no particular breed; those most in request are*
*the Irish and small Scotch. The sheep are chiefly the polled, long-*
*woolled kind. The horses are strong but ill shaped.'*
JB.

**Burley on the Hill** In 1606 it is said that Sir George Villiers, father of the 1st
Duke of Buckingham, died in this building – his nearby family mansion having
been destroyed during the Parliamentary War. EB and JB.

**Empingham, Main Street** A stone-built inn which was
originally a 17th century courthouse standing in the
centre of the village.
2007 the house was advertised as an hotel.
Landlady. 1842 P; 1912 K.

**Morcott, Stamford Road** 1740 built as a hunting
lodge for the 8th Earl of Exeter of Burghley
House. The inn was a coaching house with a smithy, which was used by a local
blacksmith until the late 1930's.
The stabling was used by 'Grittar', who won the Grand National in 1982 at 7-1 odds
ridden by Mr. Dick Saunders, a Northamptonshire farmer, who was the oldest rider of a
National winner at the age of 48, he was also one of the two amateur jockeys to have won
the National in recent years; Grittar was trained by a local farmer and permit holder[1], Mr.
F. Gilman. Landlord and R. W. Kershaw-Dalby – local historian. 1861-1888 Uppingham K; 1941 K.

**Stretton** The old stone-built White Horse was originally part of an estate which
included several farms bought by Mr. Jackson Stops, owner of the Northampton land
agents Jackson Stops and Staff. He gave the estate to his son who was killed during
the Second World War, after which the farms and buildings were rented out; one
included the White Horse public house which was taken on by the Burnham family.

1942 Mr. Jackson Stops asked that the pub be reopened as the 'Jackson Stops', a new sign was made bearing his coat of arms. It became a popular meeting place for airmen based locally at Woolfox and Cottesmore.

**Notes**

1. An amateur racehorse trainer.

# SHROPSHIRE

*'The fishermen here use a kind of canoe made of osiers covered
with hides, which they work with a paddle. It is so light, that on
quitting the river the fisherman carries it home upon his back
with the one end over his head as if it were a large basket . . .
and in every considerable town in the county malting is carried
on to a large extent.'*
JB.

*The Norman name was 'Salopescira', hence Salop as a synonym
for the county – and its official name from 1974 to 1980. The
Old English name was 'Shrewsburyshire'; thus most main roads
leading to the capital, Shrewsbury, became Salop Road.*

**Bishop's Castle, The Square** Situated on the corner with Salop Street. The
White Horse was listed as trading between 1809 and 1871. Farmers and their
wives came by pony and trap from all round the area, most of the inn-yards
provided stabling. Hard-drinking and hard dealing would take place in the inns
themselves – it is said that many a horse could find its own way home if the
occupants of the cart were in no condition to guide it!

In 1851 the licensee was T. Bluck, an innholder and grocer, who also owned
a carrier's wagon [see Shrewsbury, Frankwell] which took heavy goods from
Craven Arms before the opening of the local railway in 1865. A dinner was held
at The White Horse to celebrate the opening of the railway and a receipt signed
by Thomas Bluck reads:

'Dinner & Ale for 40 at 2/6 each. Settled Nov. 3 1865'.

1870 Mrs. Mary Bluck, his wife, was listed as licensee and by 1881 the site had
become a butcher's shop; 1891 Mr. Lamb was listed as the butcher and he later
married a Miss Mutton!

The premises have since been a pottery and finally a private house.

JP. 1822-1842 P; 1868 S.

**Bridgnorth, 5 or 7 St. Mary's Street** This street dates from the 12th century
and formed part of the new town; the brick facings conceal timber framed
buildings on narrow fronted, long 'burgage plots'' with side passages.

CFG. 1822-1828/9 P.

**Castle Pulverbach, Longden Road** The White Horse inn stands close to the
ancient castle earthworks.

An original building on the site of the inn was thought to date from the 12th or 13th century, the present building is partly 14th and 15th century; extensive alterations were made in the early 17th century and it was a farmhouse until the 19th century.

1821 the first licence was granted; the house was owned by the brewers W.T. Southam and the licensee was William Cox; the rateable value was £20. The condition of the premises was described as fair and consisted of a kitchen, a bar, 2 parlours, a club room, 6 bedrooms, a back kitchen and a cellar; there was stabling for 10 horses. The Police Station was 3 miles away.

The inn has three cruck[2] trusses, suggesting that there could once have been two houses standing at right angles, there is also an exposed timber frame of square and rectangular panels to the ground floor and an inglenook fireplace.

1896 the house had a full 7 day licence.

In 1994 an extension was made to the building.

By 2008 the inn sign stood on the opposite side of the road.

Inscribed on a piece of wood inside the front door is the following poem:

*Cathercott upon the hill,*
*Wilderly down in the dale.*
*Churton for pretty girls*
*Pulverbatch for good ale.*

ShA. 1868 S; 1879 PO; 1900-1937 K.

**Clun, The Square** Situated in the old market square, this Grade II listed building dating from c.1780 was once a coaching inn and post house.

1896 the house, whose rateable value was £13, was recorded as having a full licence. The landlord, who was also the owner, was Job Greaves. The accommodation and the condition of the buildings were described as 'good' and there was stabling for 6 horses.

ShA and Landlord. 1842 P; 1868 Posting House PO; 1895 *R.S.O.*-1837 K.

**Dawley, 4 Finger Road** 1773 originally built as two cottages before becoming a coaching inn.

1805 the first licence was granted.

It is said to be haunted by both male and female ghosts. 1879 PO; 1900-1937 K.

**Heath Hill, Dawley Road** A coaching inn which has been trading since c.1721.

1882 the inn is shown on the Ordnance Survey map.

1896 listed as an alehouse with a 7 day licence; the rateable value was £18 5s. 0d. The house which had 4 rooms downstairs and 3 up was described as in good condition, but the stabling for 2 horses required a new manger. The majority of the customers worked at the mines and the ironworks. The Police Station was about half a mile away.

One bedroom is named 'Motley Hall' because ghosts of a motley crew of happy children have been seen there by many people.

ShA and the Landlady who had held the licence since 1973. 1895 *R.S.O.* K.

**Ironbridge[3], Lincoln Hill** The actual location is at the junction of Church Road on the east side of Lincoln[4] Hill. The top of this hill where the inn stands is very high, James Bell's Gazetteer described the area as being 'immediately under the abrupt height yawning caverns disclose themselves, the entrances into the limestone quarries, from hence ever and anon wagons drawn by horses and laden with the produce of the mine are seen to issue'.

1735 a tenant was known to have been in occupation. Beer was brewed here and delivered to other public houses.

c.1780 undermining took place beneath the White Horse Inn; horses and perhaps donkeys were used to draw the wagons from these workings. Donkeys were in fact used to haul the limestone wagons and sledges along the subterranean galleries and each day they were brought to the surface for light, warmth and food. These extensive caverns with their supporting limestone pillars later became an attraction to visitors; it is said that some were large enough to hold a church.

By 1800 the house had a full licence.

1849, 29th September – a Tithe Plan by G.W. Cooke shows three single shafts set in a line from the Swan Hotel (by the River Severn) to the White Horse Hotel; by 1981 the remains of these shafts could still be seen together with banks of limekilns behind the White Horse Inn.

1882 the Ordnance Survey map marks the White Horse incorrectly on the west side of Lincoln Hill, opposite a pound for stray animals together with a slaughter house.

1901 the owner was listed as Miss Smith of Wimbledon and the landlady, Jane Edwards had kept the house for 22 years. The house, which had no stabling, consisted of 3 rooms upstairs and 3 down and had a rateable value of £21. The condition of the building was described as 'fair'.

During this same year there were signs of subsidence and on 31st December Mrs. Page's donkey disappeared with its stable, never to be seen again.

In 1970 a Fire Insurance sign could still be seen on the wall of the inn.

c.2004 the public bar was refurbished. ShA, SCM, JB and Landlady. 1868 S; 1895 *R.S.O.*-1937 K.

**Ludlow, Castle Street** 1822-1842 P; 1868 S.

**Newport, 1 and 3 High Street / St. Mary Street** The White Horse Inn stood on the site of what is now the New Market Hall. The inn traded from at least 1610, and is listed in 1681. The building was demolished in c1856 when the builder, Mr. Williams, purloined several objects for his own house including the fascia board with the words 'William Gregari 1610' which he had placed on the front of his large house at Numbers 1 and 3 High Street. In c.1991 this building, with the help of a grant from English Heritage, became the Council Offices with the proviso that all features of the building be retained – which included the fascia board-giving the impression that the building is dated 1610 when it fact the date refers to the former White Horse Inn. Miss Octavia Maclean, Local Historian. 1822-1842 P.

## Oswestry

**23 *Church Street*** Between 1828 and 1877 the various Trade Directories give either Cross Street or Church Street for the location of this inn. Solicitors documents dated 1934 gives the location as Salop Road but Kelly's Directory of the same year gives Church Street. Cross Street runs into Church Street so this is probably the same inn.

An article on Oswestry dated 1885 reads:

> 'our modern rulers have so chopped and changed the names of the thoroughfares, that we question whether any member of the Corporation, from the Mayor down to the Bellman, could, off-hand, say where one street began and another ended'.

A stone set in a wall in the rear [of the house] fixes the date of erection as 1697. Its sign, a large white stone horse, was once fixed to the front of the brick building. The inn is situated near the site of the New Gate, one of the old town gates; over the arch of which was the figure of a galloping horse with an oak sprig in its mouth – the crest of the Fitz Alan family – who were lords of Oswestry from the 12th to 14th century, the date 1570 is above the horse's head when Henry Fitz Alan was the holder of this title.

1724 the White Horse Inn was occupied by the Edwards family, and thereby hangs a dramatic tale:

> In 1784 Margaret, the landlord's daughter married Richard Coleman, an excise officer, who became the landlord two years later. He was sub-tenant of a field owned by Thomas Phipps, a solicitor, whose father had been a freeman, churchwarden and mayor of Oswestry.

Richard Coleman having mowed the field and removed the hay was accused by Phipps of having stolen the hay instead of leaving it stacked; a bitter quarrel ensued.

Some two years later in 1789 a writ was served by Phipps upon Coleman; an offer was made to drop this writ, but another action would be brought against Coleman for assaulting Phipps' teenage son. In June of that year the solicitor, Thomas Phipps, his son and the office clerk, William Thomas, swore an affidavit that in their presence Coleman had signed a note agreeing to the trespass of carrying the hay off the field. With a further false affidavit against Coleman, documents were lodged in the Court of Exchequer. Coleman insisted that both the Note and Warrant of Attorney were forgeries. The Clerk confessed that he had not seen Coleman sign either document, whereupon the two Phipps were 'apprehended in the cellar of their own house, armed with a brace of pistols properly loaded'. The son immediately escaped, but ten days later was retaken in Cardiganshire some 60 miles away.

In July 1789 the three were tried at the Shropshire Assizes – William Thomas was acquitted but the two lawyers, Thomas Phipps aged forty-six and his son aged twenty, were hung for forgery on the 5th September at Old Heath near Shrewsbury. It is said that a Dr. Milward E. Dovaston hung the skeleton of Phipps from the ceiling of his surgery.

After the trial Richard Coleman gave up the White Horse Inn and left the town.

1803 the premises were put up for auction and were thought to have been taken over by Coleman's brother in law – the pumpman John Edwards for a rent of £30 per annum.

1823 various legal papers concerning mortgages of The White Horse Inn mention messuages and dwelling houses, but give no address until 1868 when Cross Street is given.

1832 during the General Election an incident occurred involving the inn's stone sign:

> 'the rejoicings of the Tory Party in this town took rather a rough turn, and although about twenty-five special constables at a cost of 10s. 6d. each were sworn on the occasion, there was something like a riot, and some of the most violent politicians relieved their feelings by smashing the windows of the Liberals', one of whom was a bookseller Mr. Samuel Roberts, who found to his amazement upon entering his drawing room, the leg of the horse from the White Horse Inn which had been thrown through his window.

At this time a boy, Shirley Brooks was living opposite the inn. Forty years later in 'Bye-Gones' of December 1875 he wrote the following:

'I see you notice the Old White Horse. It gives me a sensation, to this hour, to think of that sign. It was opposite our house. One day I saw a log tied by a rope, outside the bars of the window on the first floor. A man's leg had been badly set, and the surgeons (probably Mr. Cartwright was one) had to break it again, in order to its being properly set. This was between 1830-33. I have seen worse things done since, but my nerves were then young, and I had a night-mare for a week.'

Between the White Horse Hotel and the Fox Inn there was formerly a small shop approached by three steps, the occupier of which in 1852 was Mr. George Weston, a druggist. The last to use these premises before it was demolished in 1872 to enlarge the hotel, was Miss Leake and her green-grocer's shop. At this time the building was refronted and the white horse plaque – still without its foreleg – was refixed to the new frontage.

1984, October – the premises were acquired by Marston's Brewery; who have kindly supplied various legal papers dating from 1823 referring to The White Horse Inn, the Sun Inn and other premises.

1998 the house was bought by the Wolverton and Dudley Brewery.

By 2001 the premises once again became a chemist shop still with the two-foot high white stone horse with a blue background on the wall.

IW, JPJ and MBe, ShA and SAN. 1822-1842 P; 1868 S; 1879 PO; 1900-1937 K.

**English Walls** 1901 'The White Horse Vaults' was listed as having a full 6 day licence with Thomas Hughes as innkeeper, the owners were the brewers Dorsett, Owen & Co. of Oswestry. The condition of the house, which had two downstairs rooms, was described as 'good'.

**Pontesbury** There are White Horse Cottages in Carver Street, possibly on the site of the public house. 1868 S; 1900 K.

**Shifnal, 8 Market Place** The White Horse was known to have been trading before 1814; it adjoined private dwellings in the town and was not far from the Police Station.

1886, 3rd December – the licence was endorsed and the landlord was fined 5s.0d. with costs of 13s. 6d. for permitting drunkenness.

1896 the house which had a rateable value of £15 6s. 0d. was recorded as having a 7 day alehouse licence, the owner was James Cheadle and the landlady, Mary Tudor. There were 4 rooms downstairs and 4 up, outside there was stabling for 2 horses. The buildings were described as being in fair condition but the ceilings needed to be lime washed.

By 2008 it was no longer trading as a public house. The site has at one time been a teashop, the White Horse café and Fennels Restaurant.

ShA. 1822-1842 P; 1868 S; 1879 PO; 1900 K.

## Shrewsbury

***Abbey Foregate*** Situated away from the town on the east bank of the river 'Near the Cold Bath'. 1780 it was listed as trading but probably closed by c.1820 when G. Beaumont advertised his removal from here to the 'Sun' in Milk Street.

The Abbey Foregate, a military depot, was erected from a design by James Wyatt (1746-1813) at an expense of £10,000. It consisted of 'two large depositories for ammunition, an armoury capable of containing 25,000 stand of arms, and neat houses for the storekeeper and the armourer. JB.

***139 Frankwell (at the corner of White Horse Passage)*** Known to have been trading since 1696.

1709, 1780 and 1786 listed as trading.

1851 the Bishop's Castle carrier went from this house, as did the Wollaston Carriers in 1871.

By 1883 it was no longer trading and became a lodging house.

c.1950 it was demolished and the site is known as White Horse Mews.

1828-1842 P; 1868 S; 1879 PO.

***4 / 7 Wenlock Road and 1 London Road (both these entrances were used for trade)*** 1508, 25th September – a deed shows this coaching inn which stood in the high Street, was conveyed by the Abbot of Lilleshall to Richard and Joan Scryven.

c.1720 according to Kelsell's Diary, Richard Lloyd of the White Horse had died, he was from the Quaker branch of the Lloyd family.

1752 the first licence was granted; on 22nd February 1759 Gabriel and Mary Jones of the White Horse witnessed a marriage at the Abbey Church.

From 1856 it was often recorded as the 'Old White Horse' which appears in the following jingle referring to local inns, remembered from childhood by an old lady resident in the Abbey Foregate district:

*'The Old White Horse has tolled the Bell,*
*And made the Peacock fly,*
*Has turned the Old Bush upside down,*
*And milked the Dun Cow dry.'*

1901 a photograph shows the Wenlock Road entrance to the Old White Horse Inn with the landlord, Mr. Benham, his wife, daughter and pet dog.

c.1902 listed as a free house and an alehouse, the class of customers being described as 'Good'; the tenant landlord, John Jones, had been there for some forty years. The owner at this time was William Hazeldine of Woodlands, Abbey Foregate. There were nine rooms, six for private use and three for the public. 'Sanitary Accommodation: W.C., 1; Urinal, 1; good'. Two rooms could accommodate six travellers and there was stabling for two horses. The rateable value was £22.

c.1930 the old building was demolished and a new one erected on the site.

Deed 15768A. 1822-1842 P; 18879 PO; 1900-1937 K.

***Castle Foregate*** The White Horse was listed as trading 1707.

(SPL 17126). ShA, LCL, SPB, JB, Hulbert's Memorials 1803 and DTr.

**Wem, 48 High Street** Wem is Old English for a swampy place.

An important coaching inn situated at the central and ancient road junction. Thought to have been standing before 1677, the year of a disastrous fire in this area of the town, the timbers from the original building survived in this now Georgian fronted building; a wattle and daub boundary wall within the building may have linked the original inn with the shop next door. Remnants of the stables survived – they were last used during the 1914-18 War for the horses of soldiers stationed in the nearly village of Bettisfield. The old garage at the back of the building has a church-like window which is a reminder that it was once a chapel – William Hazlett's father was the incumbent Unitarian minister.

A gravestone in the churchyard opposite reads:

'In respectful memory of Thomas Griffiths, late of the White Horse Inn who died 22nd March 1786 aged 41 years.'

1805 Posting House WCO; 1842 P; 1868 S; 1879 PO; 1898-1937 K. Hotel.

**Worthen** The Old White Horse House was a brewery and stables. This Grade II listed building has the date 1812-14 carved in a beam in the cellar.

1881 the landlord William Vaughan and a Mr. Evans etched their names onto one of the windowpanes.

1896 the house, owned by the brewer, C.H. Kynaston of Wem, had a full 7 day licence and the rateable value was £26. The manager was still William Vaughan, who kept a 'good and clean' house; there was stabling for 12 horses.

1966 one of the players in The Association Football World Cup match – won by England with 4 goals against West Germany's 2 – brought the trophy here to display it to the local people. This same year the inn ceased to trade and became a private house – 'White Horse House'. For a short time White Horse Furniture may have used the stabling which became White Horse Cottage.

ShA and Mrs McLoughlin, owner. 1895-1937 K.

**Wrockwardine Wood, New Road** The White Horse Tavern which was known to have been trading before 1800 was a coaching inn; the long stone building with walls fourteen inches thick is thought to be the same date as the adjoining house, where there is a stone dated 1794; the cellars run beneath both properties.

1840 a local friendly society met here regularly.

1896 listed as an alehouse with a 7 day licence, the majority of the clientele were miners; the rateable value was £22 10s. 0d. and it was owned and occupied by James Wade. There were 3 rooms downstairs and 3 up. The house was described as being in fair condition, but the stabling for 4 horses needed a new manger, the cratch[5] repairing, cleaning and lime washing; at one time there were also pigsties at the back. The house was about one mile from the Police Station.

1922 the new Wrockwardine Wood Bowling Club had a green on the White Horse field.

By 1974 there was a licensed clubhouse on the White Horse field where the Wrockwardine Wood Football Club played, it was thought to have been one of the earliest professional clubs in the country.

ShA. 1826 L; 1868 S; 1879 Wellington PO; 1900 Wellington – 1937 K.

**Notes**

1. Tenure of land or tenement in a town or city in England, which originally involved a fixed money rent. In Scotland – the tenure of land direct from the crown in Scottish Royal burghs in return for watching and warding. Col.

2. A curved wooden timber supporting the end of a roof.

3. 'Here is the celebrated iron bridge erected in 1779, consisting of one arch of 100 feet in span and 40 feet in height. It was cast in the works here, the whole weight of iron being 378 tons. Its appearance is admirable, adding a sublime feature to the scenery of that delightful glen.' JB. (See Madeley-Market)

4. The name of this area is thought to have been derived either from the local limekilns or from 'links' [Linc-oln], when extra horses or ponies were needed to be linked to the trucks being pulled from Coalbrookdale, up the hill to Church Road and on to the foundry.

5. A rack for holding fodder for cattle, etc. Col.

# SOMERSET

## Bath, North-East Somerset & North Somerset

*'Orchards are numerous, and those that have a northern aspect and
are sheltered from the westerly winds are generally very productive.
Along the northern base of the Mendip hills the fruit yields a strong
and very palatable cider, and in the vale of Taunton-Dean cider is
made of the very best quality.'*

JB.

*In 1830 the taverns of Somerset used 'public house checks' sometimes
known as tokens. These were made of brass or copper and generally
gave the value and name of the landlord and his house; they were
widely used until the turn of the twentieth century.*
'Somerset Public House Tokens' by S.C. Minnitt. 1985.

## Bath

***4 / 5 Northampton Street*** c.1848 a token issued by T. Kift
gives 'White Horse Northampton Street'.
 c.1992 renamed The Dark Horse.

***14 Somerset Buildings*** 'White Horse Cellar'. c.1856
a brass token issued by G.B. Lovell gives 'White
Horse Cellar Bath'.
 MDG. 1842 P; 1861-1883 K; 1897 PO; 1914-1939 K.
 *It is not clear to which house these dates refer.*

***Walcot Street (without North Gate)*** From the *Bath Journal* :

> 1752, 13th July – 'Tomorrow being Tuesday the 14th will be played
> for at backsword at Mr. James Griffiths at the White Horse without
> North Gate, Bath, a Pair of Silver Buckles. He that breaks most
> heads and saves his own will be entitled to the Prize. To begin at one
> o'clock and continue till sunset.'

> *In the same edition* :
> 'to let or for sale: a well-accustomed messuage with the backside, stables,
> out-houses and appurtenances thereunto belonging, commonly called
> or known by the name of the White Horse in Walcot-Street in the
> parish of St. Michael in the City of Bath.'

*Six months later* :

1753, 29th January – 'last Thursday was buried at St. Michael's Church in this city Mr James Griffith who kept the White Horse Inn. His pall was supported by six shoemakers, and six carried the corpse, all were decently dressed, they wore new gloves and aprons agreeable to the deceased's desire.'

The new St. Michael's Church was built in 1744 with a handsome dome. It was of the Doric Order. JB.

**Bayford** The inn was listed as the White Horse from at least 1813. Mrs Lovell, who was born in 1912, wrote to the author in 2002, "I understand there was a coaching inn at the top end of the village and I am wondering if that was called The White Horse, it is now a farm house."

1831 listed as trading in both the Somerset Directory and the Post Office Directory under 'Stoke Trister, Bayford'; also in 1861, 1866 and 1872.

There were three landlords listed at the White Horse in eleven years.

1875 it is thought the sign was changed to The Unicorn. However, both the White Horse and The Unicorn were listed at this date in Bayford, therefore it has been difficult to establish the correct location.

Mrs Violet Lovell, who moved to The Unicorn with her parents when she was about 17 years old, described her life while living at the inn:

'It was news to me that the Unicorn at Bayford was called the White Horse years ago. My Mum and Dad took the Unicorn in 1929 together with a small holding for his milking cows. We also had a cob which was a very light grey in colour and loving horses it gave me great pleasure to help look after him.

My Dad used to make cider in the stable and I used to help grind the apples. We had two blacksmiths in the village, one across the road and one at the entrance to Stoke Lane, the house is still called The Forge; the one from across the road sometimes brought a horse to shoe in our yard so that he could have a sly pint of cider unknown to his wife!

We had a lovely well in the back yard which never went dry; before the mains water came through the village – if the neighbours' pumps ran dry – they came to us with their buckets and drew water from our well.' JSk.

**Bradford-on-Tone, Regent Street** Situated opposite the Church. Mortgage documents record that the tavern was trading from 1691 to 1950.

1755 the White Horse was mentioned in a marriage settlement.

1780 the local men challenged the 'Rest of the World' to meet them here for duels with backsword, sword and dagger – no one kept the appointment so they fought each other!

1843, 18th September – St. Giles Day Revel; an extract from the diary of The Reverend William Burridge, vicar and magistrate, reads:

> 'Not a head of stock to keep up the name of a fair. The White Horse . . . was open all night . . . Tom and I out until past twelve to keep the churchyard free from whores and drunken fellows.'

Due to fire damage the tavern was rebuilt in the 19th century on its original forecourt.

1866-1875 the licensee was listed as a coach builder and from 1883-1889 as a wheelwright.

1998, 24th August – a sale notice describes this two-storey house as a beautiful country pub in the heart of the village, with an adjoining village shop and post office. Mr. R. McCann - landlord, JSk, SRO and L&M.

**Brewham, Kingsettle Hill (north side)** 1620 an inn was recorded in the parish, its name is not known.

1720 the White Horse was listed as trading, but by 1723 the building had been demolished. JSk.

**Bridgwater, Penel Orlieu / Pig Cross / St. Mary Street** Listed as trading from 1754.

1861 Mrs. Elizabeth Spurway is recorded as landlady at The White Horse, St. Mary Street; at the same time James Alway is listed as 'grocer & provision dealer in this street'. It was he who bought the inn for £710 in December 1869, an old photograph of about this date shows the Roman-tiled roof and crumbling plaster wall covered in posters, its sign hanging out over the pavement; Mr. Alway had it demolished at a cost of £200 and built new premises which became The White Horse Stores.

1875 the inn was not listed, however in 1880 Mr. L. Llewellyn is given as landlord, and oddly enough in 1931, the premises at 13/16 Penel Orlieu was occupied by Mr. Llewellyn, a chemist.

Penel Orlieu – 'the name theories are legion . . . not a very nice area in the old days; a rabbit warren of streets and alleys, the pig market and all sorts of other fascinating sidelines of life!' JSk, DW and ABM.

**Bruton, Coombe Street** The house stood just above the present Royal Oak. By 1798 The White Horse is not mentioned in the Tithe Rent Book. JSk.

**Cannington** Situated due north of Cannington on the road to Combwich; recorded as trading from c.1700-1743. SRO.

**Chantry** The building is thought to have dated from c.1700 and was possibly known as The White Hart; by 1861 it was trading as The White Horse until 1939. For a few years from c.1870 it is thought to have been named The George.

**Chard, 15 Silver Street** Streets named Silver refer to a running river. ' . . . a stream of water in the streets being divertable at pleasure into the English or Bristol Channel'. The English Channel is known as the Silver Streak.
1640 listed as trading.
1791 the landlord recorded as a victualler.
1820 an extract from the Council Minutes reads:

'Opinion of Council that White Horse Inn, from bad characters which are allowed to frequent it, ought to be Watched.'

1822, 12th April – Borough accounts record a payment of £35 19 11½ to Mr. C. Locock for labour and materials for repairing the road at the White Horse Inn, left unfinished by Mr. Stembridge two years earlier.
1841 the inn was shown on John Wood's Map on the north side of Silver Street standing in a large yard. On 3rd May – Volume 31 of the *Chard Union Gazette* records:

'Free Public House, or Inn, for Sale. Wm. Gregory instructed to sell by Public Auction, 13th May at 6 pm – old established inn called the White Horse in borough of Chard, with extensive stabling, brewhouse, cellars, barn, outhouses, garden. Premises now in occupation of Mrs Ann Stone, the owner. Desirably situated, in immediate neighbourhood of extensive lace manufacturers within short distance of terminus of Chard Canal. Wm. East, Solicitor, Chard.'

1842 The landlord, Abel Oram, was the owner of a woollen mill – the inn stood near to the town's millstream.
1865, 14th October – The following advertisement appeared in a local paper:

'White Horse Inn, Chard – J. Woodberry, Fancy Cane, Rush and Kitchen Chair manufacturer – Tables Bedsteads etc. constantly on sale.'

James Woodberry was landlord from 1859 to 1866.
1881 the landlord, James Hill also owned Holyrood Mill.
c.1985 re-named The Stumble Inn.
1991 re-named The Old White Horse Inn.
1997 the premises were closed.

1999 the building was demolished except for the façade, the interior being converted into apartments. During these building works a narrow well was found on the south side. RWC and SRO.

**Dulcote** The White Horse was situated west of the bridge between two mills.

1696 John West, an inn-holder of Wells is recorded as leasing the alehouse beside the west mill.

1785 the White Horse Inn and garden are listed as the property of the Vicars of Wells.

1829, 6th July – an indenture records the Reverend Thomas Coney of Dover in Kent as the owner. A map of this date shows the White Horse Inn, court and garden as 'leasehold for lives', meaning the land was leased to a tenant and his heirs over the course of several specified generations. One condition of the lease states that the inn may not serve 'refreshments, beer, wine, spirits or other liquors on Sundays to any person or persons whatever except strangers who may be travelling and requiring such refreshment'.

1861 Charles Clement Tudway re-leased the inn to Ann Churchouse and Charles Berryman, with the permission of the owner, the Reverend Coney.

By 1889, and again in 1914, the house was listed in trade directories as a public house, not an inn. Richard Pointing was the publican, he and his family owned a butcher's business in Wells, he was also a dealer; they leased the premises from the Tudway family.

1913 the Pointing family bought the premises in an auction, details included the following description:

> 'The dwelling house is built of stone with tiled roof; parlour, tap-room, cellar, kitchen, larder, W.C., 4 bedrooms. The outbuildings are stone – stables for 7 horses, pigsty, trap-house, garden, meal-house and store. The annual rent = £18.'

1914 Mrs. Sarah Pointing was listed as the proprietor of the White Horse public house, by 1919 she was listed as a private resident in Dulcote, the house having been renamed The White House. 1861 SD and PO; 1883-1914 K.

**Dulverton, 23 High Street** The premises became a private house, The Old White Horse, later to become an antique shop. 1842 P; 1883 K; 1897 PO; 1914 K.

**Dundry, Northwick Road**

**Exford** Situated beside the upper reaches of the River Exe, 'The Exmoor White Horse Inn' was originally a 16th century alehouse.

Deeds exist for the years 1707 and 1785 and mortgage documents for 1856.

By 1883 it was listed in Kelly's Directory as a Commercial, Posting and Family Hotel.

1897 the Post Office Directory showed White Horse & Crown.

1955 the inn was advertised for sale.

2007 the hotel organised salmon and course fishing, expeditions to Exmoor on foot and horseback and the Exmoor Stag Hounds met in the courtyard. SRO.

**Frome, 10 Portway** There are deeds existing between 1846 and 1861. By 1875 it had become The Old White Horse; four years later it was re-named The Bird in Hand.

1939 listed as trading in Kelly's Directory as The White Horse; during the Second World War, Field Marshall Montgomery had his D-Day H.Q. at the nearby Portway House Hotel – maybe he and his staff visited their local!

Situated between a veterinary surgery and the Post Office, the premises became a butcher's shop selling game from Longleat. WRO.

**Glastonbury, North Lode Street** 1717 SRO.

**Ham** 1726 listed as trading; in 1750 it was recorded as being of bad repute. From 1859-1875 the landlord was a farmer and butcher.

1931 listed as trading before becoming a private residence. SRO. Some directories give Creech St. Michael.

**Haselbury Plucknett, North Street** Behind the inn was a field – Winter Common Field – divided into strips of various sizes which were allocated to the villagers; it is thought that the field was probably the remains of mediæval strip farming.

At the end of the 18th century the White Horse Inn was situated in the very narrow Swan Hill between Haselbury House and the Swan Inn. ' . . . at the Three Cross Roads . . . the coachman guided the team of horses to the right . . . pulled the team up outside the White Horse Inn. There the coach stood for a few minutes, the horses scuffing their hooves in the gravel as bags and postal packets were thrown from the roof.'

1840 the inn is shown on the Tithe Map, it stood in an acre of orchard and garden, possibly a Tudor building in poor state of repair which was demolished as it did not appear on 19th century maps. The business moved to North Street opposite the blacksmith's shop.

1924, 18th September – the premises were put up for auction as part of the Haselbury Estate, owned by the Portman family who had been Lords of the Manor.

c.1950 Mr. Wheatley, the landlord of the White Horse operated a village haulage business to transport cattle to various markets, the Crewkerne market having closed.

c.1980 renamed Haselbury Inn, a restaurant.

In 1815 Haselbury Plucknett had a thriving horse hair industry supplying brushmakers, upholsterers and tailors; the horse hair was also used for long fishing line for cod in Scotland, the making of gas meters and for surgery.

MR and HPl. 1861 SD; 1883 K; 1897 PO; 1914-1939 K.

**Hungerford, Abbey Road** Originally two cottages standing beside the Washford River. At one time it had been a toll house; the toll money was collected through an alcove in the two foot thick stone walls. The original staircase beside the 'bottom fire' in the main bar is partly bricked up – the front of the bar used to be the back wall of the pub.

c.1709 alterations to the building took place when the stone walls were faced with brick, this is probably when it became a public house.

1715 Hugh Moor of Nettlecombe became the owner.

By 1730 John Bindon of Langford Budville as owner, he and his wife gave the inn to their son James in 1782.

1814 James sold the property to Thomas Bond of nearby Washford Mills.

1835 John Bond of Williton became the owner of the inn and nearby Foulbridge Orchard.

1861 and 1866 the White Horse Inn is listed in trade directories with Mrs. Elizabeth Melhuish/Milhuish in occupation.

In the 1930s Rachel Reckitt, a metalworker, blacksmith and woodworker, made the intricate metal work inn sign.

2004, July – the house was selected as 'Pub of the Month'.

SRO – DD/CH/109/12; Tim Weaver, Landlady's son, and JSk.

**Ilchester, Church Street** 1813-1822 listed as trading; by 1840 the White Horse was no longer listed. JSk.

### Ilminster

In early 1800 a new road was built through Andover, Wincanton, and Ilminster, joining the old road at Honiton. This considerably shortened the journey from London to Exeter and put Ilminster on the map:

*'Now cider is golden and beer it is brown,*
*So lift up your tankards and don't put them down*
*Till you've drunk to the taverns of Ilminster Town,*
*Singing barrels of cider and ale!*

*A Dolphin called George was swimming one day*
*When he met a White Horse and they started to play,*
*They frolicked and splashed in the bright Rising Sun,*
*Until they were joined by a snowy white Swan,*
*'I'm New Inn this sea, can I join in the fun?'*
*Singing barrels of cider and ale!'*

JBa and NH.

### 9 Bay Hill (a continuation of East Street)

A coaching inn.

1764 there are deeds and leases from this date.

Five undated checks were issued by the landlord of the White Horse, East Street:

'White Horse Ilminster' – a tulip
head circle around a roughly cut figure 1.

A 3d. check 'White Horse E St Ilminster'.

A 3d. check 'White Ilminster' – a standing horse facing right.

A 6d. check 'White Ilminster' – a standing horse facing right.

A 6d. check 'White Horse Ilminster'.

An aluminium rectangular check stamped 'White Horse' believed to have been used during the Second World War.

1830 recorded as a 'licence post'.

In 1998 there was a change of ownership.

A Grade II listed building. 1883 K, 1897 PO; 1914-1939 K; 1960 JSk.

***Ditton Street*** "Ditton Street and Bay Hill are distinct, and there must have been two houses trading simultaneously as the White Horse, one in each address. I can find no directory or other evidence to confirm the Ditton Street one, but I have no doubt that the photograph is correctly captioned." David Bromwich of SRO and HIT.

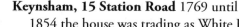

**Keynsham, 15 Station Road** 1769 until
1854 the house was trading as White Horse.

By 1897 it had been renamed The Pioneer; other names have been used including McGuinnis, The Boars Head and Fontelles. SRO and Landlord.

**Langport, The Hill** Near the junction with North Street, the building dates from c.1692.

1717 listed as The Nag's Head.

By 1726 the house was trading as the White Horse, becoming a private house by 1779. SRO and JSk.

**Mark, Church Street** Originally a wooden fish house known as High Hall. Twice a year between the 12th and 14th centuries, the Benedictine Abbots of Glastonbury and Somerton used to stop here for a drink and rest when collecting their rents and tithes from landowners and tenants. They had rowed their boats through the 'rhines'[1] to the River Yeo on their way to East Brent. Nowadays they are commemorated by 'The Abbots' Causeway, which is 1½ miles long and below sea level – it ends at the White Horse Inn.

The present building dates from c.1680 and was originally two cottages; the front door to one has been bricked up, the other is the entrance to the pub. Within can be found pews from the village Church installed in 1899, together with a history of this house.

This information was provided by William Luff who was born in this house during the tenancy of his grandmother – the family had held the tenancy for one hundred years. 1897-1939 Northwick K; 1942 JSk.

**Martock, 47 North Street** c.1980 the house ceased to trade and became a private house. 1936 JSk; 1939 K.

**Milverton** 1685 recorded as trading.

The *Sherborne Mercury* published the following notices:

> 1761, 30th November – 'To be sold in fee . . . court Farm and Court Place . . . for a sale whereof, a survey will be held at the White Horse . . . on Friday the first day of January next.'

> 1763, 17th January – 'To be sold, a large quantity of timber trees . . . for which purpose attendance will be given at the White Horse . . . on Friday fourth day of February next in the afternoon . . .' dated 12th January 1763.

> 1766, 9th June – the *Sherborne Mercury* reported that the White Horse had been sold as a private house and that the George Inn had become the most commodious inn in the town and was to be Let. However on 21st September 1767 a notice was published that a Survey concerning a tenement called Clipwell would be held at the White Horse Inn on Friday, 2nd October at two o'clock in the afternoon.

ECL. 1685 SRO; 1775 JSk.

**Minehead** 1842 P; 1883 K.

**Over Stowey, White Horse Farm** 1779 licensed as an alehouse for the last time – having served travellers on the Bridgwater road.

By 1785 it had ceased to trade, later to become a farm house.

**Queen Camel** 1772 SRO.

**Rumwell** Listed as trading from 1768-1805.

The house was sold in 1806. JSk.

**Shepton Mallet, Town Street** 1813 listed as trading. JSk; 1842 P.

**South Cheriton, Cabbage Lane** This building, dating from c.1800, was originally two cottages and a Post Office.

1866 listed as trading.

1921 the house was advertised for sale.

1958 listed as trading.

2002 refurbished and the skittle alley converted into a restaurant. WRO. 1939 K.

**Stogumber, High Street** 1748 listed as trading.

1828 a meeting of the newly formed Stogumber Union Friendly Society took place here.

The building was extended during the 19th century to include a former Market House.

1861-1875 William Dore was the innkeeper, he was also a farmer.

1868 a skittles alley was added.

2007 this Grade II village inn was still trading. SRO. 1861 SD; 1883 K; 1897 PO; 1914-1939 K.

**Stogursey** Listed as trading in 1673 and 1779 when Samuel Bisse was landlord. 'The pubs here seem to have changed their names with reckless abandon in the 18th century.' JSk. and SRO.

# Taunton

*Extra Portam (area of East Reach / Eastreech)* c.1579 recorded as trading.

1752, 30th October – The *Sherborne Mercury* published the following notice:

'To be Lett at Lady Day next, for a term of seven, fourteen, or twenty one years. All that well-accustomed inn, with a very good brewhouse, stabling and all other conveniences thereunto belonging . . . now in the occupation of Mr. Jacob Standerwick . . . '

c.1785 recorded as trading.

**East Street** 1674-c.1750.

**Paul Street** 1793-1798.

**St. James Street** Situated at the junction with North Street.

1811, 31st January – the *Taunton Courier* records that the landlord, William Nation, missed his way in the dark and was drowned in the River Tone.

The inn continued to trade until at least 1853. SRO and ECL.

**Twerton, 42 Shophouse Road**

The building was originally three cottages.

Between 1851 and 1899 the house was listed as trading. WRO and JSk. 1923-1939 K.

**Washford, Abbey Road** *See Hungerford.*

**Wellington** 1863, September – the *Wellington Weekly News* records the funeral of Mr. Stradling, proprietor of The White Horse. He had been a veterinary surgeon for many years and a member of the Ancient Order of Foresters. He was buried with military honours being a member of the Volunteer Rifle Corps.

c.1884 a token for 3d. was issued by John Hickey at 'White Horse Inn Wellington' – *the 3d. surrounded by a wreath.*

1912 the pub ceased to trade and had been demolished by 1915. Shops were built on the site. 1883 L; 1897 PO.

## Wells

**Priory Road, The Sherston Hotel** The property is said to have once been part of the Manor of St. John which was owned by the Sherston family.

1665-1724 it was listed as The Black Bull and later became The White Horse.

By 1854 it had become The Sherston Arms, then The Railway Hotel until the railway was closed.

**23 Sadler Street** 1691, 1709 and 1718 the inn was known to have been trading as White Horse; it was situated on the main road to Shepton Mallet.

By 1830 this coaching house had been renumbered 19-21.

At sometime the house was renamed The White Hart.

By 2007 No. 23 had become Armadine Picture Framing; the cobbled yard could still be seen in its original form. SRO. 1822 P; 1897 PO.

**Wincanton, 4 High Street / Market Place** This inn may have originally been known as The Hart.

1558 inherited by the heirs of Henry Williams.

Before 1655 Robert Vining was innkeeper, his family ran the inn for over 70 years; they were followed by the Deane family who were here for 61 years in total.

1733 listed as a coaching house, seventeen coaches a day passed through Wincanton which was about midway on the London to Exeter run. The house was built by Nathanial Ireson of Wincanton who in 1721 supervised the building of Stourhead for Henry Hoare.

It is said that he designed his buildings to have no guttering on the front elevations.

1737 the inn was rebuilt by George Deane, whose initials are on the keystone.

1800, 7th April – an advertisement in the Salisbury Journal reads 'White Horse, Wincanton; well-accustomed; with malthouse adjoining.'

The following year on 31st August the inn was again mentioned Thomas (sic.) Deane having improved the house thanked his patrons and advertised that 'Having no concern in the posting business, his attention shall ever be to accommodate those who chuse to call on him in the most decent style.'

c.1816 a trade card was printed.

1842-1885 the licensee Samuel Sly, opened the house as a wine and spirit store.

1927 Harry Ridout was recorded as the hotelier, innkeeper and wine and spirit merchant.

1931 the brewers Hall & Woodhouse Ltd. sold the premises to Mr Ridout as a free house.

1938 a photograph shows a porch had been built over the main entrance, this had been removed by the end of the century.

1976 the public house ceased to trade.

c.1980 this listed building was reopened as an hotel.

Occasionaly the voices of three happy unseen children are said to be heard on the stairs. Landlord, BCo and Hall & Woodhouse – brewery.

**Wiveliscombe, North Street** 1775-1906 listed as trading. The original coaching yard and stables were converted into properties to become a garden shop with flats above called White Horse Mews; the archway was used to reach the cattle market until 1960. JSk. 1822 P; 1861 SD and PO; 1883 K; 1897 PO.

**Yeovil, 10 St. Michael's Avenue / New Prospect Place** There is a nine-pin skittle alley in this pub. Mike Crocker. 1936 JSk; 1939 K.

## Notes

1. A large open ditch or drain.

# STAFFORDSHIRE

*The Stafford Knot originated from the thirteenth century,
Humphrey Earl of Stafford took it as his personal badge and
used it for his retainers' and servants' livery; it was not part of
the family armorial bearings.*

*In 1707 the 1st Battalion of The South Staffordshire Regiment
was sent abroad and spent about fifty years in the West Indies,
forgotten and neglected; in 1758 it is recorded that the men were
dressed in rags and had not any hats; a year later at Guadeloupe
they were using muskets of a half-century old pattern.*

*In 1788 the officers were suffering from tremendous hardship
having had no pay for the last seven years. To commemorate
these privations the Regiment in 1936 received official sanction
to incorporate a piece of brown canvas signifying 'sackcloth and
ashes' in their uniform.*
ECT.

**Burslem, The Potteries** In 1571 the total number of beds for travellers in
'Boslem' was 4, with stabling for 4 horses.

At a very early period this place was distinguished for the variety and excellence
of the clays in its vicinity and in the 17th century it was the principal place in
England for the manufacture of earthenware. Josiah Wedgwood was born here
in 1730 and George Stubbs was born in Liverpool in 1724, both collaborated on
the solid green jasper plaque of 19 white horse reliefs made at the nearby Etruria
Factory in 1788. PRO and JB. 1860-1900 K.

**Burton upon Trent, 158 High Street** In 1571
the total number of beds for travellers in 'Burton'
was 41, with stabling for 82 horses.

'Good stabling' was offered by this hostelry.

1875, October – a line drawing shows the
White Horse was trading in the High Street at
this date. The inn was closed between 1880
and 1911 and later a branch of the National
Westminster Bank was built on the site. PRO.
1828/9 P; 1860 K.

**Cheadle, Mill Road** In 1571 the total number of beds for travellers in this town
was 34, with stabling for 69 horses. PRO.

From the *Staffordshire Advertiser*:

> 1871, 23rd December – Sale at auction, White Horse Inn, Cheadle
> Mill, described as 'newly built and old established'.
> 1878, 12th January – To be let, White Horse Inn, Mill Road.

1822-1842 P; 1860 K; 1880 PO; 1900-1912 K.

**Checkley, 20 Church Lane** The house with a brick façade still has
its original stone foundations, but the cellar has been filled
in. This hostelry was conveniently situated on the narrow
winding lane between Uttoxeter and Newcastle-under-
Lyme, one of the major cross-country routes of England.
Local innkeepers and farmers would provide stabling
and hire out fresh horses for travellers, as there were no
coaching inns for some thirty miles.

Extracts from the Church Wardens' accounts:

> 1636 The innkeeper, Katherine Overton, was paid 2s. 6d. "for a great
> Panne and fire to brew two strykes¹ of mault and to seethe the patches, and
> for her labour and maides in carrying ye liquor and size to ye Church."
>
> Extensive repairs were being carried out to St. Mary and All Saints
> across the lane; the workmen depended on Katherine Overton to supply
> ale for refreshment and it appears she mixed water and size for the mortar.
>
> 1660 'Spent at Overtons of a workeman and tow or three of the
> nighbors about the greate bell 1s. 0d.' 1661 Feb. 3 – 'Paid to Catherine
> Overton for bread for ye communion 2s. 0d.'
>
> 1662 at a vestry meeting held at the inn, when new officers were chosen,
> one church warden paid Katherine Overton 9d. and another warden
> paid 1s. 3d. for refreshments for 'our neighbours'.
>
> This was the year the Hearth Tax was levied at a rate of 2s. 0d. for each
> hearth except those in cottages.
>
> 1666 Katherine Overton, by this time a widow, had 2 hearths to be
> taxed, the rectory had 7 hearths, one other house in the village had
> 2 hearths, the rest had only one – these may have been cottages and
> therefore exempt from paying any tax.

In the late 18th century Mr. W.C. Oulton spent a night in this 17th century hostelry;
due entirely to the note in his 'The Travellers' Guide' published in 1805, I was able
to trace the history which I have been unable to find in trade directories.

1792 first listed as White Horse with Lydia Beardmore as innkeeper.

By 1799 Mrs. Beardmore, had become a widow and was probably helped by her twenty eight year old son, William.

1813, 19th April. The last recorded vestry meeting took place at the White Horse Inn. Future meetings were held at the Poor House.

1818 William Beardmore, now forty seven, was listed as a victualler and blacksmith. Four years later a turnpike road, which became the A50, was built to by-pass the village – this of course affected the custom of the White Horse which within a few years ceased to trade as an inn, as by 1834 Mr. Beardmore was listed solely as a Blacksmith. Three years later the Parish Rate Evaluation book recorded that he had divided the premises into two dwellings. The property included the hostelry measuring 45 by 15 foot, a small pantry 9 by 8 foot, a blacksmith's shop and shoeing house 50 by 16 foot, an unoccupied back house of 13 by 18 foot and a croft[2] measuring 3 roods 10 perches.

The 1841 Census gives Wm. Beardmore, now aged seventy, as the blacksmith and a 10 year old girl, Charlotte, living in the house. The house at the back was inhabited by an agricultural labourer, William Lymer, his wife and four children.

The 1851 census gives William Ward, as blacksmith and farmer, owner and occupier of the premises, including the house at the back – he had previously lived in a cottage next to the smithy.

By the 1861 census he had moved and set up the smithy at Manor Farm. The occupants of the White Horse were recorded as Thomas Davis, an agricultural labourer, his wife and seven children; also his cousin, George Ward, a retired farmer.

It was known as White Horse Cottage until c.2001, to become the White House.

Mrs. F. J. Johnson, local historian, kindly provided the above information and the following references: Sta. Ref. D113/A/PC/1; Sta. Ref. D113/A/PV/2 and 3; Sta. Ref. D113/A/PP/1-18. WDG.

**Cheslyn Hay, 8 Mount Pleasant** The site was originally occupied by two cottages standing at the junction of a trackway used for taking the pit ponies to and from the Plant Colliery. This trackway was known locally as New Horse Road.

1965 Wolverhampton and Dudley Breweries Limited purchased the site when the Landywood housing estate was to be built on Coalpit Field, the plans of which included a public house. In August the following year the brewery was given a provisional grant to build the pub, which was opened on 16th August 1968. There is a large bar, lounge and off licence attached.

By August 2008 the landlord had been here for 40 years.

The pub has its own football and pool teams. PEv and Walsall Local History Centre.

**Fazeley, Atherstone Street** 1933, 6th January – a deed describes this substantial property at the time of delicensing:

'All that messuage dwellinghouse or tenement formerly used as a public house and known as the White Horse . . . together with the yard, garden, brewhouse, stable, shopping outbuildings . . . also all that dwellinghouse . . . Butchers Shop, garden, Slaughterhouse and Room over the same and together with the Fasting Pen, Stables, Piggeries and appurtenances adjoining the same.'

1933 the house ceased to trade and was later demolished due to road widening. 1860 K-1880 PO listed under Tamworth; 1900-1912 K.

**Hanley, The Potteries** 1822-1828/9 P.

**Longton, Anchor Terrace, Anchor Road** 1857, 31ˢᵗ January – advertised for sale by auction; described as a beerhouse with a spacious yard, stable and brewhouse.
In 1867 it was advertised to be let on 5ᵗʰ January and 7ᵗʰ September. SA. 1880 PO; 1900-1936 K.

**Newcastle under Lyme, Penkhull Street** In 1571 the total number of beds for travellers in 'Newcastle' was 24, with stabling for 38 horses. PRO and KU. 1822-1842 P.

**Rugeley, Bow Street** In 1571 the total number of beds for travellers in 'Rudgley' was 23, with stabling for 54 horses.
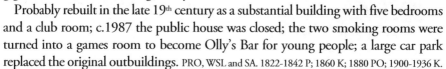
Situated at the Marl Pits, one of the oldest areas on the south side of Rugeley, adjoining the London Road. 1788 the inn is recorded as being owned by the Earl of Uxbridge – *who lost his leg at the battle of Waterloo in 1815.*
1854, 7ᵗʰ October – For Sale by Auction and described as 'old established freehold public house near the Market Place and Sheep Fair with stable, gig house, other outbuildings and yard.'
Probably rebuilt in the late 19ᵗʰ century as a substantial building with five bedrooms and a club room; c.1987 the public house was closed; the two smoking rooms were turned into a games room to become Olly's Bar for young people; a large car park replaced the original outbuildings. PRO, WSL and SA. 1822-1842 P; 1860 K; 1880 PO; 1900-1936 K.

**Silverdale, Newcastle Street** 1941 K.

**Stafford** 1828 P.

**Stanton, 99 Woodlands Road** The Bass Brewery Minute Book records:

1887, 29th November – the Board agreed to 'purchase 1 rod of land adjoining the inn . . . for £90.' Six months later on 19th May a cheque for £80 was paid.

1912, 19th October – 'a small plot of ground opposite the White Horse' was bought by the Company for £45.

1922, 3rd April - land adjoining the White Horse Inn comprising 8 acres 3 rods 6 poles was bought by Bass for £475; the remainder, measuring 2 acres 1 rod 31 poles, was sold to Halls Colliery Ltd. for £25, on condition that a boundary fence was erected and maintained. Three weeks later the deeds had been received, but it was discovered that the area of the land sold to Halls Colliery measured 2 acres 2 rods 7 poles, so that Bass had bought land measuring only 6 acres 0 rods 39 poles, instead of 8 acres 3 rods 6 poles.

1926, 3rd May – 'Alterations were made to this property amounting to £725.'

**Tamworth, 1 Lichfield Street** Situated at the cross roads on the corner with Silver Street, opposite the White Lion Hotel. The building was demolished in 1966 and the site, near a tower block of flats, became an open space due to the widening of Lichfield Street.

RSu. 1822-1828 P; 1860 K; 1880 PO; 1900-1940 K.

**Tunstall, 143 Brownshill Road** In the title deeds of this beerhouse dated 1789 'White Horse' is not mentioned.

1856, 20th September – from the *Staffordshire Advertiser* :

'A rare opportunity. To be let and entered upon at Michaelmas, one of the oldest and best beerhouses in the Potteries situated on a good thoroughfare and in the midst of potteries, coal, iron and brick works . . . '

1902 a tenancy agreement was made with Norris's Brewery; by the end of the 19th century the whole adjoining area had been redeveloped to accommodate H.&.R. Johnson's Tile Works. SA., Landlord and R.E. Blacknell.

**Uttoxeter, Bridge Street** In 1571 the total number of beds for travellers in 'Uttoxiter' was 57, with stabling for 143 horses.

1828 listed as trading in Pigot's Directory in the Market Place where there was once an ancient market cross with 24 steps.

The Bass Brewery Directors' Minute Book records the following concerning the White Horse:

> 1898, 18th July – land adjoining the above house to be sold for £600.
> 5th September – conveyance of two cottages and the land adjoining the White Horse, £600 to be received.

> 26th September – land adjoining sold for £600. That the £600 had been received.

> 1925, 16th November – It was agreed to demolish property in order to open up the frontage of the inn, giving a contribution of £500.

> 1928, 23rd April – 'Resolved to transfer the Roof of Salt's Bottling Stores, Wetmor Road, to the yard of the White Horse at an estimated cost of £191 15s. 0d.'

c.1960 to make way for new town planning, the building was demolished and rebuilt in twentieth century style.

2004 the roof and the upper floor were badly damaged by a fire.

PRO. 1842 P; 1860 K; 1880 PO; 1900-1936 K.

**Notes**

1. To level off grain in a measure.

2. A small enclosed plot of land, adjoining a house, worked by the occupier and his family.

# SUFFOLK

*'At the period of the Roman invasion this county formed part
of the territory inhabited by the Iceni, one of the most powerful
of the native tribes, from whom the Iknield street, or road
of the Icenic, derived its name. Unmounted horses appear
on the majority of their coins, which may have had a sacred
significance. This may account for the prevalence of White Horse
Inns in this area.*

*Buck-wheat is sown upon poor sands, chiefly for feeding poultry.
Carrots have been cultivated in the 'Sandlings' from time
immemorial and were wont to be sent by sea to the London
market. They are now grown principally as food for horses.*

*Victuallers' records for public houses in Suffolk give a sample of
how, over a hundred years, the advent of coaches travelling from
stage to stage caused a revolution in accommodation for both
man and horse. Figures provided by the Excise Officer for the
years 1571 and 1685, were based on the number of 'Spare Beds'
and 'Standings for Horses'[1].*

JB and PRO W30/48/49/50.

**Badingham, Woodbridge Road** A Grade II listed
traditional coaching inn.

1939 July – the house was bought by
Adnams the brewers.

2007 There are old beams, flagstones,
horse brasses, jugs dated 1450 and a settle
dated from c.1700.

Landlord and Adnams & Co. Ltd. 1846-1853 PO; 1874-1900
Framlingham W; 1912-1937 K.

**Badwell Ash, The Street** In 1685 the total number
of beds for travellers in this village was 4, with
stabling for 8 horses.

The inn is thought to date from the 16th
century.

In 2007 the stables at the rear of the
inn still had the mangers in situ; there is
also a large barn, probably on the site of the

original coach house. The inn is said to be haunted by a little old lady and a gentleman, and a coach has been heard passing the inn at midnight on New Year's Eve. PRO, Landlady and Landlord. 1870 PO; 1885-1900 WPO; 1912-1937 K.

**Beccles, 29 New Market** In 1571 the total number of beds for travellers in this town were 26, with stabling for 52 horses; one hundred years later the numbers had increased to 122 beds and the stabling for 159 horses.

2007 the White Horse was still trading.

PRO. 1823-1830 P; 1846 PO; 1850 S; 1870 PO; 1885-1900 WPO; 1912-1937 K.

**Beyton, Bury Road** In 1685 the total number of beds for travellers in 'Byton' were 14, with stabling for 18 horses.

Situated on The Green, the date 1603 can be seen on the chimney stack; a well to an underground spring is now inside the building in the hallway. There is a large garden and a patio.

By 2007 a barn had been converted into six en-suite bedrooms. PRO. c.1727; 1846 PO; 1870 PO; 1885 WPO; 1900 PO; 1912-1937 K.

**Bradfield St. George, The Green** In 1685 the total number of beds for travellers in 'Bradfield' were 14, with stabling for 12 horses.

c. 1960 this tavern became a private house. White Horse Lane is an old bridle path, which runs from the corner of the village green to where this house stands.

PRO and Mrs. Holden, the one-time Postmistress. 1874-1885 W; 1900 WPO; 1912 K.

**Brandon, White Horse Street** In 1685 the total number of beds for travellers in this town were 28, with stabling for 56 horses.

Thought to have been a coaching inn. The present building dates from 1930 having been built on the site of the former inn.

1995 the premises were extended and 'modified'.

PRO and Landlord. 1830 P; 1846 Town Street 1870 PO; 1885 WDG; 1900-1912 W; 1937 K.

**Bungay, Staithe[2] Road** In 1571 the total number of beds for travellers in 'Bungie' were 33, with stabling for 66 horses; one hundred years later the numbers had increased to 60 beds and the stabling for 90 horses.

The building probably dates from the 15th century. 1757 it is recorded in a Copy Lease; at this time it would have been a flourishing business being situated near the Staith – the busy navigation centre on the River Waveney – used by the crews on the wherries which travelled daily between Bungay and Yarmouth, and workers from the busy Maltings nearby which had traded from the end of the 18th century.

During the 19th century it was run by the same family - Emma Codling running it single-handed for forty-five years from 1917 until 1962 when aged eighty eight, her son Francis took it over.

1978, June – the inn was closed, trade having declined from the 1920s due to the arrival of the railway and motor transport which superseded the wherry trade.

c.1980 it became a private house.

PRO, CR. 1823 P; 1830 Staithe P; 1846 PO; 1870 POT; 1874-1885 W; 1900 WPO; 1912-1937 K.

**Bures St. Mary, Cuckoo Hill** This tavern ceased to trade between 1900 and 1910, it became a private residence, White Horse House. 1870 PO; 1885-1900 W.

**Bury St. Edmunds, Butter Market** In 1685 the total number of beds for travellers in this town was 265, with stabling for 530 horses.

The original building will have been of an earlier date and thought to have been rebuilt after a fire in 1609.

By 1735 the Masonic Lodge No. 78 was meeting at the "White Horse Inn on the Cornhill", after only eight years the Lodge was erased in 1739. The ornate sign bore several Masonic signs including the Sun, Moon, and Square, the Level, Plumb Rule and Compass, also Crossed Pens.

Sometime after, an ancestor of Charles H. Bullen purchased the Inn, the family to remain as owners for the next one hundred and fifty years.

1758 John Pate, a local coachman, became landlord; by 1765 the coaching business had grown, running three times a week between Bury and London – to the Kings Arms in Leadenhall.

Some years later it was superseded by other coaching inns and its trade declined – by 1868 the landlord had become bankrupt.

1875 the inn ceased to trade and this was the last date it appeared in the rate book. The Bullen family converted the premises into an upholstery shop 'The Old Established Decorating and Furnishing House on Market Hill'; Charles H. Bullen was an expert in the restoration of 18th and 19th century furniture – he was also the local auctioneer. He would arrange for 'furniture and objets d'art to have careful packing to any part of the Kingdom, or of the world'.

1937 listed as White Horse in Kelly's Directory.

By 1996 it had become 'Palmer's Restaurant', or 'Purdy's'.

PRO, BSE, JRB and BN. 1824-1850 P; 1853 Corn Market PO; 1870 POT.

**Capel St. Mary, London Road** An architect, who inspected the property suggested that the building dated from 1402, due to the material used in the walls – horse hair, etc.

1670 the will of Richard Partridge showed that he bought the inn from Mr Sicklemore of Ipswich and, with Mr Henry Libbis as occupier, had left the business to his son John. A condition was made that he, John Partridge, gave an annual sum of £5, in half yearly instalments to his wife, Anne, for the rest of her life – the money was to be handed to her in the south porch of Capel Church.

1821 it was mentioned in an agreement which also involved a number of public houses in Ipswich:

'The inn was a well-known staging post for the London, Norwich and east coast coaches, and the stable where the horses were changed, or put up for the night, lay along the roadside adjoining the turn-in to the yard at the side of the inn. In those early days, the front door of the inn opened out onto the edge of the London Road. The journey to London at that time took about ten hours and the cost was three pence per mile.'

The last stage coach called here in 1854 and the stables being of little use were converted into Stable Cottages each with its own front garden; an attractive row of six dwelling houses, with corbelling beneath the eaves running the full length of the building; these were demolished in 1969 and were replaced by two chalet bungalows named Aysgarth and Ostler House.

This Grade I listed public house is supposed to have a ghost, although the landlady hasn't seen it. It is thought that a young girl called Martha, believed to have worked in the White Horse, was drowned in the nearby Castle moat.

IRO, BWr and Landlady. 1846-1853 PO; 1874-1912 W; 1937 K.

**Cavendish, High Street** In 1685 the total number of beds for travellers in this village was 6, with stabling for 12 horses.

The building stood between the present day Bull Inn and the Post Office. It had a frontage of 103 feet; the original property consisted of stabling, a brew house, granary yard, gardens and other outbuildings. The village cricket field must have been nearby as it is known that horses and pony traps were left at the White Horse during matches.

1877 John Churchyard the owner died, leaving the business to his daughter, Mrs Ann Coe.

1908 the licence was refused and compensation was awarded by the Suffolk Licensing Committee. A year later Mrs Coe sold almost half of the premises to a harness maker, who was also an insurance agent; this part of the site became two houses. She sold the remaining frontage of 59 feet to a builder who in 1922 built a house named Brockholm; he had a workshop in the building which had been the skittle yard.

1964 the house was renamed Stormont.

PRO, 'Cavendish, Its People and Its Heritage' 2002 – and Mrs. Margaret Godwin, Postmistress.
1846 PO; 1874-1885 W; 1900 WPO.

**Corton, 47 The Street** In 1685 the total number of beds in this village was 1, with stabling for 4 horses.

The oldest surviving inn of the town – it originally stood in the old main street, north of Baker's Score.

1658 an innkeeper died at Corton – thought to have been from the White Horse.

1712 Elizabeth Plantin is recorded as landlady, after she died in 1750 the Mewse family were landlords until 1803 when it was sold for £450.

Described by The Rev. John Pridden in his list of Suffolk inns of 1786 as 'a decent public house, very pleasant to the sea'.

1846 the premises were sold with a paddock.

By 1862 the house had been abandoned due to the erosion of the nearby cliff and the new White Horse was built on the old paddock. On 8th November the original building and its land was sold to a local farmer who converted it into two farm cottages. These were sold in 1879 for £52.

In 1880 a conveyance describes the property as a 'freehold messuage, tenement or Cottage, then divided into two dwellings with the barn, stable, outhouse, edifices, buildings, yards, garden, lands and hereditaments[3]'. The cottages were depicted in a painting by William Marjoram in 1888.

1872 the new public house was sold for £800.

Over the years the White Horse has been known for the 'hospitality it provided for the many seamen wrecked over the years on Corton beach. Mrs Elizabeth Smith, landlady from about 1890 until the 1920s, was especially well known for her kindness to them'. During the 1939 war, bodies recovered from a shipwreck were laid out in the lounge downstairs; in the nearby passage the air is still always chilled to this day.

The landlady tells of her experience in January 2000 while upstairs on a very hot day she experienced a sudden chill in the room. The curtain, with no window behind it, moved and the hair on the back of her head stood up; she called the girl up from below, who had precisely the same experience. Moving back along the upstairs passage the air was chilled – the rest of the house was still hot.

1981 the landlord strangled his wife with the iron flex – a crise de passion – he was sentenced to fourteen years in prison.

c.1995 a conservatory was added to the south side of the inn and the former vegetable garden became a children's play area and car park.

<small>PRO, MSo and Mrs. Barbara Worrall, Landlady. 1853 PO; 1870 PO; 1874-1885 W; 1900 Lowestoft WPO; 1912-1937 K.</small>

**Earl Stonham, A140 Norwich Road** The building was originally two cottages, it was converted to an inn in c.1866.

By 2002 it had become a private house named White Horse Cottage.

<small>1870 POT; 1885 W.</small>

**East Bergholt** A coaching inn. 1992 renovated to become a private house – the outhouses having been removed. <small>1846 PO; 1853-1870 PO; 1874-1900 W; 1912-1937 K.</small>

**Easton, The Street** A 16th century coaching inn situated on the village green next to the church.

The ghost of a man has been seen in the area between the bar and the kitchen.

<small>1846-1853 PO; 1870 Wickham Market PO; 1885 2 listed W; 1874-1900 Framlingham WPO; 1912-1937 K.</small>

**Edwardstone, Mill Green** In 1685 the total number of beds for travellers in 'Ediestone' was 1, with stabling for 2 horses.

The inn may originally have been attached to a mill or brewery, as one of the three staircases leads to the loft where the old floorboards still show that barley was dried there for the maltings, possibly to supply The Fleece, the nearby busy coaching inn.

By 2007 the pub had been a family business for over 20 years and had been listed in the CAMRA Good Beer Guide for 16 years. There were 2 self-catering cottages and its own adjacent caravan and camping site with 30 pitches.

<small>PRO. 1870 Boxford PO; 1885-1900 Boxford WPO; 1912-1937 K.</small>

**Exning, 23 Church Street** According to the Landlord, the inn is 300 years old and was a coaching inn.

During the Second World War the White Horse was used as an officer's mess, it was open 24 hours a day to accommodate the Group Three Bomber Command of American and Canadian servicemen.

Between 1923 and 1990 the inn was run by the Welford family, they then sold the premises and emigrated to the U.S.A. but returned 12 years later to find the pub about to close to become a private residence.

2005 the Welford family had moved in and were once again running their pub.

1846-1870 PO; 1874 W; 1900 Newmarket W; 1912-1937 K.

**Felixstowe, 33 Church Road** The original long, low building was of red brick and tiles with a wooden lean-to and half-hatch doors at one end, it stood in White Horse Lane and was one of the few buildings in this area; the road was little better than a cart track, although it was the main road to Felixstowe Ferry.

1712 a license was granted at the 'Whitehorse'.

The landlord for most of the 19th century was George Hall, who was also a boot and shoemaker.

The stables were used for the horses of those people attending church services and vestry meetings. This was also a meeting place for the local donkey boys and their charges.

1904 a new building was constructed at the back of the original inn.

c.1960 an additional third bar was added.

LPT. 1853 Felixstow PO; 1874-1900 W; 1912-1937 K.

**Finningham, Station Road** In 1685 the total number of beds for travellers in this village was 2, with stabling for 4 horses.

The following details of this Grade II listed 15th century building are given in the Listed Building Register dated 29.7.1955:

'Public house. C15, floor and stack inserted late C16 or early C17, extended late C17, further extended and raised C18, altered C19. Timber frame, plastered with some panelled pargetting[4]. Plain tiled roofs. Originally 4 bays, a small 2 bay open hall with storeyed lower bay to right and smoke bay or storeyed solar bay to left; a parlour added to left and a cross gabled bay added to front right. Now all 2 storeys.

Main range has entrance to right into former lower bay of hall, a 6 fielded panelled architraved door[5], to left a 3-light glazing bar casement.

First floor 4-light leaded metal frame casement. Rebuilt ridge stack at original left end. C17 parlour to left has an early C129 2 storey canted bay window to front with 2:3:2 part opening metal frame leaded casements, transomed[6] on ground floor, taller roof has a hip to left, end wall ground floor brick casing.

To front right C18 gable fronted bay extends forward, inner return plaster incised as ashlar[7] over brick base, gabled front transomed 4-light casement on ground floor, 2-light leaded casement on first floor, exposed plates[8] and purlins[9]. Right return rebuilt external stack with off sets in a C20 outshut, pantiled[10] lean-to behind. Original range re-roofed with a hip to rear right.

To rear catslide roofs over C18 and C19 brick lean-to outshuts, an external stack at service end.

Interior: hall has original open truss posts with traces of moulded shafts to arched braces, towards upper end inserted chamfered[11] storey posts to chamfered cross axial binding beam, frame largely concealed at upper end. Remains of a panelled screen to cross passage, a service doorway from cross passage with a chamfered 4 centred arched head. Parlour has a stop chamfered axial binding beam. C18 fireplace with a lugged architrave[12], Greek key pattern to mantelpiece. First floor some exposed studding[13] and arched braces in walling, raised eaves over hall, re-roofed.'

PRO. 1846 PO; 1853 PO; 1874-1900 W; 1912-1937 K.

**Framlingham, 27 Well Close Square** In 1571 the total number of beds for travellers in this town was 9, with stabling for 12 horses; a hundred years later the numbers had increased to 51 beds and the stabling for 58 horses.

1632 Nicholas Sheen was the owner, he was also a Churchwarden.

1832 the inn and stabling for twenty six horses was bought by James Brunning at auction for £900; he lived here, carrying on a very good business. His son continued the business until c.1929. PRO, MLK and PJS. 1823-1830 P; 1846 PO; 1850 S; 1885-1900 W; 1912-1937 K; 1960 JM.

**Great Finborough, The High Road** In 1685 the total number of beds for travellers in this village was 1, with stabling for 2 horses. At one time there were a number of businesses in the village – maltings, a sawmill, hosiery factory and fish and chip shops, a blacksmith, thatcher and wheelwright.

This was a 'closed' village with one main land owner, the Woolastons family, who were Lords of the Manor, owned The White Horse Inn. The original building fronted on to Church Lane, facing the Green, it was one of the inns used for Petty Sessions for Hirings[14].

In 1794 the estates in Great Finborough were sold, the inn was listed in the deeds of sale. Mr. Pettiward bought the main Manor, thus becoming the Squire – it is recorded that he banned the inn from opening on Sundays.

1844 The White Horse Inn is shown on the tithe map as the only public house in the village; it had a six day licence, brewing its own beer using the water from the well on the Green. By this date the ownership of the inn had passed to the widow of Squire Pettiward, who had remarried to become Lady Jane Seymore Hotham.

1876 the inn was demolished; an Estate Office was erected on the site, a substantial building of ornamental brickwork with the date 1880 above the main door.

The landlord Francis Mitson, moved across the road to Chestnut Horse Farm and transferred the licence from The White Horse to the Chestnut Horse Inn.

1936 a sale of the whole estate took place; the Estate Office on the site of the old inn was converted into three private houses named White Horse Cottages.

PRO, Mrs. Mary Williams, SCA and The Postmistress.

**Great Waldingfield, Lavenham Road** In 1685 the total number of beds for travellers in 'Little Waldingfield' was 3, with stabling for 6 horses.

This Tudor building is said to be five hundred years old – built as a coaching inn with a forge at the back. PRO, PE. 1885 W; 1900 WPO; 1912-1937 Little Waldingfield K.

**Hadleigh, High Street (on the corner with Duke Street)** 1558 Thomas Alabaster acknowledges ownership of this freehold messuage – from the Manor of Toppesfield court roll.

1648 John Alabaster surrenders one messuage called The White Horse to Anne Holgrave – from the minute book of the Manor of Toppesfield.

In 1685 the total of beds for travellers in this town was 49, with stabling for 98 horses.

1709-1794 title deeds exist for these dates.

1853 listed as trading by the Post Office.

By 1855 the building had been demolished and a Police Station built on the site at a cost of £1,100.

During the 1970s the building was converted into Hadleigh Library.

PRO, SUF. and Mrs. Sue Andrews, Hon. Archivist.

**Halesworth, Chediston Street** In 1571 the total number of beds for travellers in this town was 14, with stabling for 20 horses; one hundred years later the numbers had increased to 27 beds and stabling for 50 horses.

1759, 1ˢᵗ March – a notice in the *Ipswich Journal* stated:

'A POST-CHAISE, with Able Horses, to any Part of England.
That Thomas Bonner, who has been carrier to Norwich twenty five Years, has set up a four-wheel Carriage to Yarmouth: Sets out every Monday Morning, at Six o'Clock, from his own House, the White Horse Tavern in Halesworth to the Buck on the Key in

Yarmouth, with a four-wheel Carriage and able Horses, to carry Goods and Passengers, by Way of Wangford, Wrenthun, & c. and returns on Tuesday Morning, at Six o'Clock, for Halesworth: Sets out on Wednesday Morning for Saxmundham. The Goods and Passengers forwarded by Tho. Shave and Richard Frewer for London; to be at the Saracen's Head within Aldgate on Saturday Afternoon. Proper Care will be taken to accommodate Passengers, by their humble Servant, Thomas Bonner.

He continues the NORWICH and BUNGAY Carriage, as usual, to the Star in the Market, Norwich.'

PRO. 1846 PO; 1870 POT; 1885 W; 1900 WPO; 1912 K.

**Haughley, New Street** The pub became a private house.
1846-1853 PO; 1874-1885 W; 1912-1937 K.

**Haverhill, High Street** In 1685 the total number of beds for travellers in 'Haverill' was 20, with stabling for 40 horses. PRO. 1874-1885 W.

**Hitcham, The Street** In 1685 the total number of beds for travellers in this village was 2, with stabling for 8 horses. For the years 1797 and 1818 there are probate documents concerning the White Horse which are held by the brewers, Greene King; they also have various indentures[15] dating from 1837-1878 regarding the ownership of the inn – that of 1838 refers to Sir Benjamin Collins Brodie, Bart. selling to Sarah Salmon for £180, a 'messuage or tenement now used as a beer shop and known as the sign of the White Horse and Blacksmith's Shop with the Yards Gardens and appurtances . . . situate (sic) at Hitcham, Suffolk' – cottages are also mentioned.

Mr. and Mrs. Gray lived at The Elms in Hitcham; in 1931 Mr. Gray was accused of murdering his wife, Lucy, by giving her arsenic as a cure for a sexually transmitted illness; she died aged fifty nine of an arsenic overdose and was buried on 28th September. The trial was held in the White Horse public house, Mr. Gray being acquitted. A play was written about the murder and trial – the narrator's final words sum up the result – "tonight, at Hitcham White Horse, justice has not been done." PRO, Mrs. Balshaw, Landlady. 1853-1870 PO; 1874-1900 W; 1912-1937 K.

**Holbrook, The Street** In 1685 the total number of beds for travellers in this village was 2, with stabling for 4 horses.

1937 listed as trading in Kelly's Directory – it became a private house.

PRO, Local Post Office.

**Hundon, Mary Lane** In 1685 the total number of beds for travellers in this village was 3, with stabling for 6 horses.

1937 listed as trading in Kelly's Directory. It became a private house named White Horse Cottage. PRO.

**Ipswich, 43 Tavern Street (on the corner with Northgate Street – originally the North Gate)** 'The Great White Horse Hotel'.

1518 listed as 'Whit Hors Inn'. It was always one of the foremost inns of the town. It is thought there was an inn on this site for many years before this date, it may have been a pilgrims' inn as Ipswich was largely a monastic town.

1520 the Duke of Norfolk was a regular guest here for at least 15 years.

1528 'The Whit Horse' paid an annual rent of 6d. for setting the signboard on town land.

1561 Queen Elizabeth I visited Ipswich and her retinue stayed here.

In 1571 the total number of beds for travellers in this town was 116, with stabling for 496 horses; by 1685 the numbers had increased to 193 beds and stabling for 630 horses.

1610 shown on John Speed's map.

1689 known as a regular coaching hostelry and terminus for coaches and Mail bound for Norwich and Great Yarmouth to and from London; nine regular coach services ran through the town daily.

1736, 14th January. King George II stopped at The Great White Horse. 'His Majesty had landed at Lowestoft that morning and arrived at Ipswich late in the evening, after a journey from Saxmundham that took four hours. At 11 o'clock the King met the town's civic dignitaries and clergy at an assembly in the great dining room on the first floor of the hotel, where all kissed his hand. He resumed his journey to London before midnight' – 'to the loyal cheers of the greater part of the population of the town'; having slept for three or four hours at Stratford, he arrived at St. James's Palace at 2 p.m. the next day.

It is not known when the stone statue of a white horse was placed above the porch; possibly when the inn became known as The Great White Horse to commemorate the Hanoverian monarch's visit.

1797, 13th November – Lord and Lady Nelson stayed here while viewing Roundwood Place at Rushmere which they bought. Lady Nelson was a frequent visitor to the hotel.

1800, 6th November - Lord Nelson, who had been appointed High Steward of Ipswich, stayed the night, together with Sir William and Lady Hamilton; they too had landed at Yarmouth, *[or possibly Lowestoft]* arriving by coach at Ipswich. The enthusiastic citizens unharnessed the horses and drew the coach through cheering crowds to the end of St. Matthew's Street which runs in to Westgate Street and then into Tavern Street.

1807, Tuesday 3rd November at three o'clock King Louis XVIII of France arrived from Yarmouth with his suite in order to change horses for the journey to London.

Nowadays it seems curious that so many travellers from abroad should have chosen this route incurring some 123 miles to reach the capital, one explanation lies in the fact pointed out by James Bell in his Gazetteer of 1843:

'The quay of Yarmouth is justly the pride and boast of the inhabitants, for it is allowed to be equal to that of Marseilles (sic), and the most extensive and finest in Europe, except the far-famed one at Seville, in Spain.'

1818 the Borough Council acquired the frontage of properties in Tavern Street, including the White Horse, in order to widen the road which was only twelve feet wide. This involved demolishing the three-gabled frontage with the large stone statue of the white horse above the main door – the statue can still be seen today in the village of Tattingstone, a few miles from Ipswich, raised on a pole in front of the White Horse public house. A smaller statue was put up over the new entrance of the 'Great White Horse'.

The broader road greatly facilitated the fourteen regular coach services which passed through daily on their way to London, Norwich and Yarmouth, all changing their horses in Tavern Street; the fresh horses were harnessed ready for the mail coach arrivals, each change over took just over ten minutes. Meanwhile sealed mail bags were loaded on to the coach from the Post Office opposite.

The stabling at the inn was for the customers' horses and carriages, mainly local gentry. ' . . . the landlord, in order to accommodate the many bagmen or commercial travellers of the time, hired all the rooms at the Royal Oak in Northgate Street for their use, so that his regular customers should not be disturbed by their boisterous behaviour'.

1819, 16th June – first mention of The White Horse being the headquarters of the 'Blues' or Tories, at parliamentary elections. The partisanship between the rival camps of The White Horse and The Crown and Anchor was always keen at every election. It continued to be the Blues' headquarters until early into the 20th century.

1822 King Louis XVIII of France took refreshment on his way to Norwich.

1834 Charles Dickens (1812-1870) stayed here while covering the Sudbury bye election for the *Morning Chronicle,* he then became a regular visitor and wrote:

'In the main street of Ipswich, on the left-hand side of the way, a short distance after you have passed through the open space fronting the Town Hall, stands an inn known far and wide by the appellation of The Great White Horse, rendered the more conspicuous by a stone statue of some rampacious animal with flowing mane and tail, distantly resembling an insane cart-horse, which is elevated above the principal door. The Great

White Horse is famous in the neighbourhood, in the same degree as a prize ox, or county paper-chronicled turnip, or unwieldy pig – for its enormous size. Never were such labyrinths of uncarpeted passages, such clusters of mouldy, badly-lighted rooms, such huge numbers of small dens for eating or sleeping in, beneath any one roof, as are collected together between the four walls of the Great White Horse at Ipswich.'

In 1924 the play, 'London Life' by Arnold Bennett and Edward Knoblock, sets the scene in Act I Scene 2 in the courtyard of the White Horse Inn at Ipswich. It was performed at the Theatre Royal, Drury Lane in London. The stage setting:

'An archway centre leads to the street. To the left centre a smaller arch leads to the bar-parlour. To the right a doorway leads to the rest of the house. The Courtyard is roofed over with glass. There are several tables right and left with chairs, also a screen right, sheltering tables from the draught.'

Major The Hon. Arthur-Riggs-Falkiner (afterwards Lord Plinlimmon) played by Mr Graham Browne – announces "So this is the world-renowned White Horse, immortalised by Dickens!"

1937 Edward VIII came to stay with Mrs. Wallis Simpson after his abdication. The solicitor conducting her divorce was an Ipswich man.

The present hotel can boast nearly 500 years of trading and is probably the oldest established business in Ipswich. Apart from some exposed timber framing, nothing of the original building remains, though a listed building entry dated 19th December 1951 states:

'Originally a 16th-17th century timber-framed building . . . The present front is of grey gault brick with a parapet and a rusticated stucco ground storey . . . Part of the original internal courtyard has been glazed over and part of it has been preserved in the present lounge . . . '

PRO, SUF, SS, LPT and CHW. 1805 'White Horse' WCO; 1830 Commercial, Posting & Family P; 1846 Family Hotel and Posting House PO; 1850 S; 1870 2 listed – possibly that at nearby Tattingstone – PO; 1885 Northgate Street W; 1900 WPO.

**Kedington, Sturmer Road** Roman or Romano-British remains of either a building or burial site have been discovered on the Newmarket road within 200 metres of the inn.

While a map of the area dated 1722 does not show a building on the site of the inn, a building with its stables and smithy – are shown on the 1774 Glebe Map, it is likely to have been trading as a coaching inn by c.1730. It stood at a road junction on the route from Newmarket to Colchester at almost the highest point of the village; the coach road between Cambridge to Bury St. Edmunds also passed nearby.

At one time there was a wheelwright's workshop next to the 18th century thatched cottage standing below the pub.

1844 John Garwood, a blacksmith, was listed as landlord – the inn was not named.

1883 a sale document describes the inn as 'an old public house with stabling and accommodation'.

From the late 19th century a friendly society 'The Ancient Shepherds' met here.

The present building on this site is listed as 19th century. Until recently Gowers farm stood to the rear of the pub.

1961 the smithy was demolished.

By 2007 the house was owned by Greene King and Co.

BER, Mr. J. Pelling – local historian – and Landlady. 1870 PO; 1885 W; 1900-1937 K.

**Kersey** In 1685 the total number of beds for travellers in 'Cersey' was 3, with stabling for 6 horses.

Thought to date from the 16th century, this old tavern was used by farm workers who met in the long passage which led to the tap room where beer was served from the barrels. The women met in a room where darts and cards were played; another room had a large open fire where a poker was put in the hot embers until it was red hot – then plunged into a tankard of beer – hence Poker Beer, Mulled or Mauld Beer, drunk on a cold winter's days.

1981 the building was described ' . . . as near a pink and white birthday cake as a pub could be.'

c.1960 the premises were refurbished.

Lorah Orris – local Postmistress, PRO and BBV. 1846-1870 PO; 1885-1900 W; 1912-1937 K.

**Kirton, 15 Bucklesham Road** In 1685 the total number of beds for travellers in 'Kitton' was 1, with stabling for 2 horses.

Situated on the main road passing through the village, which at one time had been a sandy bramble lane. The original building was thought had been a coaching inn dating from c.1695.

1765 the inn was licensed – at this date the building was possibly two cottages, together with the gardens; the new road was built directly in front of the building, thus separating the gardens to the other side of the road which were later used for housing. Originally another two cottages stood on the car park. Stable tethering rings can still be seen on the walls of the outbuildings.

2002 Mr. Fred Pitt had been landlord for twenty five years. Over the previous eighty years only four landlords had managed the inn.

PRO, Mr. Pitt and Ms. Alison Hayden. 1874-1900 W; 1912-1937 K.

**Lavenham, Water Street** Writing in 1946, Mr. L. P. Thompson pointed out that an ancient water course still ran beneath the floors of the inn.

1824-1850 P; 1870 POT; 1974-1885 W; 1900 WPO; 1912-1937 K.

**Laxfield** The Old White Horse, a farmhouse dating from 1520. A Grade II listed private house. JSS. 1846-1853 PO; 1874 W; 1912-1937 K.

**Leiston, Station Road** Dating from the early 18th century, the White Horse Hotel has always been the centre of life for local people and for some early inn keepers relying on the Black Economy[16], an important secondary income to smuggling.

1937 listed as a Family and Commercial Hotel in Kelly's Directory. 1853 PO; 1870 PO; 1885 W; 1898 *R.S.O.* K.

**Long Melford, 14 Southgate Street** The building has a 'Victorian red brick frontage on top of earlier terraced buildings'. It became a private house but is still listed as the White Horse on the electoral roll. Postmistress. 1937 K.

**Mendham** A general misconception may have arisen that the local public house, the 'Sir Alfred Munnings' was at one time known as The White Pony; as there is a wrought iron sign of a pony outside; this does not depict a grey pony but Charlotte Gray's brown pony. The artist was born in the nearby mill in 1878 and his early works were painted in this vicinity – these included 'The Pass to the Orchard' and 'The Poppy Field' – both depicting a grey pony.

**Otley** 1846-1853 PO; 1874 W.

**Rendham, Bruisyard Road** In 1685 the total number of beds for travellers in 'Rentam' was 3, with stabling for 4 horses.

According to the deeds the building is thought to date from 1723.

PRO and Landlord. 1853 PO; 1870 PO; 1885-1900 W; 1912-1937 K.

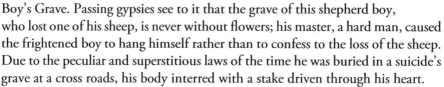

**Rickinghall Superior, The Street** In 1571 the total number of beds for travellers in 'Rickinghall' was 5, with stabling for 7 horses; by 1685 the numbers had decreased to 1 bed and stabling for 3 horses.

In 1605 Nicholas Fowle reported an affray which took place at the Crown and The White Horse, no doubt affrays like this would have caused loss of custom.

The present building dates from 1660. There is an early 19th Century stable block and also a kitchen / brewhouse.

1830 listed under Botesdale in Pigot's Directory and again in the 1853 Post Office Directory. PRO, SUF Ref. TM 0475 and AKF. 1874-1885 W; 1912-1937 K.

**Risby, Old Newmarket Road** Until c.1982 this house was listed under Great Saxham. Situated on the A14 road – previously the A45.

According to folklore – still common knowledge in 1946 – at a nearby cross roads a mile or so from The White Horse is a grassy mound known as the Boy's Grave. Passing gypsies see to it that the grave of this shepherd boy, who lost one of his sheep, is never without flowers; his master, a hard man, caused the frightened boy to hang himself rather than to confess to the loss of the sheep. Due to the peculiar and superstitious laws of the time he was buried in a suicide's grave at a cross roads, his body interred with a stake driven through his heart.

1783, Friday 15th January – *The Bury Post* reported:

> 'that a local man was attacked by a highwayman within a few yards of The White Horse. The villain laid hold of his horse's bridle, and ordered him to stop, upon which he struck at him with a stick he had in his hand, but this not disengaging him from his hold, he dismounted and compelled him to desist, when the robber applied his mouth to a whistle he had fixed to his coat, which brought forth his companion from the hedge, who immediately fired a pistol; at the same instant the rider had remounted his horse and galloping off happily received no injury.'

Seven months later on 23rd August *The Gentlemen's Magazine* reported:

> 'About nine last Monday evening an uncommon and beautiful meteor suddenly burst forth from the elements in the N.E." *Apparently this meteor was seen some half an hour earlier over Salisbury.* "It remained about half a minute in one station, affording a tremulous light not unlike the moon emerging from a cloud; then proceeded in a very regular and swift horizontal motion through the East, where dividing into several glowing balls of light, it disappeared." *Meanwhile, passing over Suffolk –* "Mr. Amyss, master of the White-horse inn, five miles from Bury, in the road

to Newmarket, was looking out of his ground-floor window, he saw a great light in the horizon, seemingly over Cavenham, and called to his family to come and see the strange light, which kept proceeding slowly directly towards his house, looked bluish, and when within a quarter of a mile plainly shed innumerable stars, each of which appeared to have a tail, seeming to pass directly over his house, and, as he thought only just clear of the chimnies. He ran to a back window; saw it keep on its course towards Great Saxham, and judged it might be about three rods *(16 or 17 feet)* in length. About one minute after he lost sight of it, he plainly heard a loud noise as of something heavy fallen down in the room overhead. He then looked at his watch, and it wanted twenty minutes of ten. He judges that the whole lasted three minutes. The course appeared to Mr. Amyss as from N.W. to S.W. nearly.'

1828, Saturday 2nd August – on their way to Bury St. Edmunds to conduct the famous trial five days later of William Corder (1804-1828) – the Lord Chief Baron Alexander, together with Mr. Justice Holroyd, stopped at The White Horse. They donned their wigs and robes before a large mirror – which later hung in the lounge bar – and coming downstairs were welcomed by the High Sheriff with the customary attendance, the procession then set off to the Shirehall at Bury. Corder was accused of murdering his lover, Maria Marten, the year before. He was 'executed amid popular execration in 1828'.

1996 the building was refurbished. Mr. and Mrs. Nathan - Landlords, LPT. and DNB.

1850 S; 1853-1870 2 listed PO; 1885-1900 W; 1912-1937 K.

**St. James, South Elmham** 1875 the White Horse was mentioned in a lease document.

1887 a list of fixtures and furniture for the White Horse, 'Southelmham St. James'.

c.1993 it became a private house, still known as White Horse.

IRO. 1846-1853 PO; 1874-1885 W; 1937 K.

**Sibton, Halesworth Road** A 16th century inn standing at the edge of the village. The bar has an inglenook fireplace, horse brasses and old settles and pews; there is an old oak servery with gleaming brass beer engines – a viewing panel shows the working cellar with its Roman floor.

2007 the inn and most of its outbuildings was Grade II listed.

RWIB. 1874 PO; 1885-1900 W; 1912-1937 K.

**Southwold, High Street (South Side)** In 1571 the total number of beds for travellers in 'Southwould' was 10, with stabling for 12 horses; one hundred years later the numbers had increased to 15 beds and stabling for 26 horses. PRO. 1846 PO.

**Stoke Ash** The building, which has been a farm house, coaching inn, public house and a roadside café dates from 1647 and is described by the local postman as a 'lovely old building'; the entrance porch resembles that of a Church. At one time there was stabling and a number of barns belonging to the inn. The open fireplace in the restaurant would have originally been used for cooking as well as heat.

The old Suffolk custom of the annual Petty Sessions for the Hiring and Retaining of Servants was held here at Michaelmas on 11th October. On this day 'the servants would assemble at The White Horse, and stand in a row . . . most of them would merely be re-engaged by their old masters. Such of them as were open to fresh engagement stood with a straw in their mouths'. Servants were handed a small sum of money to bind the contract, then a riotous party followed. In 1765 two of these sessions were held here, but this proved to be an unpopular arrangement.

Several ghosts are busy around the inn, one of a lady who restricts her activities to opening a window when it has just been shut – so long as a little blue plaque in one of the bedrooms is left in place she will behave herself. A former landlord, Mr. Symonds, saw the ghostly figure of a lady standing in the hallway on his way to the bathroom in the middle of the night, when he returned she had gone. There is the ghost of a man who committed suicide by jumping from one of the windows in the upstairs restaurant; there are many tales of windows opening and shutting by themselves and also of Napoleonic prisoners in uniform stomping around.

The Bar has an atmosphere of antiquity with the attraction of the Parish-boundary running along the centre of the present day bar counter.

LPT. 1846 PO; 1853 also Thorndon PO; 1874-1885 W; 1912-1937 K.

**Stoke-by-Nayland, School Street** 1550-1599 a beer house is recorded in the Parish Survey.

Between 1844 and 1912 although three public houses were listed, the sign of the White Horse does not appear.

1952, May – a sketch was made of a gaming dial in the bar; it seems that the numbers on the dial were not correct, as opposite numbers were supposed to add up to 13. In c.1954 when the house closed, the dial was bought by Mr. Robin Green, a local historian.

The premises became a private residence – 'White Horse Cottage'.

BER. Former Reference No. HA 533/4/2.

**Stone Street, near Hadleigh** The house was known locally as The Donkey because the sign looked more like a donkey than a white horse and it could be identified amongst the many local White Horse inns.

By 2007 the house had closed to become an antique shop.

1823-1830 P; 1870 Ipswich PO; 1937 K.

**Stowmarket, Stowupland Street** In 1685 the total number of beds for travellers in 'Stomarkett' was 41, with stabling for 82 horses.

The Petty Sessions for the Hirings of Servants were held here.

c.1880 a drawing of the entrance to the White Horse yard shows the maltings buildings in the background.

c.1903 the house closed. A photograph of a charabanc outing showed the façade of the building had been altered to a shop front and by the end of the 20th century another photograph showed an undertaker's sign above the door. PRO. 1823-1850 P; 1870 2 listed – possibly Great Finborough POT; 1874-1885 Station Road W; 1900 WPO.

**Stratford St. Mary** 1755 recorded in a deed as The Bird in Hand.

1774 by this date it had become the White Horse.

1837, 10th May – a sale agreement between William Baker, twine spinner, and William Back, miller, states 'All that messuage or tenement sometime since an Inn or Public house called or known by the name or sign of the Bird in hand afterwards called or known by the name or sign of the White Horse with the rights members and appurtenances thereto belonging situate lying and being in Stratford Saint Mary . . . Also all those two tightles (*sic*) or parcels of land formerly and now used and occupied with the said messuage or tenement . . . Together with all singular houses outhouses Buildings walls ways paths passages waters watercourses . . . unto the said William Baker his heirs or assigns the rent of one peppercorn only upon the last day of the said Term . . . '

It appears that by 1837 it was no longer trading. SUF Ref. HD 21/360/63.

## Sudbury

*Ballingdon Hill* In 1685 the total number of beds for travellers in 'Ballington' was 8, with stabling for 16 horses.

Founded c.1793. Leased by the Mauldon family in the early 1800s.

1874 listed in White's Directory under Ballingdon Street with Mrs. A.M. Mauldon as landlady.

1880 a small brewery was built behind the public house which was later expanded.

1863 listed under Ballingdon Cum Brundon, the landlord being a brewer, spirit merchant and victualler.

1895 listed as trading with A.M. Mauldon & Son as licensees.

1901 the building was enlarged after being partially burnt down.

1958 the premises were taken over by Greene King to be closed two years later. A proviso was stipulated that 'no alcoholic drink be sold on this site in the future'. A small stone plaque of a white horse still appears over the main door of the building; a figure of the Invicta white horse was engraved on the glass of the main door – now disappeared.

By 2001 it was occupied by Travel and Leisure Advisory Services Ltd.

PRO, GrK and BHS. 1830 P; 1846 PO; 1848 Ballingdon Cum Brundon W; 1850 S; 1870 PO; 1885 3 listed – possibly North Street and Great Waldingfield W.

***North Street*** In 1685 the total number of beds for travellers in Sudbury was 42, with stabling for 84 horses.

1744, on or about 19th November a mortgage is recorded on 'that Messuage, Tenement or Inn known by the name of the White Horse'. In c.1782 the artist, Robert Pollard (1755-1838), depicted a white horse standing above the door of the White Horse Inn in Fetter Lane, London – a similar figure stands on the flat area above the bow window of the White Horse Hotel.

1804 August – these substantial premises were sold for £1,250 which included a brewhouse, granary, malt kiln, stables, other outbuildings and a shop. The mill house had stabling, a well, and banking premises.

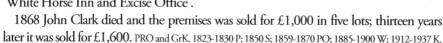

1819, July. John Clark of the White Horse to have 'the right and privilege of using the ram water cistern'.

1846 listed in the Post Office Directory as the 'White Horse Inn and Excise Office'.

1868 John Clark died and the premises was sold for £1,000 in five lots; thirteen years later it was sold for £1,600. PRO and GrK. 1823-1830 P; 1850 S; 1859-1870 PO; 1885-1900 W; 1912-1937 K.

**Swefling** In 1685 the total number of beds for travellers in 'Swefland' was 3, with stabling for 1 horse. PRO. 1853-1870 PO; 1885-1900 W; 1912-1937 K.

**Syleham** A closely timbered Grade II listed building – originally thatched, was built as a farmhouse in the second half of the 16th century; the roof is supported by a pair of green posts, typical of the Elizabethan period. At sometime the house was divided into two; the second entrance door still remains.

The premises consisted of three rooms – one with a jug and bottle counter; a public bar, a small 'snug', and a small room where drink was stored, these were connected by serving hatches. It is believed there was a cellar, now filled in.

Another public house in the village was 'The Black Horse' which was thought to have been an ale house, while The White Horse served spirits and was a venue for local landowners and farmers to talk business.

At one time both these public houses were owned by the Sparrow family, who still own the Black Horse; the brewers, Lacons, took over the White Horse, followed by the Norwich Brewery until the business closed in 1965 to become a private house named 'The Old White Horse'. The original globe bearing the inn's name still hangs over the door. 1846 PO; 1874-1900 W; 1990 WPO; 1912-1937 K.

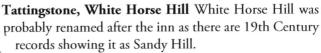

**Tattingstone, White Horse Hill** White Horse Hill was probably renamed after the inn as there are 19th Century records showing it as Sandy Hill.

The inn is said to date from c.1680, probably an old coaching inn on the main road from Manningtree to Ipswich. There was a smithy attached to the inn throughout the 19th Century, the blacksmith holding the pub licence.

In 1685 the total number of beds for travellers in 'Tatison' was 3, with stabling for 3 horses.

c.1836 the large stone figure of a white horse was brought from the Great White Horse at Ipswich. As many as three or four of the merrier young men of the village have been known to clamber up on the horse's back.

There is an old hand pump out side which at one time supplied the house with water from the well. Inside there is an inglenook fireplace and old beams.

2007 the inn is a meeting place for a number of clubs which includes the local football team and a car owners; club for Wolseleys and Morris Oxfords. There is a campsite in the field to the rear.

PRO, Mrs. S. Hardy, local historian. 1846-1853 PO; 1874-1885 W; 1912-1937 K.

**Thelnetham, Hopton Road** At one time this was a bakery and a public house.

c.2002 the house was sold to a brewery and run by a succession of managers.

2005 work was done to restore the pub organised by a local business man who was a regular customer.

1846-1853 PO; 1874-1885 W; 1912-1937 K.

**Westleton, Darsham Road** In 1685 the total number of beds for travellers in 'Weselton' was 1, with stabling for 1 horse.

The inn is situated by the village green and was once a thatched beerhouse with wattle and plaster walls.

1870 listed under Saxmundham by the Post Office.

1873, 14th June – a For Sale notice in the Suffolk Chronicle describes the premises as 'a beerhouse, together with two cottages, a barn and stables'.

1898 the buildings were demolished and replaced with a large brick building.

1937 K and listed as an inn and brewery.

1972, 18th August – an article in the *Suffolk Mercury* reported that until this year the sign only advertised 'Adnams Ales', then a new 'White Horse' sign was painted by Mr. David Barber of the Southwold sign writing firm. The landlord, Mr. Maurice Eve, maintained that the only true white horse had been a racehorse and that its trainer docked its tail so that no other horse could be passed off as the real white horse. *The racehorse, Montblanc, carried a full tail – see Alton, Hampshire.* PRO. SRO. and SM 18th August 1972 and Adnams & Co. Ltd.

**West Row, 57 Beeches Road** The building dates from c.1650 and has been a coach house and a mortuary before becoming a public house.

A horticulturist, Mrs. Reeves, grew flowers in the four and a half acres of land near the White Horse, when the lease ran out she bought the house as well as the land.

2007 a drinking pub only, some of the pub's custom comes from nearby RAF Mildenhall and RAF Lakenheath; the Mad Mutts Bikers' Club met here every month.

It is said there are 4 ghosts – 'the bar patron' who has been seen standing at the back bar or walking around the room; upstairs two cavaliers have been seen and the ghost of a woman. L. Deadman, Landlord. 1846-1853 PO; 1874-1885 W; 1870 Mildenhall PO; 1885-1912 Mildenhall W; 1960 JM.

**Wetheringsett Green, Pitman's Corner** In 1685 the total number of beds for travellers in 'Wethersett' was 1, with stabling for 4 horses.

2002 the Postmistress, Miss Winifred Margaret Smith, M.B.E., aged 91, remembers the wooden settles and beer barrels with taps being used in c.1931.

1992 the premises became a private house. PRO. 1846-1853 PO; 1900 WPO; 1912-1937 K.

**Whepstead, Rede Road** According to the landlady 'the building has very old beams'.

2007 the pub was still trading. 1846-1853 PO; 1885-1900 W; 1912-1937 K.

**Wickhambrook, Church Road** In 1685 the total number of beds in 'Wickhambrooke' was 4, with stabling for 8 horses.

Situated opposite the church, it is believed that a tunnel ran under the road from the inn to the rectory.

On holidays and festive occasions the landlord employed a man to lead or wheelbarrow the drunks to the triangle of grass at the entrance to Wash Lane and leave them there.

c.1972 the public house closed and was converted into a doctor's surgery and residence which was renamed White House; later the surgery was relocated in the village, the house continued to be a private dwelling.

PRO. 1846-1853 PO; 1874-1885 W; 1886 W; 1912 K.

**Withersfield, Hollow Hill** In 1685 the total number of beds for travellers in this village was 2, with stabling for 4 horses.

The building dates from the 15th century and was a coaching inn; it is situated on the old Colchester to Cambridge garrison road. The cricketer, Sir Donald Bradman (1908-2001) – knighted in 1949 - was a customer here.

By 2007 the cart lodge had been converted into ensuite bedrooms. PRO and Landlord. 1846-1853 PO; 1883-1900 W; 1912-1937 K.

**Woodbridge, Market Place** In 1571 the total number of beds for travellers in this town was 34, with stabling for 68 horses; one hundred years later the numbers had increased to 57 beds and stabling for 212 horses.

PRO. 1824 P; 1846 PO; 1850 P; 1870 POT;1874 Market Hill – 1885 Market Hill W.

## Notes

1. These later became 'Guest beds' and 'Stabling for horses'.

2. a wharf; a structure for shipping coal; an embankment. Old English 'stæth' – a bank; Old Norse 'stŏth' – a landing-stage. C.

3. Any kind of property capable of being inherited. Col.

4. Plaster or mortar used to cover walls. Plasterwork with incised ornamental patterns.

5. Moulding around a doorway or window.

6. A horizontal member across a window.

7. A block of hewn stone with straight edges for use in building.

8. Thin flat sheets of a substance such as metal or glass.

9. Horizontal beams that provide intermediate support for the common rafters of a roof construction.

10. A roofing tile with an S-shaped cross section, laid so that the downward curve of one tile overlaps he upward curve of the adjoining tile.

11. Narrow flat surface at the corner of a beam or post.

12. A projected piece by which something is connected, lifted or supported.

13. Supporting framework of a wall or partition.

14. A statute fair, virtually the same as a mop fair, formerly an annual event in most market towns in England and Wales at Martinmas (11th November), when men and maids stood in rows to be inspected by those seeking servants, farm workers and the like. B2.

15. Any deed, contract, or sealed agreement between two or more parties. Col.

16. The portion of the income of a nation that remains illegally undeclared because of payments in kind or for tax avoidance.

# SURREY

The following song immortalised in 'Oklahoma', the American musical by Richard Rogers and Oscar Hammerstein, was first performed in London in 1943:

*'You will sit behind a team of snow white horses,*
*In the slickest gig you ever see!*
*Chicks and ducks and geese better scurry*
*When I take you out in the surrey,*
*When I take you out in the surrey with the fringe on top!*
*Watch that fringe and see how it flutters*
*When I drive them high steppin' strutters.'*

A surrey is a light four-wheeled horse drawn carriage having two or four seats; originally made in Surrey and called a 'surrey cart'.

**Dorking, High Street** An 18th century building with parts dating from the 15th and 16th centuries.

The Knights Templars had a building on this site which passed to the Knights of St. John in 1278. At one time it was known as Cross House – there being a Maltese cross of the Order of St. John of Jerusalem on the building. In 1540, during the Reformation, the Order lost its properties in this county.

Richard Keverne in his 'Tales of Old Inns' recorded the legend that there were 'underground passages beneath the building; the cellars are cut in sandstone and from them a curious passage of well-worn steps descends to end in the shaft of an ancient well. Its walls are covered with carved names and initials and dates, many of them of the early years of the eighteenth century. The head of the well is covered by a slab in the courtyard'.

1750 the inn was established as the White Horse.

1823 listed as a posting inn in Pigot's Directory.

1837 the Pickwick Papers published Charles Dickens's description of the White Horse under the name of the 'Marquis of Granby' – this inn was known as the great posting house of the town.

1871 Kelly's Directory described it as a family commercial hotel and posting house.

1939 Richard Keverne records the inn as claiming to have the biggest apple tree in the South of England in its garden. 'It is of great age, yet still yields an average of about twenty-eight bushels of Lord Derbys each season'.

An early inn signboard showed the building with the sign and a stage coach in the foreground; in 1955 a replica of this sign was presented to The White Horse Tavern, Newport, Rhode Island, U.S.A. which was never used and is now mislaid. 1988 the sign was repainted but no longer showing the old signboard or the stagecoach in the foreground.

By 2005 it was known as The Macdonald White Horse Hotel.

SUR, RK and PNC. 1839 P; 1826 L; 1845 2 listed one possibly Shere – 1859 PO; 1867-1938 K.

**Egham** 1623 a deed held by the Flintshire Record Office relating to various properties recorded the following information about the White Horse:

> A 'messuage or tenemente with the appurtenances being a Comman Inne sometime called or knowne by the name of the White Horsse and sithence called or knowne by the name of the syne of the Swann scituate and being in Eggham in the Countie of Surrey late in the tenure or occupation of . . . Browning or els his assignee'.

FRO Ref: D/PT/719.

**Epsom, 63 Dorking Road** This town did not appear to have been of any great extent until the discovery in 1618 of a medicinal spa, the water became famous for its laxative qualities – *hence 'Epsom Salts'*. It then became a fashionable resort and rapidly increased in buildings and population.

The original timber-framed building dates from the late 17th century and was famous during the early days of the spa; it was originally named the New Inn.

1667 the house is recorded in Samuel Pepys' diary when he visited Epsom to take the waters. 1672 the White Horse was mentioned in the court circular – alas the date is unknown.

In 1780 The Derby Stakes was run here for the first time, over a distance of 1½ miles; it is said that the 12th Earl of Derby won the privilege of using his name by tossing a coin with Sir Charles Bunbury.

Five weather boarded cottages used to stand to the side and rear of the inn. Opposite lay the way to the spa – along White Horse Drive.

1830 a Victorian frontage was added with a parapet around the front and sides.

LHP and JB. 1845 Esher PO; 1867-1938 K.

**Esher, High Street** There were eight inns here during the 18th century due to the village being a principal coach stage midway between London and Portsmouth; travelling by coach from the City of London took two hours; passengers sitting inside paid 5s. 6d., those on the outside paid 3s. 6d. Naval couriers changed their horses here.

Pre 1770 the Blake family had been landlords of the tavern.

1797, 28th June – the Poor Rate claimed £16 from David Remnant of the White Horse.

c.1800 the premises were sold to the Moore family, brewers, of nearby Moore Place.

Although a small building, the words 'Loose Boxes & Coach House' were painted in large letters across the double doors adjoining the building, indicating accommodation for private carriages.

1913 the business ceased to trade and the house was converted into Esher Police Station; minor offenders, such as drunkards, had earlier been locked up in a stable on Cato's Hill.

c.1960 the building was demolished to become a car park for the railway station.

SUR, AMi, LHu, IS and PJM. 1839 P; 1845 2 listed, one possibly Epsom, perhaps confused with the White Lion Inn – 1859 PO; 1867-1913 K.

## Farnham

The hops here 'bring invariably higher prices than any other in the kingdom, and for more than a century they have formed the stable trade of the town.' The local engineering firm of Elliott's – whose family built a machine for bagging hops – had been granted a licence for the White Horse alehouse in 1860.

***11 / 12 West Street*** 1720 recorded as The White Horse by Nigel Temple in his book 'Farnham Buildings and People', but in 1727 and 1756 it was listed as the White Lion up until 1770 when the building was replaced after demolition.

***47 / 49 / 50 & 50A West Street*** Pre-1779 the property was described as a cottage, barn and garden.

1809 the house, divided into two tenements with a workshop, was sold by Isaac Holloway, a wheelwright, to G.C. Knight.

1842 John and James Knight are recorded as owning the beerhouse and workshop.

1849, 1st March – George Elliott, blacksmith, of the White Horse Beershop applied for an alehouse licence 'but all which applications refused'. However an alehouse licence was granted to him on 1st March 1860.

1872 the public house was valued at £17.

1892 listed in the Petty Sessional Returns of licensed houses as serving the labouring classes.

1891 the house with a large blacksmith's shop and cottage were put up for sale by Farnham Brewery.

c.1901 the building was destroyed by fire and the shell was bought by a local engineer, George Elliott, a member of whose family had his name incised on the wall of 50, West Street – the building in which John Henry Knight's petrol-driven car was built.

1904 the inn had accommodation for six persons, sanitary facilities and stabling for eight horses – with a Poor Law Assessment of £18 it offered bread, cheese and minerals, sleeping accommodation for two persons and no stabling.

1905, 5th June – a licence renewal was refused and the premises ceased to trade as a public house on 19th December.

By 1913 cottages had been built on the site and in 1914, before the outbreak of war, 49, 50 and 50A West Street – one of which had been the St. John's Ambulance Station – were offered for sale with other properties belonging to the Elliott family.

The shop premises became Heath and Wiltshire's Garage whose petrol pumps were still standing in 1994 by which time the building was empty and a demolition application had been made. SHC, JHF, JPa. 1845 Guildford – 1859 PO; 1867-1899 K.

## Guildford

Originally the High Street and Spital Street were separate but running into each other; only the High Street is shown on later editions of the Ordnance Survey map of Guildford, so Spital Street appears to have become part of the High Street – though in 1874 the White Horse was still listed as in Spital Street.

***253 High Street*** There are deeds dated 1696, but part of the present building dates from the early 18th century. The 1739 map of Guildford shows White Horse Yard adjoining a large field beside the London Road known as White Horse Field, where it was claimed soldiers were billeted during the 1770s.

1895 Moses Puttock set up a livery stable named White Horse next to the inn, which became a motor garage in the early 1900s.

1904 listed as trading in the Petty Sessional Returns of licensed houses.

The inn sign shows a three dimensional knight riding a white horse which dates from 1970.

By 2006 the house was known as The Guildford Hotel.

Pub Brochure. 1839 P; 1845-1887 PO; 1913-1938 K.

***1 Spital Street*** 1851 PO; 1867 K; 1871 PO; 1867-1874 K. SHC Ref.1483.

**Hascombe, School Lane / The Street** Dating from the 16th century; by 1727 the building was the subject of a complex marriage settlement between Sir John Frederick and Barbara Kinnersley when a number of cottages and the mill were included in family trusts which continued throughout the 18th century.

c.1800 one of the cottages had become the New Inn.
By 1821 the inn had become the White Horse when
it was sold by the trustees to the brewer, John Hall
Grinham, for £1,425. Three years later the brewer
died and his executors conveyed the inn to Charles
Whitbourne in trust for John Gardner, a grocer, to
whom in 1826 the premises were sold with other
properties. He later became a brewer and from then
on the business was heavily mortgaged by its various
owners until 1851 when it was sold to the sitting
tenant for £630. Conveyances took place in 1855, 1858 and 1867.

Mark Sturley in the second part of his book 'The Breweries and Public Houses of
Guildford' describes The White Horse in 1990 as an excellent place to seek refreshment.

It is a listed building, the sign is thought to have been painted by Gertrude Jekyll.

MSt. 1845-1887 PO; 1899-1938 K.

**Haslemere, 22 High Street** In 1805 it was the town's
principle inn. By 1887 it was advertised as:

'White Horse Family & Commercial Hotel,
Haslemere. This Hotel is entirely under new
management, and will be found perfectly
quiet and comfortable. It is within easy
distance of Hindhead and Blackdown. Good
Stabling and Posting House.'

The inn was patronised by the author, Hilaire Belloc (1870-1953).

The building has been described as 'typical of the first half of the eighteenth
century, still generally built with the low entrance archway[1] common before
coaches began to carry outside passengers'.

AER and WCO. 1826 L; 1839 P; 1845-1887 PO; 1898 *R.S.O.* K; 1938 K.

**Headley** By 1904 the property was no longer listed under White Horse in the
Petty Sessional Returns. 1870 K; 1871 PO.

**Old Woking, The High Street** Until the 19th century
the village street was known as Town Street – to
become The High Street.

Once standing on the corner with Broadmead
Road, this half-timbered house was thought to
date from the 16th century.

c.1800 Vestry meetings were sometimes held
here as well as the other public houses to emphasise

secularity; by 1820 ten per cent of the parish were 'so poor as to need assistance'; ten years later 246 persons received Poor Relief – thirteen per cent of the parish.

1892 listed as trading in the Petty Sessional Returns of licensed houses.

c.1900 a photograph of the inn shows a distinctive banner sign hanging out over the street.

1904 the inn was listed in the Petty Sessional Returns as a tied house with poor trade, it offered food, three bedrooms, sanitary facilities and stabling for four horses.

1920 part of the property – *possibly the stables* – was demolished for road widening; three years later Mr. Conway West had converted the remaining hotel building into a garage. SHC, AC and IaW. 1845-1859 PO; 1870-1913 K.

**Reigate, Albert Road** Marked on the 1ˢᵗ Edition of the Ordnance Survey Map of Reigate as White Horse Inn, later editions show an unidentified building on the site. Listed by Kelly in 1870.

**Ripley** A busy coaching town with a number of hostelries, the High Street being on the old London to Portsmouth road.

1778, 1st June – a For Sale notice in the *Salisbury Journal* read 'White Horse at Ripley in Surrey occupied by Mrs. Hawkins'.

1779, 10th May – another For Sale notice was published in the same newspaper:

'. . . all that freehold messuage, tenement or Inn called the White Horse, situate at Ripley, together with the stables, coach-house, garden, meadow land and premises thereto belonging and in the renting of Mrs Hawkins'.

1791, September – the White Horse was mentioned in the Gentleman's Magazine. The inn was listed as trading until 1853. MSt. 1839 P; 1845-1851 PO.

**Shere, Shere Lane** The building, set in the centre of the village, dates from 1425; it is claimed that the house was converted from a hop drying building but it is more likely that it began as a farmhouse from c.1500 when it was called Cripps with two acres of ground called 'Marysse' – a hop garden.

Documentary records of the building date from 1562; it was constructed as a two bay open hall with two cross wings and a number of subsequent alterations and additions.

By1664 the house had become an inn with a brewhouse.

By c.1720 a cellar had been added, later to be boarded up.

c.1830 a malthouse was built next door.

c.1923 a timber extension was added.

1955 casks of brandy of this date were discovered in the 'unknown' cellar, perhaps confirming legends of local smuggling activities.

2007 there were wooden stocks outside the inn and it was said that timbers from Nelson's Ship, Victory, formed beams in the restaurant. 1845 Dorking-1887 PO; 1899-1938 K.

**Stoke in Guildford / Stoke-next-Guildford, High Street** 1696, 22nd April – probate of the will of Thomas Browne, cooper, which was included in the deeds of Flutter's Brewhouse, Guildford who owned the White Horse.

1782 this licensed house was in the ownership of Francis Skurray, a brewer of Guildford, with his son and grandson.

1904 listed as a free house of good trade with a Poor Law Assessment of £94 – offering food and sleeping accommodation with 'good' sanitation – but no stabling.

1823 P; the Post Office Directory of 1895 lists two White Horse inns under Guildford; one would certainly have been Stoke which was one mile north of Guildford. SHC Ref. 1483/4/13.

**Sunbury-on-Thames, Thames Street** Standing on the corner with The Avenue – once a track running north.

1720 Samuel Slater bought the property as an alehouse.

1729 The Register of Innkeepers and Alehouse keepers records the White Horse as being the oldest alehouse in the town with William Diddlefold as licensee; he had six children some of whom were born here – the eldest son, William, also became an alehouse keeper in Sunbury.

1750, 16th April – the Court Book for Kempton Manor gives details of a property, consisting of three new cottages, 'which to the west abutted on the White Horse alehouse'. These cottages were bought by a brewer.

1751 George Slater, son of Samuel Slater, gave a fifth share of the White Horse to his mother, Hester Perry, who had remarried. Three years later George's brother, Isaac, sold his fifth share to Mr. Diddlefold, who moved away in 1764 having lived there for at least thirty four years.

1773 William Hopton was landlord for the next thirty years – showing the business to have a very strong trade.

The White Horse was paid two guineas a year by the local council to make its 'convenience' available to the public – to meet the need of the increasing numbers of people taking a day's outing from London.

By 1834 'Sunburry' is described as lying along the bank of the Thames and containing a great number of ornamental villas and handsome seats.

1845, 20th May – the premises were sold, together with adjoining property. A plan showed 'a rectangular building with a splayed bay at the extreme west end of Thames Street frontage'.

1876 Charles Durrant Hodgson, a member of a local brewery family, became landlord in July – in October he bought the property for £38 7s. 6d.

1882 the landlord was fined 10/-[2] with eight shillings costs for 'permitting drunkenness in his licensed premises' – four years later he was convicted for opening just before one o'clock on a Sunday.

1886 by this date the premises were owned by Hodgson's Kingston Brewery. The landlord, William George Court was described as 'holding the White Horse Hotel' – it is unlikely that there was any residential accommodation.

1923 a piece of land to the rear of the public house was sold.

1926-1937 Richard Charles Winter was listed as landlord – he was the father of Fred Winter, the famous steeplechase jockey; Fred would visit his father whenever he was racing at nearby Kempton Park.

By 1988 the brewers, Courage, had bought Hodgson's Brewery which included the White Horse. SHC, SBC, JB and KH. 1823-1839 P; 1845-1898 PO; 1914-1933 K.

**Notes**

1. From c. 1760 – the accession of George III – outside seating on coaches became more common. The archways of inns were heightened to accommodate taller coaches.

2. Ten shillings, e.g. 50p.

# SUSSEX

*An act of Henry VII (1504) directed that for convenience the county court should be held at Lewes as well as at Chichester, and this apparently gave rise to the division of Sussex into East and West parts, each of which is an administrative county.*

# EAST SUSSEX

*If you wake at midnight, and hear a horse's feet,*
*Don't go drawing back the blind, or looking in the street,*
*Them that asks no questions isn't told a lie.*
*Watch the wall, my darling, while the Gentlemen go by!*
*Five and twenty ponies,*
*Trotting through the dark –*
*Brandy for the Parson,*
*'Baccy for the Clerk;*
*Laces for a lady; letters for a spy,*
*And watch the wall, my darling, while the Gentlemen go by!*

'A Smuggler's Song' from Puck of Pook's Hill
by Rudyard Kipling (1865-1936), published in 1906.
*The author having moved to Bateman's near Burwash in 1902.*

**Bodle Street Green, North Road** The original White Horse inn, a brick building dating from the 16th or 17th century, stood at the apex of the triangle known as The Green – due to a later absentee landlord the Green was built over, hence officially there is no longer 'a green'. The road, then named Back Lane, was a poor area with large families living in small cottages.

1838 the house was owned by Mr. John Griffiths and was listed as an inn keeping business with Mrs Philadelphia Oxley as innkeeper, she kept a very good house and knew how to manage the 'regulars'.

c.1850 the present White Horse was built by Mrs Oxley's son, Richard on a site some 50 yards south of the original inn; upon the death of his mother nine years later, he became innkeeper and blacksmith when it became a very rough inn.

An old inhabitant of Bodle Street Green learned from his grandmother that:

Young men had nothing else to do except to go to the White Horse where there was always fighting. "Richard's brother Will 'was a devil for a fight' but not unless he was forced into it – he was six foot two and as thick as a barrel. The two brothers were standing at the bar one night when Jim the well digger framed up to Will. Will hit him once and his head went through the panel of the door; he got up bleeding like a stuck pig and the landlord washed the back of his head and tidied him up. Then he came up to Will and said – "Look 'ee' never hit me like that again or else". Will hit out again but Jim dodged him and was out the door like a shot from a gun, and so Will's fist went through the other panel. The landlord fetched two bits of board and nailed them across to keep the draught out. They were there for years was them boards."

Cocky Wren, the post office receiver, was so frightened of a fight with the well digger that he would not deliver letters down Back Lane.

According to Philadelphia Oxley's will, proved in 1859, the property included:

'public house, blacksmith shop & 2 small plots of land (1 & ¼ acres) in occupation of Ric Oxley & George Burgess. The saleable value being £500 & the annual rents £26.'

1862 a receipt signed by Mrs. Oxley's daughter, Elizabeth reads:

'Received £5 of Ric Oxley to give up one room to hire in the house of the White Horse Inn, left to me by my late mother & that I have no other claim upon the said Ric Oxley'.

1886 the Beerhouse Licences for Warbleton show Richard Oxley as holding an off-licence for wine and beer. During this year the premises were sold to the Star Brewery of Eastbourne.

By 1887 Edward Pankhurst was listed as innkeeper and wheelwright. The following year the brewery allowed a mortgage for the freehold premises of the White Horse Inn for £800 and interest at 4½ per cent.

1891 the census showed that Richard Oxley had retired; he died the following year aged seventy six.

1901 Edward Pankhurst died aged forty four; later in 1910 his brother, Alfred, became innkeeper and carried on the wheelwright business until he became bankrupt – the bailiffs arrived and he plied them with drink, while his tools were removed to safety in Longview Cottages.

c.1920 Fred Brett, or 'Fred the fox', was landlord; his daughter, Nancy lived in the pub for forty three years, taking over from her father in 1951 after she married Harry Wells the previous year:

Fred was the first person to own a motor car in the village. He did not always close the pub on time and one night, hearing that the police were coming, he removed all the money from the till and supplied free drinks. The chief inspector, entering by the back door, was knocked out by Mr. Brett with his one good arm – the other was blown off in the First World War. In court he defended himself, his case being that it was a private gathering of friends and family and he was only guarding his property against intruders. When asked what weapon he used he held up his other arm. The case was dismissed.

The distinctive white horse, painted in the 1920s on the red newly retiled roof was originally copied from a picture of Mr. William Cardwell's horse – he was listed as owner and licensee in 1915 – he wanted the polished elm counter for his coffin. The sign writer made a mistake so that the hindquarters were lower than the front, this was left until it faded and was blacked out during the Second World War as it was a sure navigator's line for the bombing of London. It was repainted when the war was over and is believed to be a portrait of a racehorse owned by a director of the Star Brewery.

At this time the White Horse was the centre of village activities, such as a collection point for all the iron railings in the district which were used for the war effort.

1972/3 the pub was closed for a year to be refurbished.

Since 1993, the pub has held musical evenings.

The original inn was demolished except for the cellar, rebuilt and named Mount Pleasant; it became the home of Mr. Christian and his family; it was thought to be the only house in the village with cellars. By the end of the twentieth century it had become Brick House. JWi, Mrs. Ayres and Mr. David Hosey, Landlord.

## Brighton

***30-31 Camelford Street*** By 2002 the old sign had been changed to a black diamond. 1845 PO;1870-1899 K; 1974 K.

***65 East Street*** A well-known hotel and posting house with a covered carriage road which ran through to extensive stabling at the rear.

Situated at the south end of the street close to the beach where the Fish Market was held, fish being weighed on the large scales which stood near the inn. At very high tides the sea would wash past The White Horse and flow down Pool Lane to Pool Valley behind the stables.

1788, 15th December – the landlord, advertised in the *Sussex Weekly Advertiser*:

> 'W. Henwood begs Leave respectfully to acquaint the Nobility, Gentry, and the Public in general, that he has greatly enlarged, and fitted up in a commodious and elegant Manner the said Inn, with a Number of exceeding good Beds; and assures those Gentlemen Travellers in the different Branches of Trade, that it will be his peculiar Care and Attention to accommodate them in the best Manner, at all Times of the Year. His Wines and Liquors of every Kind are of the choicest Quality, and a good Larder. Where by a constant Attention to the Duties of his Profession, he hopes to merit their Patronage and Protection.'

> N.B. Neat Post-Chaises and good Horses; likewise good Stabling.'

Under Mr. Henwood's management the White Horse further developed its business, especially as a venue for important meetings. On Wednesday, 26th August 1789 the opening was held here of the first Lodge of The Ancient and Honourable Society of Free and Accepted Masons of England with the H.R.H. The Duke of Cumberland as Grand Master was present.

1799 listed as trading.

1812 after the county election when Sir Godfrey Webster, the Liberal candidate, had won one of the most expensive County elections ever, his carriage stopped outside The White Horse where the Liberal committee room of Capt. George Pechell, R.N. was celebrating; the landlord, William Allen, 'came out and taking his watch from his fob, held it up and said "Look here Sir Godfrey, this is paid for – paid for!" Sir Godfrey bowed, and the procession moved on. *There had been a rumour that Sir Godfrey had not paid his tavern bills.*

1825 alterations took place to incorporate a private house lying between the hotel and The Rising Sun – a tavern which had been occupied by Mr. Willis, a fisherman.

1829, 28th May and 1837, 24th June – the White Horse was included in the sale particulars for the Newhaven Brewery.

c.1835 'It is a spacious and convenient establishment, and it is resorted to by very genteel company. Mr. Hodd, the proprietor – *possibly Mr. Allen's son-in-law* – is indefatigable in attending to the wants of his numerous customers, and by this means his house is always full of company. We may likewise add that his charges are extremely moderate.'

1839, 4th April – The White Horse Tavern with The Tap, Stabling and Appurtances was put up for sale, as part of the Newhaven Brewery Sale, with an annual rent of £200.

It was described as a well established, flourishing business offering the best accommodation in the most fashionable part of Brighton; the well placed Tap was noted for serving 'Newhaven Tipper'. The particulars included:

'The situation of this favourite Hotel possesses all the advantages of being immediately connected with the Sea, the Steine, the Grand Junction Parade, the East Cliff and the Baths. It commands an extensive view of the Sea, and directly in front of the House is the fashionable Drive and Promenade, and attractive and constantly exhilarating scene of gaiety and pleasure.

The Hotel includes good Cellaring, a spacious Coffee Room, Bar and Parlour, detached Kitchen and Offices, three sitting Rooms and Twelve Sleeping Chambers; a Tap with necessary conveniences, Stabling for Twenty Horses, Yard and other Premises.'

c.1845 during the County Elections when Mr Curteis, a candidate, and his committee were sitting at the hotel, 'the counter-election cries outside so exasperated the cook that she threw a bowl of scalding water from the window upon the crowd beneath', shouting "No red herring soup" which landed on the keeper of the temporary prison known as the Black Hole situated to the north east of the Market – one Barber Harmer who limped back to the prison was silenced for the remainder of the election.

Mr. Curteis had said that red herrings were food for the poor; during this election red herrings hung from trees along the road between Brighton and Chichester. The carriage of his opponent, Sir Godfrey Webster, was decorated with red herrings.

1869 this considerable site, together with The Rising Sun was bought by Brill's Baths Company to build their large swimming pool and other baths. The demolition took twelve days and employed one hundred and five men. An old cellar connected to the White Horse was discovered, completely filled with empty wine bottles – van loads of them were removed:

> " . . . if the consumption of beer was enormous at The Rising Sun,
> that of wine at The White Horse was not a whit behind it."

BH, KPC, ESR, AMS 6300/4/5 and JGB. 1805 WCO; 1823/4 P; 1826 L; 1839 P; 1845 PO; 1870 K. Hotel.

**Edward Street** Situated in what was once a poor area of the town.

1870 listed as trading. K.

1871 recorded in a lease of licensed premises and a property register.

1876 the business was sold by Kidd and Hotblack Ltd. to Keeping and Co.

In 1892 a conveyance records it was sold to Tamplin and Son.

1896-1966 the house was recorded in the property register.

Schedules of title deeds for 1900-1959 are held by the East Sussex Record Office together with a house specifications book for 1909 and repairs books from 1912-1962.

**Ditchling, 16 West Street** This distinctive white-painted building dates back to the 16th Century. The history of St. Margaret's Church records that the local pathways, twittens[1] and lanes were well-used smugglers routes, *which would probably account for a tunnel running from the cellar at the White Horse, under the road and churchyard, to the church;* during 1774 it was reported that four hundred men and two hundred pack horses passed through the village.

1794 regular London to Brighton coaches stopped in Ditchling to change horses before ascending Ditchling Bostal[2], the steep track, over the Downs; this had levelled passing places for the horses to rest at intervals all the way to the top.

1856 the premises were owned by the Bear Brewery, Lewes.

1898 the house was bought by Southdown and East Grinstead Breweries Ltd.

1915 the hotel was put up for sale.

1920 the title deeds are held to this date.

1992 the house ceased to be a hotel. ESR. 1839 P; 1845 PO; 1870-1899 K; 1974 C&L.

**Hartfield** 1851 PO; 1870 K.

**Holtye, Holtye Road / Holtye Common** The building is said to date from the 13th century.

1835 the Post Office Directory records William Kenward as landlord; he was a farmer of 'Holty' Common. The pub is situated next to the golf course.

Landlord and ESR. 1899-1913 Cowden K.

**Hurst Green, Silverhill** 1888, an assignment of a lease of the White Horse Inn.

By 2006 two restaurateurs from London had bought the pub and refurbished it. ESR. 1845 PO.

**Lewes, St. John Street** According to the *Bath Journal* of 18th February 1744, smugglers armed with pistols and blunderbusses rode openly through Lewes; they called for a bottle of wine at the White Horse.

The following are held by East Sussex Record Office:

1896 a property register lists the premises as being purchased by Tamplin and Son Ltd., Brighton.

1899 house valuation; 1900 schedule of title deeds; 1909 house specifications book; 1912 repair book; 1959 schedule of title deeds; 1962 repair book and a 1965 house valuation.

**Robertsbridge, 61 High Street** Until 1567, when George Padyham was landlord, the house was called The Tonne and was a copyhold[3] tenement of Robertsbridge Manor; it was renamed The George and described as a messuage with stable, milkhouse and garden of half an acre in several parcels; recorded as catering for a 'select clientele'.

The inn was rebuilt over the stone wine cellar, rather than being enlarged, with a 14th – early 15th century spiral staircase. In about the 15th or early 16th century 'a detached two bay stable block with deep foot-braces' was built in the rear yard, which was continuously altered over the years.

Between 1630 and 1700 the medieval front of the building was demolished and was replaced by 'a lofty building which covered the entire frontage of the plot' and consisted of a thick brick wall with plinths[4] and poor quality tile-hung timber-framing. A balcony was added over the front door for important people to make pronouncements. *This balcony could have been used by the Earl of Leicester who held the Lordship of Robertsbridge before the Sidney family moved to Penshurst Place.*

A four storeyed rear kitchen wing was added, with its ground floor being sunk below the level of the main building. The kitchen had an inglenook fireplace almost 10ft. wide, a storage area and its own staircase leading to the staff accommodation above. In the main building a fine staircase was constructed from the ground floor to the attic, leading to a large meeting room on the third storey and the attic on the fourth, where the roof span was carried on an aisled construction.

1660 Justice Courts were held here in the meeting room.

*1662 John Padiham was assessed for seven flues. This was the year the Hearth Tax was levied at a rate of 2s. 0d. for each hearth except those in cottages.*

1667 the rate book records that he paid £9 for The George and Brewhouse.

1711 John Lulham was assessed for rates at £7 for the George, £8 for Lordsbrook – a meadow, £22 for his farm, £1 for Adam's House and £3 for Pottery field; in 1713 he paid rather less – £4 for the George, £3 for Lordsbrook, £11 for his farm and £2 for the Pottery field.

Title Deeds of the 18th century and of 1807 record that this property was called The One Star or The Star.

By 1864 the house was known as the White Horse.

1938 listed in Kelly's Directory.

c.1960 it ceased to be a public house.

By 1972 the balcony had been taken down and its doorway converted into a window; four of the original windows had been bricked up.

By 1997 this imposing Georgian building with the original cellars had been named Robertsbridge House and was owned by A.W. Gore & Co., Licensed Property Valuers.

The 17th century brewhouse became Langham Cottage; at one time a stream ran along its western wall, and possibly provided water power. DMa, 1972 Periods C & D, *1972 Periods A – D.*

**Rottingdean, High Street / Marine Drive** Originally called The King of Prussia, the inn is situated at the cliff edge, and in close proximity to the windmill (built 1802). This has made it a hub of activity, drawing in both travellers and sailors alike. It was also one of the oldest rendezvous for smugglers in Sussex who used the large cellars, which ran beneath the roadway, to store their contraband.

In the 18th Century the inn was a famous venue for bull-baiting and cock fighting. The Inn was also the perfect venue for auctioning off the wreckage of ships that had been driven into the unfriendly shore below. Some ships had almost certainly been lured to the cliff edge by wreckers, who had at one time

been active along this coast. The inn was used as a halt for those travelling along the Dover Road, and later as a terminus for horse-drawn buses.

1822, 12th April – *The Brighton Gazette* reported an inquest which was held here on Mr. Briggs, a hatter of Brighton, and Mr. Knowles of Cowfold who were killed three days earlier when their horse and cart fell over the cliff near Roedean Gap between Rottingdean and Brighton. Mr. Samuel Sutton, landlord, reported that the two men had called in to The White Horse at nine o'clock in the evening and had stayed for ten minutes, they were both sober:

> 'Just before they got up into the cart, Mr. Briggs expressed a wish to drive, saying that he knew the road better than his friend; but Mr. Knowles objected, remarking, that he knew the horse better than Mr. Briggs. It was rather dark, the moon not being then up.'

At ten o'clock that night, three labourers on their way to Black Rock Bottom found a man's hat lying about six feet from the cliff edge, looking over they saw a man and a horse lying on the beach so they went down on to the beach by Green-way Gap; the first thing they found was the wheel of a cart lying near the sea and then found the bodies of the men. When the landlord learned about the accident at half past eleven that night, he sent a horse and cart for the bodies to bring them to the inn.

At the inquest, the Coroner read a letter from Henry Sayers, an ostler at the Kings Head Inn, Brighton which stated that he:

> "well knew Mr. Knowles's mare, which was always unquiet in harness; that she had run away with him no less than three times, and that she was in the habit of gibbing[5]."

The cliff was eighty two feet high and the road was only three yards from where the accident happened. Both men wore watches – one had stopped at twenty five minutes past nine and the other at ten minutes to ten:

> 'An immense concourse of people have visited the spot during the week, particularly on Sunday, when every description of vehicle appeared to be in requisition for that purpose. The road was literally thronged during the greatest part of the day, which was one of the finest we have had this spring.

> The very large sum of £511 10d. was deposited last Saturday night, in the Savings Bank, from more than 60 depositors: it is an undeniable proof of the industry and better condition of our labouring classes.'

1833 bought by Vallance Catt and Company, Brighton.

By 1899 it had been sold to Tamplin and Son Ltd., Brighton who demolished the original Inn in 1934 to make way for the current building; it remained in their ownership until 1963. BH, SWA and A-S. 1839 P; 1845 Brighton PO; 1851 PO; 1870 Hotel and Posting House – 1898 K.

**Salehurst** Listed in Kelly's Trade Directory of 1877 by which time The Star in Robertsbridge had become the White Horse – a confusion may lie here.

Originally established for the workers of the local estate and was named The Old Eight Bells until c. 1975, when it became 'Salehurst Halt Pub'. It was owned by Whitbread before becoming a free house.

Salehurst Halt was on the Kent and East Sussex Railway line which was opened in 1900 and ran from Robertsbridge through to Tenterden; it partially closed in 1954 and was shut down in 1961, but has reopened between Tenterden and Bodiam as a standard gauge, single track, branch line with passing places at the five stations.

**Notes**

1. A narrow thoroughfare, or alley. PC.

2. A path up the side of a steep hill. PC.

3. A tenure less than freehold of land in England evidenced by a copy of the Court roll. Col.

4. The lowest part of the wall of a building that appears above ground level, especially one that is formed of a course of stone or brick. Col.

5. A most objectionable habit where the horse refuses to move forward and, in some cases, runs backwards. RSS.

# WEST SUSSEX

The 32 year old Horace Walpole MP, writing on 26th August 1749 from his newly acquired house, Strawberry Hill in Twickenham, to his close friend and contemporary George Montague MP – tells of the 'distresses' he and John Chute (1701-1776) have encountered while travelling in Sussex:

*"If you love good roads, conveniences, good inns, plenty of postilions and horses, be so kind as never to go into Sussex. We thought ourselves in the northest part of England; the whole country has a Saxon air, and the inhabitants are savage . . . Coaches grow there no more than balm and spices; we were forced to drop our post-chaise, that resembled nothing so much as Harlequin's[1] calash[2], which was occasionally a chaise or a baker's cart. We journeyed over Alpine mountains, drenched in clouds, and thought of Harlequin again, when he was driving the chariot of the sun through the morning clouds."*

'The Letters of Horace Walpole', Vol. II, (London, 1840).

**Angmering** 1687 the probate inventory for William Adams of West Preston in Rustington records that he held the Lease of the White Horse at Angmering for £30. JPe and WSR - EP1/29/164/007.

**Arundel** On 27th October 1571 and 15th April 1580 the exact position of this alehouse with its garden was given as 'abutting S. on land of Henry, Earl of Arundel, N. on a tenement of Philip Stopwell, E. on the moat of Arundel Castle and W. on the High Street'. A Bond in £100 of the latter date 'for the peaceable possession of the property' refers to Mary, wife of Adam Chamber, a tailor, and lease for seven years at an annual rent of £4 granted to the occupier, Francis Cradle, on 16th June 1575.

1591, 5th November – a declaration was made for a change of use of the property in respect of 'Messuage, barn, garden and land called the whithorse (sic.) in Arundel, late the inheritance of Adam Chambers'.

**49 Tarrant Street** Standing on the south side, west of Brewery Hill junction.
   1845 2 listed PO – one possibly Bury.
   1910 listed in Pikes Directory.
   JPe and WSR – Catalogue of Lavington Archives Nos. 151-155.

*Coaches from Arundel ran three times a week to The White Horse, Fetter Lane, London.*

**Billingshurst** The Court Rolls of Bassetts Fee Manor in Billingshurs, 1604-1642 records the White Horse in 1618. *This is a large document written in Latin!*
JPe and WSR – 6077.

**Bognor Regis, Bersted Street** 1842 the tithe map shows a White Horse Inn on the site of what became the Queen Victoria Inn which is no longer trading.

**Bury** A coaching inn. The original White Horse sign can be seen on the side of the building.
In c.1997 it was renamed the 'Squire and Horse', the sign shows a rider on a white horse. 1851 PO.

## Chichester

*Near Eastgate* 1900 it was known as the Swan Inn. It has also been named the White Swan and the White Horse but the dates of these signs are not clear.

*Northgate* 1701 the house was named 'The Queen's Head' and later it was 'The Bear'.
By c.1763 the name was recorded as The White Horse.
1930 it was destroyed by fire. RMo and JPe. 1823/4 P; 1851 PO; 1870-1913 K.

*61 South Street* 1416 built as the Chichester Law Courts – one of the oldest inns in the city with tunnels leading to the cathedral.
1533 according to a local historian the inn is mentioned in *Sussex Notes and Queries*, 14, p.57.
1680 it became a coaching inn on the Brighton run.
The decorative wrought iron sign which hangs out over the street shows two white horses, one a small weathervane mounted on a hooped tendril of vine leaves carrying a gilded bunch of grapes, the other hanging below, is the painted signboard of a rampant white horse; this being the crest of the Fitzalan family from nearby Arundel Castle – for five hundred years the home of the Dukes of Norfolk.

1974 the house was renovated under a Preservation Act.
2006 became Prezzo Restaurant.
1823-1839 P; 1845-1851 West Pallant PO; 1870-1899 Hotel 'Family and Commercial and Cyclists' K. 1974 renovated under a Preservation Act.

*3 Westgate* 1649 a Parliamentary survey refers to 'a piece of ground where lately was the house of the White Horse Inn'.

1682, 26th December – several young men were arrested for drinking 'treasonable toasts', presumably in connection with the aftermath of the Civil War – King Charles II was to die in 1685.

By 1909 the house had become the private residence of a solicitor, George A Tyacke. c. 1960 the building was demolished.

1851 West Pallant PO; 1870-1899 K. A-S; CDH; JPe, MJC; RMo, and WSR.

**Chilgrove, 1 The High Street** Situated at the foot of the South Downs.

This former coaching inn is thought to date back to 1765, though there is a stone dated 1720 which can be found at the front of the house. The original parts of the building have low beams and open fire places. There is also a flint-walled garden and a small courtyard. Deeds for the public house show that accommodation was provided for travellers, who had their horses shod at the nearby smithy before climbing the steep hill to the top of the Downs.

1817, 12ᵗʰ May – the house was advertised for sale.

Since c.1980 the signboard depicted a tabby cat, 'Lucy', painted by the local ironmonger's artist wife; although the inn remains The White Horse, Mr Barry Phillips, the landlord here for thirty years had asked for a sign of a white horse, but the artist said she was unable to paint horses; the landlord suggested she painted something else, so she painted her cat.

2008, May – the author speaking to the on-site builder discovered that the sign depicting the cat was unknown to him, but the present sign was of a white horse. By July 2008 the sign was to be of a small fish – this 17 bedroom building under new management, with eight Jacuzzis, would be known as The Fish House. The new car park, presumably taking the place of the garden and courtyard, was 'massive'. The stone dated 1720 will be reinstalled and flood-lit. Mr Barry Phillips had retired in 1996.

After World War II some transactions between the old established brewers for purchasing public houses, which included this house, were decided by cutting packs of playing cards. SWA. 1877-1899 K. Mr. Barry Phillips.

**Easebourne, Easebourne Street** A 16ᵗʰ century coaching inn.
1839 Midhurst P; 1845 Petworth – 1851 PO; 1870 Midhurst – 1899 K.

**Fittleworth** 1594 a Chichester Consistory Church Court Deposition, written in Old English, mentions the White Horse. *This may refer to Petworth.* JPe and WSR – Ep1/11/7.

**Graffham, Heyshott Road** A 17th century inn which was originally a farm house and dairy situated on the Petworth to Guildford Road, from which milk was delivered until the late 1930s.

A picture hangs in the pub of 'Granny Pratt', she was a very important person in the history of the house, the date 1800 is written on the back – it is said that the picture must never be moved. She used to become very annoyed with the village children who teased her by chanting ditties, such as 'Granny Pratt looks like a rat'.

Landlord and Bar Maid. 1851 PO. 1879-1899 K.

## Horsham

*Carfax* The market area of Horsham where this timber-framed building stands, but is no longer an inn.

1698-1706 listed as trading.

1706 an inventory of Thomas Rade, Inholder (sic.).

1760 Henry Waller – of a notable Sussex inn-keeping family – was listed as occupier of the 'late White Horse'.

1765 Henry Waller died and an abstract of title referred to the 'White Lyon'.

JPe, WSR - EP1/29/106/306, SAS mss (ND75) and AHu.

*15 Springfield Road (formerly No. 6)* c.1816 a chapel – later known as the White Horse – was built on the opposite side of the road from the Malt and Shovel.

1888, 26th March – an Indenture of Conveyance referred to:

> " . . . a piece of ground whereon is erected a brick building formerly used as a Meeting House for the exercise of Divine Worship but now converted into a messuage formerly used as two dwelling-houses but subsequently converted into one dwelling-house and now and for many years past used as a Beer-house known as The White Horse."

Trust Deed dated 1st April 1912. RBB.

**Hurstpierpoint, Albourne Road** A coaching inn, which has been trading since 1591.

The oldest part of the building has become the restaurant; the rest of the building was added between 1800 and 1850.

At one time, the local customers were crossword enthusiasts and the sill of one bay window was full of dictionaries. 1839 P; 1845 Hurst PO; 1870 3 listed – 1899 K.

**Lindfield, 22 The High Street** The village was mentioned in a Saxon charter of 765 AD as Lindefeldia – 'open land with

lime trees'. The High Street follows an ancient
north-south track which existed for thousands
of years before the Romans built a major road
a mile west of the village. King Edward III
granted Lindfield a royal charter to hold a
market and two annual eight day fairs –
the August fair becoming one of the largest
sheep sales in Sussex.

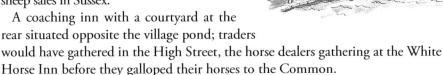

A coaching inn with a courtyard at the
rear situated opposite the village pond; traders
would have gathered in the High Street, the horse dealers gathering at the White
Horse Inn before they galloped their horses to the Common.

1757 is the date on the corner stone of this brick house.

It is reputed that there are smugglers' tunnels near All Saints Church and on
a dark night in 1782, smugglers are reputed to have passed through the village
leading three hundred horses carrying contraband.

1791 the Old Series Ordnance Survey map shows the building.

1844 the site is shown as a house and garden owned by the Mason family.

1874 the Ordnance Survey map marks the building as a public house.

1901 Tamplin's Brewery of Brighton owned the premises, Mr J. Dennett being
the landlord.

1906, 15th January – Alice and George Creswell moved into the White Horse;
they celebrated their golden wedding here in July 1938. During their tenancy they
had a marquee in the garden where they served meals to shepherds, farmers and
dealers who came to the Lindfield fairs to sell cattle, sheep and horses. Mrs Creswell
nursed and visited sick people and was named the Good Samaritan of Lindfield;
she would also dash to the pond to rescue people who had fallen in. Her daughter
and son-in-law took over the inn after Mr Creswell's death and were there in 1959.

By April 2006 new owners had turned it into a pub and Thai restaurant.

Richard Bryant – Volunteer Archivist, HH, the Landlord and Lindfield Brochure.


**Manningsheath** Situated two miles from Horsham on the turnpike road which
ran through Henfield to Brighton.

1794, Wednesday 20th August – the White Horse was auctioned in one lot at the
Star on Horsham Common 'between the hours of four and six in the afternoon'.
The Particulars and Conditions for Sale record 'a Messuage or Public House known
by the Sign of The White Horse . . . and Two Tenements (one of them entirely new
built) adjoining . . . with the Outhouses Gardens and Appurtenances thereto . . . '
The bids ranged from £300 to £410.

1912, 1st April – A Trust Deed gives the following details:

'. . . messuage or Public House formerly known as The White Horse
but now and for many years past known as The Dun Horse situate

at or near Manningsheath in the Parish of Nuthurst in the County of Sussex with the stable and buildings garden orchard backside and premises together with the shop and other buildings forming part of the said messuage or Public House and also those two messuages or tenements on the north or north-easterly side of and adjoining the said Public House.' RBB.

Presumably this change of name was because two White Horse inns were merely two miles apart – the other being in Horsham.

The present pub has two bars, beamed ceilings, and there are stained glass windows at the front which advertise an old Brighton brewery.

Horsham Museum MS SP7.

**Maplehurst, Park Lane** A coaching inn on the old east-west route from Cowfold to Billingshurst before this part of the A272 was built.

1826 Built for workers on the Burrell Estate at nearby Knepp Castle.

1912 The following description appears in the Trust Deed of The Rock Brewery, Brighton Limited:

'... messuage tenement or inn ... and land thereto adjoining and belonging and known as The White Horse inn situate at Maplehurst ...'

The present pub is said to have the widest bar in Sussex. Other features include a wood block floor, exposed brick work, an old Victorian stove and an original wood beam in the bar.

2007 awarded CAMRA Pub of the Year for the Surrey and Sussex region.

Landlord. 1870-1913 Nuthurst K; 1974 C&L.

**Mare Hill** The original building is said to date from the thirteenth or fourteenth centuries.

2008 The pub is owned by the Fuller's Brewery, and offers a landscaped car park and helicopter access. 1845 Petworth PO; 1877-1899 3 listed K. – possibly referring to Nutbourne and West Chiltington. An R.S.O. Hotel.

**Midhurst** 1610 12th January – the property is mentioned in a Quitclaim[3] for £50; on 7th October a Conveyance reads:

"Messuage, wherein Thomas Bailie lately dwelt, being parcel of his burgage[4] sometimes called the 'Whte [sic] Horse, and now divided into two tenements or dwellinghouses . . . the other now in occ. of John Kelinge alias Blisse, with his skilling[5] and little stable situate on the N. end and E. side of the barn . . . and so much of the gate belonging the said burgage as lies by a straight line from S. end of the house of office unto S. end of the said little stable between it and the pale[6] which divides the said burgage from the burgage sometimes called Brewers but now the sign of the Rose, together with all chambers, rooms, gardens, etc., belonging in as large and ample manner as said Thomas Bailie lately held and used the same, and together with free ingress etc. with horses and carts only for the carriage away of the dung and soil as need shall require through the said gate . . . " Cowdray 4465 and 4466.

1660 by this date deeds refer to this dwelling house as 'lately an inn called the White Horse'.

1695/6 and 1740 deeds refer to 'a messuage standing near the west corner of the land on the north side of the White Horse inn.

1785 An article in Sussex Notes and Queries No. 13 by E. M. Gardner entitled 'Midhurst Court Rolls' describes the property as a 'One Burgage tenement held by Sir Charles Mill of Mottisford in the County of Southampton – formerley the White Horse Inn, now a private tenement.

Three years later in 1788 a deed records the White Horse as 'now a private tenement' with the stables, outbuildings, yard or backside and garden belonging . . . having the street leading from out of the North Street towards the Market Place on W., and Hog Lane on N – formerly the Beast Market' Cowdray 4098-4114.

1816, 19th and 20th August – a conveyance describes the property as 'heretofore called the White Horse inn'. WSR and Cowdray 2833-2850 and 3116-3141.

**Nutbourne** 1623 According to an archivist from the West Sussex Record Office, the White Horse inn is mentioned in a Church Court Deposition of 1623.

This inn is possibly listed under Marehill and West Chiltington.

**Oving** 1817, 12th May – the house was advertised for sale in the *Sussex Weekly Advertiser*.

It is possible that the White Horse once stood on the site of The Tangmere Hotel which was demolished in 1943/44 in order to extend the runway of nearby Tangmere aerodrome. The population of Tangmere in 1831 was a mere 197 – the White Horse probably stood in a large area which had several small taverns.

The aerodrome opened in April 1918, but closed in November; it re-opened in 1925 in anticipation of the Second World War when runways, blast protection pens and a perimeter track were built.

Mr. Travers Johnson, Tangmere Aviation Museum, WSR and JPe. 1851 Oving PO.

**Petworth** 1845 2 listed PO; 1870 2 listed – 1877 K. *Possibly the same as Fittleworth and Easebourne.*

**Roffey (situated on the corner of Crawley Road and Leechpool Lane)** In August 1614 the Horsham Carriers 'who lieth at the White Horse In Southwark *(see London)* reported a strange and monstrous Serpent or Dragon in Sussex, two miles from Horsam, in a Woode called St. Leonard's Forrest reputed to be nine feet, or rather more, in length. He will cast his venom about four rod[7] from him, as by woeful experience it was proved on the bodies of a man and a woman coming that way, who afterwards were found dead, being poisoned and very much swelled, but not preyed upon'. *This alarming incident will most likely have taken place on the road which is now the A264.*

1874, 12th March – an Indenture of Conveyance gives the following details:

' . . . messuage formerly in two tenements standing thereon and formerly used as a Grocer's shop and Beerhouse but now used as a Beerhouse only and known as The White Horse.'

Known as 'The Donkey' or 'The Pony' by the locals.
c.1971 the house was demolished and the site became eight houses and a surgery.
Landlord of The White Horse, Maplehurst, RBB and CH.

**Rogate, East Street** Situated on an old coaching route, at one time the inn had a smithy attached. It was a favourite of the poet Edward Thomas (1878-1917).

2006 the inn, which backs onto the village recreation ground, has a darts team and holds quiz nights and folk evenings. 1845-1875 Petersfield PO; 1895-1899 Petersfield K.

**Rudgwick** 1620 recorded as trading in the House of Lords Record Office, no. 33412, Box 2a.

**Slaugham** Trading from the 18th century and possibly earlier, the inn was used for the Manor Courts and also for Vestry meetings until 1780 when the auditor refused to allow any expenses for a meeting place.

1879 the White Horse inn is shown on Sheet 14 of the Ordnance Survey map. The inn with its outbuildings was situated in front of the church adjacent to the churchyard entrance and the lych-gate which was built in 1903. The innkeeper provided essential tethering for the horses of church-goers until the motor car arrived and fewer horses were used.

1922 in order to improve the view of the church Colonel Warren, a local resident, funded the demolition of the inn, its barn and sheds and the re-laying of the electricity and telephone lines underground. At this time the village shop opposite the White Horse was converted into a public house named The Chequers and Mr. King, who had been the landlord of the White Horse, moved in with his family. Mr. Brian Funnell – local historian.

**South Bersted, 39 Chichester Road** 1817, 12th May – advertised For Sale in the *Sussex Weekly Advertiser*.

1842 the tithe map shows this site as three tenements with gardens.

1845 and 1899 the pub was listed under Bognor.

By 2003 it was named 'White Horse and The Trax II'. 1845 P; 1851 PO; 1870-1899 K.

**South Harting** 1851 PO. *This was a misprint for The White Hart.*

**Steyning, 23 High Street** Witches were burned in this ancient town.

1613 William Holland of Steyning bought this large timber-framed posting house with stables at the back, situated on a corner opposite the Old Mint. The stage coach for Brighton stopped here.

*The Shoreham Herald* of February 1955 records the following:

> 1614 the owner of the White Horse Inn, William Holland, 'bequeathed the property to his trustees on condition that the sum of £5 should be paid yearly to the vicar and churchwardens out of the rent, to succour and relieve the poor people who may be inhabitants in the town of Steyning.' This became part of the Steyning Charities, run by the parish Church, which produced about £15 annually to be distributed within ten days of Christmas amongst the town's poor of over forty five years of age. Mr. Holland also endowed the local boys' Grammar School.

1651, August – there is a tradition that King Charles II, having escaped after the Battle of Worcester, stopped at the White Horse for refreshment before departing from Shoreham on board a coal vessel for Fecamp in France.

1706 H.M. Justices of the Peace held their Court in this house.

Every Michaelmas all the men of the Borough were summoned here to their social headquarters for the Borough Court-leet[8] to vote for the two Members.

1790 the inn was re-fronted and a music gallery and assembly room were added. By 1806 there were Barracks in Steyning, and this large extension was used as an Officers' Mess; after the defeat of Napoleon in 1815 the Police Court was held here until about 1834 when the law forbade licensed premises to be used as witnesses were often drunk – Customs and Excise also met here.

By 1835 the inn was owned by the Duke of Norfolk, who sold it in 1869 to the Rev. John Goring. Mr. A.H. King, a licensee at the time, who had to pay more than his annual rent to repair the pump twice, maintained 'he had something to come from the owner'.

1894 The Parish Council took over the running of the Charity.

Charles Bateman, the ostler at the inn, was the grandfather of John Bateman Hunt, local historian and village Postmaster in 2002; he recounts that hay and straw for the coach horses were provided by the landlord and that "The horses from the White Horse were used to pull the Steyning Fire Engine with my grandfather driving them, or holding them back, as they got very excited!"

These may have been greys as it was vital the team could be seen at night when most fires started, due to the open hearths which heated the houses.

1949 on the night of the Steyning Fire Brigade's annual supper, this substantial hotel caught fire and burned down. The present inn is red brick, and consists of the old kitchen, coach house and more modern stabling – converted into a function room; a ring for tying up horses could be seen in the ladies toilet; the site of the original building is now a car park and gardens.

By 1955 the property was owned by the Portsmouth United Brewery Co. who still abided by the terms of the 1614 will, but in the 1960s it seems that the Charity was closed due to this relief no longer being required.

c.1993 the sign showed a rampant white horse, which was to be replaced by a horse pulling a cart.

2004 this Grade II listed building was refurbished.

c.2006 the house won both the West Sussex in Bloom and the Steying in Bloom competitions for the mass of hanging baskets and tubs.

JPe, SAC Vol 43 p.72, FD&EC and JB. 1805 WCO; 1851 PO; 1898 *R.S.O. Hotel* K; 1974 C&L.

**Storrington, 2 The Square** In 1535 it is said to have started to trade under the sign of White Horse.

Sir Arnold Bax (1883-1953), the composer and celtic revivalist, was resident here – where a commemorative plaque can be seen. Some twenty five years after his death in c.1978 a letter arrived at Christmas addressed to his executors, the landlady locked the letter in his old room until after the holiday. When she picked up the letter to send it on, it had been smudged and looked as though it had been trodden on. Several landlords over the years have experienced a Phantom Nudger, this is not an unusual type of phenomenon – an invisible spectre that gives its victim a sharp push or nudge – one landlady was picked off her feet – footsteps have sometimes been heard together with a feeling of depression and a sharp fall in temperature.

The building has been extended to incorporate two large bars, a restaurant and several bedrooms. JPe, WSP, RL and Landlady. 1839 P; 1851 PO; 1877 K.

**Sutton** A village with flint and stone buildings, lying at the foot of the South Downs.

The deeds dating from 1746 confirm that this has always been the village ale house.

During the Second World War members of the French Resistance were billeted nearby; one of their pastimes was playing darts at the White Horse where they spent most mornings until closing time, to the annoyance of their landladies who had prepared lunch for them.

1990 the premises were transformed from a pub to an inn with six bedrooms and bathrooms, keeping the old structure with sash windows and 'higgledy' floor levels. The bar has painted half clad walls and wide planked scrubbed pine flooring. The inn brochure and 'French Resistance in Sussex' by Barbara Bertram 1995. 1851 PO.

**Westbourne, The Square** The old stables, still with the original doors, are used as cellars.

2005 The pub, which had been owned by George Gale and Co., was bought by Fuller's Brewery.

1851 PO; 1870-1877 Emsworth K.

**West Chiltington** 1616 Charles Johnson was landlord.

The reference PRO A/01/2396 has not been recognised by the National Archives at Kew, and the West Sussex Record Office can find no reference to Charles Johnson in Pulborough, Nutbourne or West Chiltington.

**West Dean** 1848 East Dean SAC Vol. 10; 1851 PO; 1870 K.

**West Tarring** 1597 one inhabitant was licensed to sell wine retail, whereas in the 1630s there were four alehouse keepers, an innkeeper and a tavern keeper.

A House called the White Horse in the early 17th Century was presumably the same as the one in the Market Place recorded between 1715 and 1770 in the Manor Court Book. VCH Vol 6 Part 1 (1980) – T.P. Hudson, editor.

## Notes

1. Alluding to Phaeton, the son of the Sun-God Helios, who tried to drive his father's solar-chariot for a day, which ended in disaster. A phaeton is a four-wheeled carriage for personal driving.

2. A light carriage with low wheels having a removable folding hood.

3. A formal renunciation of any claim against a person or of a right to land. Col.

4. In England tenure of land or tenement in a town or city, which originally involved a fixed money rent. Col.

5. A shed or outhouse especially a lean-to, a penthouse. SOE.

6. A wooden post or strip used as an upright member in a fence. Col.

7. A unit of length equal to 5½ yards – about 5 metres.

8. A special kind of manorial court that some lords were entitled to hold. Col.

# TYNE & WEAR

*This metropolitan county is located around the mouths of the rivers Tyne and Wear. Coming into existence in 1974, it was formed from areas of County Durham and Northumberland. However, many organisations, in particular wildlife and biological recording groups and sporting organisations, do not regard Tyne and Wear as a county, preferring instead to refer to the historical boundaries.*

**Hetton-le-Hole** In 1801 the population was recorded as 212 but by 1831 it had increased to 5,887 due to the 'sinking of a colliery'. 'The most extensive colliery railway runs from this place to Sunderland.' JB. 1848 listed under Houghton-le-Spring S; 1855 listed in a local trade directory.

**Newcastle, Groat Market** This small public house, well known for its three dimensional sign – a white horse facing out over Groat Market, stood between The Black Boy and The Lord Chancellor (No. 31). It had a simple front bar with a half-rectangular bar counter which was altered in 1929 to a grander semi-octagonal counter attached to the back corner of the bar.

1933 alterations were made by the new owners, Newcastle Breweries, who replaced the original simple interior.

By 1989 it had been demolished together with The Lord Chancellor, to become Thomson House and Maceys. LFP. 1828/9 P; 1849-1876 S; 1897-1938 K.

**South Shields, 178 Quarry Lane** The two-storeyed house, of Georgian design, has a main double door with carved stone decorations above; built in 1957, its name was taken from the white horse painted on the rock face on the Cleadon Hills, which can be seen from the house. In the bars and sitting room there are murals depicting local landmarks – including the white horse – which were painted at the time by a local artist, Mr. Alf O'Brien.

1997, 26th May – John Mackay, who was the landlord from 1978 to 1990, returned as landlord – he planned to install a 'horse park' with hitching rails outside the pub. SSL, SG of 11.3.1959.

**West Boldon** 'White Mare Pool Inn'. 1857 the 6" Ordnance Survey map gives an area called 'White Mere Pool' to the south of Wardley, east of Heworth and west of Boldon.

1897 the Post Office Directory and the 25" Ordnance Survey map shows this area as 'White Mare Pool' with a 'White Mare Pool Hotel':

' . . . On the road between Newcastle-upon Tyne and Sunderland there existed, before its drainage by a railway cutting, a large pond known as The White Mare Pool; and the public-house close by, which goes by the name, has for a sign the figure of a white mare carrying an old man, portraits, as tradition says, of two unlucky companions who were drowned in the pool many, many years ago.'

By 1919 the hotel had changed its name.

By 1991 there was a 'White Mare Pool' filling station with a 'White Mare Pool' roundabout nearby.

By 2000 the surrounding area had been turned into a golf course and the public house had been re-named The Green.

T&W and NQ 500 Dec.17 1864 – 'The White Mare of Whitestonecliff'. 1834 P; 1848-1876 S.

# WARWICKSHIRE

*'My White Horse shall bite the Bear,*
*And make the Angel fly;*
*Shall turn the Ship her bottom up,*
*And drink the Three Cups dry.'*

The above lines were written by a Warwickshire innkeeper about the neighbouring rival taverns. As a result the other houses lost their custom and the fellow made a considerable fortune.
LH.

Wheat is sown once in four or six years in the rich loams, and much of the barley is made into malt, which is chiefly consumed within the county.
JB.

**Atherstone, 127 Long Street** A coaching inn. 1801 listed in the Alehouse Inn Keepers Recognizance Books. WCR. 1822-1842 P; 1851-1940 K. Hotel.

**Baddesley Ensor, White Horse Lane (to become Newlands Road)** This was a public house in a mining village; the building was demolished after standing empty for a long time before new houses were built on the site. The illustration was done from a photograph taken in 1993. 'A Walk Around With Reg Day' – Local Historian.

## Bedworth

*19 Mill Street* A listed building in a conservation area. c.1995 the owner having died, it was bought by the council who planned to demolish the house for road widening, however this was rescinded and the building was sold to Bansal Trading Co. who renamed it 'JB's White Horse' – the initials of the owner – after refurbishment. 1842 P; 1851-1940 K

*24 Roadway (redeveloped as Park Road)* 1940 Kelly's directory lists Albert Bond Cotter as licensee.

**Chapel End** The Local Studies Librarian wrote 'as to the White Horse at Chapel End, I cannot find any further details, but Chapel End is a couple of miles from Nuneaton, and was a separate place.'

1899, September – an article in *The Nuneaton Chronicle*, 'Pubs of the Past', records: 'The White Horse, Chapel End, where now there is only an off licence.' *This must be a different site to Tuttle Hill, Nuneaton.*

**Coleshill, Coventry Road** The original site of the White Horse is now a private house, which is almost opposite The George and Dragon public house at No. 154. 1851-1940 K.

**Curdworth, Kingsbury Road** c.1714 the notice board in the entrance records that there has been a building on this site since Georgian times when the owners combined roles as publican and farmer.

In 1845 the owner, John Lucas was listed as a victualler and a blacksmith.

1896 Kelly's Directory lists William James Lucas as licensee, he was also a farmer. The Lucas family ran the White Horse until the early 20th century.

Landlord and J. Emberton – local historian. 1872-1940 K.

**Ettington, 52 Banbury Road** There were three farm cottages on this site in the 12th century; by the 17th century two dwellings had been converted into a tavern and in c.1903 another cottage was added.

The house had three 'spirits' – an old man who died upstairs, a dog and Mr. Townley who occasionally pushes people about when he runs about in the bar and kitchen area.

**Henley-in-Arden, 130 The High Street** An hotel. By 2003 the building had been converted into flats, keeping the Georgian façade, and had been renamed 'The White House'. 1828/9-1842 P; 1851-1912 K.

**Leamington Spa, 4-6 Clarendon Avenue** It was for many years a staging post with a stable yard, coaches and horses entered through the archway, which years later became the entrance to the lounge bar. At one time the building became a lemonade factory stretching the full length of the building.

1948 the Leamington and Warwick Magic Society was established, the founder member Steward Williams wrote in his history of the society:

'with the first meetings taking place at the White Horse Hotel, where the landlord allowed us the use of the Smoke Room for our business and where, business concluded, we would regale the customers with close-up magic. After advertising for new members, and receiving far more responses than they had expected, the landlord of the White Horse immediately made the upstairs function room available to us for our regular meetings.'

2008 the pub was a popular place for the young, referred to as 'The Leamington Youth Club', with musical evenings most nights. There are two bars, the smallest in the front being the oldest. For many years there has been a weather-beaten stone statue of a lady in the back yard. 1842 P; 1900-1940 K.

**Lowsonford, Lapworth Road** Dates from the 15th century. Became a private house, situated opposite the Fleur de Lys. 1912 K.

**Nuneaton, Tuttle Hill** The *Nuneaton Chronicle* of September 1899 reports that 'at Nuneaton Fair, which in the good old times lasted for three days, tents used to be erected in the Market Place, and beer was sold from them. A branch of oak hung outside indicating to the thirsty that there they could be refreshed, and the tents themselves going under the name of bower houses.'

Originally this inn was a coaching house.

The *Nuneaton Diary* for January 1835 records that a poling booth – for the Northern Division of Warwick - was set up for the general election outside the White Horse, Tuttle Hill. 'But little sensation produced here from the dismissal of the Whig Ministry and the assumption to power of the Duke of Wellington. A seeming apathy or doubt with all parties.'

*1835-1841 The Duke of Wellington was leader of conservative opposition in the House of Lords.*

The 1851 census for The White Horse, 132 Bar Green lists Thomas Cox, Victualler, aged 62; his wife Mary, aged 69; their son Charles, 22 and daughter in law Sarah 21 – both shown as servants – and a grand daughter, Mary, aged 4. *It seems that Bar Green was a local name for this area around Tuttle Hill and Punch Bowl Bridge.*

c.1990 the name was changed to 'The Crazy Horse', a Chinese and Cantonese restaurant and nightclub.

By 2003 the name had become 'The Red Ruby'. Nuneaton Sorting Office, NL, NC, ND and DNB. 1822-1842 P; 1850-1880 Barpool WCC; 1851 Barpool – 1940 K.

**Rowington, Old Warwick Road** A coaching Inn. The building, thought to have once been thatched, now has a tiled roof. The house was renamed The Cock Inn, but by 2003 it was known as The Cock Horse. 1872 K.

**Rugby** *This could be an error as no White Horse Inn has been located in the vicinity of Rugby c.1900.* 1900 K.

**Sherbourne, Vicarage Lane** Situated just off the A46, midway between Warwick and Stratford-upon-Avon. At one time this was the home to the Canon of St. Mary's Church, Warwick. Its present name (The Old Rectory Hotel) is a reference to this. The building is Grade II listed and dates back to the 17th century when it was a timber-framed house encased in brick. An additional wing was built during the 18th century. It has steeply pitched tiled roofs and gable ends; the front door is pilastered[1] with a hood on console[2] brackets; inside there is an old chamfered ceiling beam.

By the middle of the 19th century it was known as the White Horse Inn. At some time it was a farmhouse named Sherbourne Corner and for sixty years during the 20th century it became a rectory and was occupied by clergy.

By 2006 it had become The Old Rectory Hotel and was described as being classically furnished with a selection of four-poster beds.

Christine Gore – owner. 1842 Warwick P; 1851 Warwick K.

**Shipston On Stour, Church Street** Was refurbished as an hotel in February 2002 and renamed 'The Falcon'. 1828/9 Excise Office-1842 Posting and Excise Office P. Hotel.

**Stratford-upon-Avon, 6 Chapel Street** Although not trading as the White Horse until 1828, the history of this building dates from 1504 when the White Bear Inn stood on this site. No history is recorded for a hundred years until 1602 when it was rebuilt by Thomas Mortiboys.

Between 1721 and 1730 it was recorded as 'together with No. 5', whilst in 1727 and c.1750 two wills described the property as 'a messuage with a gateway and chambers the same adjoining', or with 'gateway and chamber above'.

1774 the licence ended and the house was sold to George Cooke, a joiner.

1806 it became a butcher's shop owned by Job Archer.

1827 Job Archer became bankrupt and a year later it was bought by Ann Court, who leased it to Job's son, Edward – he converted it back to a public house and opened it as The White Horse.

1850 following Ann Court's death, again the license ended.

From 1853 to1890 the three owners are listed as butchers; possibly combining as publicans, as in 1872 Kelly's Directory lists the building as a White Horse Inn.

Shakespeare Birthplace Trust. 1842 P; 1872 K.

**Warwick** There is an undated ¼d token issued by John Jackson of Warwick – *a horse. It is possible that this referred to the White Horse at Sherbourne.* GCW No. 185.

## Notes

1. A square or rectangular column or pillar. SOE.

2. An ornament in any material which projects about half its height or less, for the purpose of carrying anything. SOE.

# WEST MIDLANDS

*The Industrial Revolution in the 18th century could be said to
have been inspired by the remarkable men – household names
to this day – who founded the Lunar Society[1]. Meeting at the
time of the full moon, in order to see their way along the dark
unpoliced roads in and around Birmingham, they met to discuss
a broad spectrum of revolutionary ideas . . .*

*The many miners and factory workers, in sometimes wretched
conditions, suffering from the excessive heat in the work places
meant that the taverns of small hamlets were in almost constant
use from early morning.*

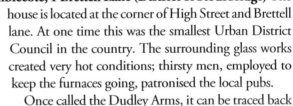

**Amblecote, 1 Brettell Lane (District of Stourbridge)** The
house is located at the corner of High Street and Brettell
lane. At one time this was the smallest Urban District
Council in the country. The surrounding glass works
created very hot conditions; thirsty men, employed to
keep the furnaces going, patronised the local pubs.

Once called the Dudley Arms, it can be traced back
to the 19th century, but was possibly trading in the
late 18th century when the road from Stourbridge to
Wolverhampton was turnpiked, it became one of
the principal hotels for the Amblecote tramway junction.

The inn is large and unusually shaped, but was once much larger, as it occupied the
adjoining buildings. The archway leading to the old stabling yard is a legacy of this.

There is a legend that a landlord turned away a starving beggar one cold winter's
night, the beggar was found dead in the road next morning – the house then
became known locally as The Starving Rascal or The Starver.

1834. Benjamin Hyde was listed as publican.

c.1880 Peter Williamson was Publican. Not long after taking over The White
Horse, he was hauled in front of local magistrates for permitting bagatelle on his
premises; his license was not endorsed.

1887 the owner of the pub (John Guest of Brettell) sold the premises to the
Holt Brewery Company.

1891 James Cook, manager for Holt's, was convicted of selling liquor to a
drunken person. He was fined 25 shillings and costs.

c.1890 Holt's rebuilt the White Horse, but they must have been disappointed
with their return of capital because they sold the building to Thomas Higgs.

1904 Thomas Higgs became the third publican to appear in front of the local
magistrate: On June 13th he was convicted of 'delivering beer to a child under 14

in a vessel that was not corked and sealed'. It is thought that the boy was probably sent for liquid refreshment by workers at the glass factory, so the conviction was perhaps harsh. Higgs was fined 25 shillings plus costs.

1906 a photograph from this year shows a window in the pub bearing the slogan 'Home Brewed Ales', suggesting that Higgs was brewing up behind the White Horse Inn. This tradition continued until Mitchell and Butlers acquired the Pub at the end of World War One.

1907 the hotel was advertised as 'offering every accommodation travellers and cyclists, and good stabling'.

1919 Charles Waldron, licensee, fined £10 plus costs by local magistrates for selling rum above the maximum price.

1942-44 the pub was popular with the American soldiers, whose base was nearby.

1974 an inn sign was designed by G. E. MacKenney, one side depicting a beggar in the snow being turned from the door and on the other side a ghost enjoying a pint in the pub. The new sign was dedicated by Ray Barlow, a former West Bromwich Albion star – he owned a newsagents close by.

1992 bought by Mark Boxley and Timothy Poole, who converted the White Horse into The Maverick Drinking House. They gave the pub an American theme with a lot of interior woodwork and decorated it with Western memorabilia, 'in a bid to capture the frontier spirit' from the Second World War: 'The bar provided the visitor with the feeling of being in a U.S. cavalry tent'. At one time the ceiling was covered in canvas but fire regulations forced them to take it down. The pub was also modernised, with the addition of a second bar, beer garden and patio.

1999 the house was purchased by Enterprise Inns.

2006 the Maverick's unique interior attracted the attention of BBC programme makers and the pub was used as a location for 'The Lightning Kid', directed by Stephen Tomkinon. It is a drama centred on a man obsessed with the Wild West.

By 2008 the name had been changed to 'The Starving Rascal', the inn sign had been repainted, though the artist attempted to keep to the original design; designs for the original inn sign were displayed in the pub. The exterior of the listed building had been painted, but the interior had not been altered.

JH, Licensees of the White Horse Inn and John Emberton. 1822 3 listed – 1842 P; 1851 2 listed – 1936 K.

**Balsall Common, Kenilworth Road** The pub was originally named The Plough and built as a two up, two down cottage selling ale or beer. It stood in the centre of the common which was thought to have been very wet and boggy. Later however, a road with a Turnpike was built.

The Plough was the venue for the annual 'Balsall Wake' which was held on the first Monday after the 15th August. Local women raced from High Cross to the inn, the prize being a quarter pound of tea; the men bowled in the field behind the inn for a duck.

One inn-keeper of The Plough, Mrs Matthews, had a husband known as 'Stocking Joe', because he sold stockings and braces around this area.

By 1854 John White was recorded as a beer retailer at the White Horse; however by 1872 Mrs Mary White was listed as the licensee.

By the end of the 20th century, The White Horse, owned by Spirit Group Ltd., had become a large building bearing no resemblance to the original.

December 2001, the identity of the pub changed, from being a Big Steak pub to a Two for One venture.

2008 the White Horse was advertised as a restaurant.

From the Solihull Local Studies Index. 1872-1940 K.

## Bilston

*92 High Street* 'The vicinity abounds with numerous forges, furnaces, steam engines, and various manufactories, the smoke of which darkens the air by day, and the flames of which serve to illuminate the night.' Snuff and beauty patch boxes were also manufactured here until c.1850, they were made of gold, enamel and painted copper. c.1960 once again a company was formed to manufacture Bilston enamelware.

1734 listed as trading with John Cooper as licensee.

1818 listed as trading in Parson's Directory with Abraham Cross as licensee.

1834 Thomas Cardell was listed as landlord, Sarah Cardell took over in 1841; the owner was James Cardell, gentleman, of Birmingham.

1851 Slater's Directory published an advertisement for John Corbett of the White Horse inn – a Wine and Spirit Merchant, also an Auctioneer and Appraiser.

1864 Samuel Butler was recorded as landlord; his son, also Samuel, succeeded him in 1873.

1875, 10th February – the Midland Counties Evening Express published the following article:

> 'Charles Taylor was summoned for being disorderly in the White Horse and for assaulting Samuel Butler . . . and James Butler, his son, on Monday, 18th January. Mr Barrow, who appeared for the complainant, was about to make an opening statement, when Mr Spooner requested him not to do so, as the defendant had no attorney. The Defendant said he had engaged Mr. Bowen, and paid him a fee of 15s. but he now refused to appear in the

case, and declined to give any reason why he did so. He wished to know if he had no remedy in the matter. Mr. Spooner said the defendant could take proceeding in the County Court if his statement was correct. Mr. Bowen shortly afterwards entered the Court, and on being informed what had occurred, intimated that he refused to appear without an 'extra fee', as he had already attended three times and the case had been adjourned. The Defendant agreed to pay an extra fee and Mr. Bowen undertook his defence . . . The whole of the summonses were dismissed.'

1882, 27th February – Samuel died and his wife Sarah Fanny Butler took over as licensee, she was also the owner.

1896 Kelly's Directory lists the inn at 42 High Street.

1898 Miss Elizabeth Evans was licensee, the Evans family ran the house for the next twenty one years – Edward Evans was the owner, his wife, Susan became licensee in 1906, she was also a brewer.

By 1912 the address was 92 High Street.

1914 listed as trading in Bennett's Business Directory.

1919 Edward Evans died after slipping in the malt room, he was aged 62. Oliver Brown, a brewer took over the licence which he held until his death in 1932.

1938 Joseph William Pridmore, the licensee also worked part time for Collins the undertakers.

By c.1970 the pub was demolished, along with the whole of this end of the High Street to make way for the construction of a new road scheme and a supermarket.

1822-1842 P; 1860 K; 1880 2 listed PO; 1900-1936 Daisy Bank K.

**24 Temple Street** 1916 licensed as a beerhouse with Edward Roberts as landlord, he remained here until 1940.

**42 Ward Street, Priestfield** One owner, Henry Williams was a fishmonger in Wellington Road, Bilston.

1873 listed as trading with Edward Jones as licensee.

1917, 6th February – Thomas Bates, the landlord, was fined 15s. for selling ale after 8.30 pm.

1919, 24th July the licence was refused by the Compensation Authority.

1921, 24th December – compensation of £915 was paid.

JB, Jack Haden and Mr. John Emberton – local historian.

*The White Horse pubs at Sedgley, Coseley and Bilston were all within 4 miles of each other.*

**Birchills, 19 Green Lane** During the 19th century this was an important metal working area of Walsall.

The pub, which had its own brewery and malthouse attached, is situated opposite Alma Street. In 1861 the owner was John Gregory, a victualler from London.

1899 Wolverhampton and Dudley Breweries bought the house.

By 1925 the premises were recorded as being old and in poor repair, there was a brewery and malthouse, both in a dilapidated condition, and a club room which was unfit for use. Cornelius Williams was the tenant who paid a rent of £60 per annum.

In 2008 the pub was closed.

J. Emberton – local historian. 1860 K; 1880 PO; 1900 - 1936 K.

## Birmingham

In 1643 Prince Rupert set fire to the town and levied a tax on its inhabitants in aid of the Royal Stuart cause. In consequence there was great patriotism and loyalty to the House of Hanover whose heraldic sign is a galloping white horse, which might account for the number of White Horse public houses at one time in this city. JAL and BPL.

*Congreve Street* Between 1770 and 1777 the street name was officially recorded as Friday Street.

This old inn dates from the time when the Colmore Estate was broken up in 1778, it was situated in a derelict parkland at the corner of two new roads that intersected what had long been a deserted rabbit-warren with a stile path. Before the end of the century the inn was described as being well appointed with ample accommodation and stabling; Benjamin Jeavons was landlord 'whom we must picture to ourselves with his three-cornered hat, his knee breeches and buckled shoes'.

With the steady increase of traffic along the new Dudley Road and the traditions of its smoke room, the White Horse soon became known as a popular market day hostelry and was the social centre for the neighbourhood for a hundred years, with many public activities. Its customers came from all walks of life and were described as men of taste and culture – James Watt's house at Harper's Hill was said to be visible across the fields.

By 1805 four important businesses had been established in Congreve Street – a merchant, a button maker, a metal dealer and an engraver. Tradesmen of this street formed an association called the 'Old Congers' who held their annual dinner at The White Horse.

The Inn's reputation continued to grow under William Holloway, who besides being the innkeeper was also a coal dealer; the nearby canal[2] would have facilitated the movement of the coal.

By 1828 Mr. Holloway had died as Pigot's Directory lists his widow, Elizabeth, as a victualler and a coal dealer at Old Wharf. 1830 the inn was sold by auction.

The sale plan gave Friday-street rather than Congreve Street. The ground rent was £7 7s. 6d. per annum and the expiry of the lease, 1867.

1834, 31ˢᵗ January – a notice advertised the Annual General Meeting of a Book Society, possibly the same as The Birmingham Shakespeare Reading Club which at one time had made the White Horse its headquarters; The Sociable Club, a Bohemian brotherhood also had its meeting here.

From 1845 the house was advertised as a 'Commercial and Market Inn'.

By 1851 the White Horse was known for its homeliness and its skittles alley. Described as a 'Birmingham house of embarkation', it was the nearest licensed house to the embarking point of the docks on the canal, which before the railway was a ready means of communication with Wolverhampton, the journey taking about two hours.

It was well known to those travelling by horse and vehicle from the countryside and was also one of the leading rendezvous for actors, both professional and amateur.

1869 the lease having expired, the landlord paid an annual rent of £135.

c.1870 the house was advertised as 'a centrally situated hotel, three minutes walk from the railway station, with stabling and a lock-up coach house'. It became famous for its 'daily Ordinary', served at half past one in an upper room and costing eighteen pence; it comprised soup, a fish, a side dish, a joint, with pastry, cheese and salad to follow. This became so popular that an additional larger room was built out into the yard at the back – some people dined here day after day for years; one colossal gentleman, who never ate less than three servings of every dish, had to sit on two chairs.

From 1872 the Railway Guards Annual Dinner was held here for over thirty years, as were Volunteer banquets and dinners of the Postal Telegraph employees.

1884 Joseph Bailey, landlord for about twenty years, died.

1889 his wife, Margaret, and two sons acquired the property as a freehold. Margaret, 'an excellent business woman of ladylike address' remained a tenant for life, the consideration money being £10,300.

1890 Margaret Bailey died, she and her husband were remembered for their free Christmas dinner, the principal dishes were a boar's head and a round of beef.

1905 the old premises were sold at public auction and purchased by brewers, Mitchells and Butlers Ltd. who submitted plans for a large new building. The justices demanded that four licenses be entirely surrendered before the alterations could be permitted.

The exterior of the new building was in Renaissance style with a clock tower on the corner. The main entrance had a fine wrought iron awning with a gate which was lowered into the basement during business hours and raised when the house was closed. In the archway, a Venetian glass mosaic pane, depicting a white horse, was from a sketch designed by the Leamington artist, Basil Nightingale. The new entrance hall, with Majolica tiles, was 12ft. wide with a fine staircase leading down to a billiard room and grill room with mosaic pavement, teak counter and oak fittings; the restaurant, 48ft. by 26ft. was fitted with oak inlaid panelling and an ornate ceiling, with an oak parquet floor; the smoke room, also of oak measured 39ft. by 22ft.

The public bar being designed by Wood and Kendrick included a brilliant-cut glass[3] partition, a Terrazo marble floor, the counter area was of fine Spanish mahogany and tile-covered walls and ceiling.

Upstairs there was the assembly room, the coffee room, the original dining room which was hardly altered, and a reading room.

The up to date kitchen was on the second floor and measured 40ft. by 23ft., it comprised the pastry department, the chef's office, and the refrigerating department.

'A Waygoods best electric passenger and goods lift has been erected, which commands every floor, in addition to two special electric service lifts.'

1907 a number of societies were recorded as meeting here, namely The True Blue Society, the Amalgamated Society of Carpenters and Joiners, and several pigeon fanciers clubs, such as the Columbarian[4] Society, the Central Counties Flying Club and the Long-Faced Tumbler Club.

c.1966 the public house was demolished and part of the site became the Central Library. The Birmingham Evening Despatch 2nd December 1907.

*Now, all that is left to remind us that Congreve Street ever existed is Congreve Passage that leads off Chamberlain Square.*

**106 Constitution Hill** 1801 listed in the Alehouse Inn Keepers Recognizance Books. 1822-1842 P; 1851 K; 1856 PO.

**80 Constitution Hill** White Horse Cellar. 1937 K.

**Lower Essex Street** 1937K.

**Steel House Lane** 1801 listed in the Alehouse Inn Keepers Recognizance Books.

**William Street** 1964 a photograph of the building shows a side yard with a low brick structure, which was almost certainly the gent's toilet, or urinal. WCR, KT.

**Brierley Hill, High Street** 1818 the pub was listed with D. Taylor as licensee.

1857 William Boyd was listed as landlord; by this time it had a concert room. The house was still trading in 1938.

J. Emberton – local historian.

**Brownhills, 10 White Horse Road** Situated on the corner of Chapel Street. The road which was cut by c.1834 did not appear to lead to any buildings; it is thought to have been used to serve loading bays on the canal.

1861 The original pub building opened, with Samuel Bickley as its first landlord.

1896 listed in Kelly's Directory with Charles Alltree as licensee.

By 2008 the road was named 'White Horse', the house had been modernised and was surrounded by buildings. WLH and J. Emberton – local historian. 1936 K.

**Coseley, 2 Upper Ettingshall Road** 1835 and 1841 listed as trading under Bond Street and Brierley Lane.

1851 listed as trading in Slater's Directory with Ebenezer Stanley as licensee, he was also listed as a shopkeeper in 1854.

c.1892 John Fellows was listed as licensee, Telia Millicent Fellows took over the licence in 1910 but only for one year.

In about 1930 a customer, later accused of murder used the White Horse as an alibi in his defence during his trial.

1961 the inn was modernised.

J. Emberton – local historian. 1860-1936 K.

## Coventry

The street pattern of central Coventry is still similar to that of the 13th century; this includes Bayley Lane and Hay Lane.

There have been four inns with the sign of the White Horse within the boundaries of the City of Coventry, possibly due to the legend of Lady Godiva (c.1040-1080) processing through the streets on a white horse – re-enacted in 1678. Col. and DNB.1822-1842 P; 1851-1940 2 listed in 1872, 1900-1912 K.

*29 Bayley Lane* Situated half way between Hay Lane and Pepper Lane.

1733 the first reference to the inn was given by James Hewitt, Viscount Lifford (1709-1789) , MP for Coventry in 1761.

1874 listed as trading.

c.1932 the house was closed.

CA, JA and AEW. 1822 P.

*49 East Street / Payn's Lane* T.W. Whitley gave a description of the street in his 'Humorous Reminiscences of Coventry Life, Coventry Coaching and Coach Roads, and Other Local Works' published in 1888:

In 1830 'a few houses existed in East-street, which not being made, its surface was in such a deplorable condition that the miller's wagons oftentimes stuck in the ruts up to their hubs . . . '

By 1845 the tavern had been renamed the Weavers Arms.

1864 Henry Pepper, the landlord, grew a pear tree in his garden which had very large pears, the largest weighing 1lb 5oz, it was 8 inches long and had a girth of 13 inches.

From 1865 the house was called the Bunch of Flowers.

1866 listed as a tavern.

By 1874 the premises were listed as White Horse.

1940, 15th November – it closed having been badly damaged by enemy action.

1941, April 8th/9th – the building was totally destroyed by enemy action. The landlord later took the licence from here to the White Horse, Henley Green.

CA, JA and AEW.

**10 Hay Lane** In 1684 it was already a large busy inn with stabling, owned by Mr. Hobson, a draper, and Mr Jesson, a wealthy man, who during the year sold the premises to Mr. Walter Gilbert. The inn, conveniently situated close to Drapers' Hall, was subsequently owned by a number of people in the cloth trade; the first recorded tenant rented the property for £38 per annum, payable in March on the feast of St. Michael and in September on the feast of the Annunciation.

The cellar was described as 'well constructed with brick shelves and wine racks'.

A year later in 1685, Mr. Gilbert sold the inn for £122 3s. 0d. to Gabriel Cotternes, a yeoman, whose family remained the owners for almost eighty years.

1719 the tenant, Thomas Comley, an apothecary of Stratford-upon-Avon, bought a share in the property for £300.

1748 the inn is shown on Bradford's Map.

1745 during the Civil War troops were garrisoned in nearby St. Mary's Hall, it is likely they would have used the inn for stabling and refreshment.

1750 a new tenant, John Evans, rented the inn for £46 per annum on a twenty one year lease; he sublet part of the inn to the Wood family.

The inn was described as having several public rooms, most of which had paper hangings and a wainscot. 'The Groom Room, where the servants would have eaten, the Kitchen and the Kitchen Parlour had a chimney and a mantle piece. There were closets in the Street Parlour and the Kitchen Parlour. A bacon crack (sic.) was provided in the Kitchen and in the Pantry was a dresser and shelves. There was also a Red Room and a Dining Room. The yard contained lead pipes and a brass cock. At this time even the water supply in London was not so advanced. This same pump was used in 1770 to earth the new lightning conductor when it was installed on the spire of St. Michael's. The stables were furnished with six stands. The yard had two entrances, at the side and the back. Both had iron gates and an inn sign. It must have been a delightful and well appointed building.'

1756 Eight soldiers were billeted here.

1759 the premises were first used for auctions.

1762 a sergeant of the King's Royal Musketeers was summoned for stealing articles from an ostler employed here.

1763 the property was sold to Charles England, a glover, wine merchant and land owner, for £520 and his tenant, James Burrows, ran the inn for about forty years.

1770 the weathercock on nearby St. Michael's Church was reguilded and the spire repaired; an order of vestry was made:

> 'that a lightning conductor be put up the steeple, and to terminate in the well of the White Horse pump. The pump here referred to is that which stands near St. Mary's Hall, and was, at the date mentioned, at the back door of the White Horse premises which extended to there.'

By 1772 the White Horse had become a staging post for the Leicester to Bristol coach run – some 118 miles.

1777 Mr. Rew began a Post Chaise Service from here to nearby Warwick.

1786 records show that the premises were still being used for auctions.

1804 two horses were suffocated when the stables caught fire.

1808 the property was sold for £535 to Thomas Wilmot, a benefactor of Coventry, whose family remained here for some sixty four years.

1874 the premises were sold to Thomas Dewes, a solicitor, for £1,550; he had his office here.

By 1923 it was known as The Council House and had been made into two premises, No. 10, the Town Clerk's Department and No. 10A which became the Local Government Officers' Social Club.

By 1979 the building had become empty. CA, AEW, CLI, BPo and SVH.

**Henley Green, Broad Park Road** Sometime after 1941 the original licence was transferred by the landlord of the White Horse in East Street, which had been destroyed by enemy action on 8th/9th April 1941.

1940 listed in Kelly's Directory under Bell Green.

1961, 13th December the evening opening of the new White Horse was covered by local television.

**Cradley, 70 High Street** The White Horse was originally two small cottages in a cobbled street with outhouses at the back, to become double fronted with two entrances and a brewery at the rear; it was very popular with local colliers who would often be seen drinking at six o'clock in the morning. 1830 Joseph Attwood, from a Cradley farming family, turned his nearby High Street butcher's shop into a beerhouse selling ale and meat; it was known as the White Horse. In 1842 he died leaving his house, slaughter house, stable, outbuildings, shop fixtures, yard and pigsty to his son William.

c.1870 listed as a beerhouse at 41 High Street. A year later the White Horse had become a grocery shop and beerhouse.

1881 William Attwood's son, John, had become the owner of the White Horse; he also owned a butcher's shop in Cradley, where he preferred to work, he therefore employed Charlotte Batham to brew the beer and by the following year her husband, Daniel, had become the owner and landlord and continued to run the house for forty years.

A sales ledger from 1897 to 1901 records that the beer from the White Horse sold right across the Black Country and the Midlands. By this time there was a licensed house in Cradley for every seventy four residents – twenty taverns and inns and forty two beerhouses.

1906, 18th September – Charlotte Batham died.

1922, 15th July – Daniel Batham Senior died and the premises were put up for sale:

> 'entrance hall, with out door department, side hall, large well fitted front bar 26 ft – 16 ft; two smoke rooms, large clubroom, three large cellars, two wine cellars with rolling way, one bed-sitting room, three bedrooms, clothes closet, living room and scullery. The outbuildings – two storeyed brewery, large malt and hop room, coach house, two stall stable with loft over all. Piggeries and usual conveniences and garden with cartway entrance from High Street right through to New Street at the rear'.

On 8th November the premises were sold by auction to Hezekiah Dunn for £3,455. His family ran the house for the next fifteen years.

c.1930 central heating was installed and the central door was blocked off.

1934 Hezekiah was the chairman of Brierley Hill and District Free Home Breweries.

1937, 2nd February – Amy Dunn had leased the old beerhouse to Arthur Joseph Batham – *perhaps the son of Charlotte and Daniel* - for £119 12s. 0d., who later bought it. By this time the premises included a smoke room, back smoke room, club room, sitting room two bedrooms, two cellars and a garage. The bar was described as 'exceedingly well made, with a mahogany back fixture 17ft x 10ft with numerous embossed and engraved panels and undershelves.' A painted and grained window board measuring 7' 9" x 2'6" advertised 'Home Brewed Ales' and 'Stout and Cyder'.

1949, 1st February – a full licence was granted and in October it was sold to Wolverhampton and Dudley Breweries.

1971, 31st January – the house closed, was subject to a Compulsory Purchase Order and demolished to be rebuilt as garden flats; by 2003 these were being demolished. JRi, J. Emberton – local historian and Mrs. Carole Neale. 1828/9 P; 1912 2 listed – 1940 K.

## Dudley

***5 / 6 New Mill Street*** 1860 listed as trading with Richard Wilkinson as licensee, he was also a vice maker and issued tokens from here.

1868, 11th July – the *Dudley Herald* reported on the anniversary of No. 6 Lodge of Loyal Good Fellows held at the house of Richard Wilkinson of the White Horse Inn in Mill Street. They also met here on 23rd January the following year.

1872. 10th February – the *Dudley Herald* advertised the White Horse for sale. It was described as an old licensed public house comprised of a front bar, tap room, smoke room, three chambers, club room, extensive cellaring, brewhouse with malt room over, a yard and a coal house. It was still in the occupation of Richard Wilkinson at a low rent of £18 per annum.

1895, August – an objection was made to the renewal of the licence to G.H. Brain because the premises were being used for betting purposes.

1938 the licence was transferred to the Hill Tavern.

1842P; 1872 PO; 1900 2 listed – 1912 K.

***Queens Cross*** 1819 William Powers was listed as licensee; he had died by 1851 as the Census lists his wife Hannah as a widow and publican. They had five children – William, a vice maker aged 20; Thomas, a blacksmith aged 18; Richard, a blacksmith aged 15; a daughter Anna aged 10 and George aged 8. They employed a servant, Maria Sloya aged 14; there were three children staying in the house, Thomas Brooks aged 8 and George and Elizabeth Brooks both aged 4.

1854 Hannah Powers was listed as a beer retailer and blacksmith and in 1860 she had become a farmer, shoeing smith and beer retailer.

***Woodside, Dudley Wood*** 1864 listed as trading with John Williams as landlord.

1867 records show that an inquest was held here. J. Emberton – local historian.

**Harborne, 2 York Street** Originally built as two cottages which were converted into the public house in 1861.

1914 listed as trading in Bennett's Business Directory.

During the latter part of the 20th century a bank robber hid in an upstairs bedroom after a bungled attempt at robbing a bank.

1997 a major refurbishment was carried out by the brewers, Ansells, but leaving the frontage untouched. The interior being dark wood and the floors bare boarding; the walls had a smoke stained effect.

The house is situated on the infamous Harborne Run, being one of a number of public houses visited by the many freshers of Birmingham University.

2003 for the sixth year the public house had been included in The Camra Good Beer Guide.

2007 Considered a "truly traditional locals pub." The landlord has a list of publicans dating back to the start. C. Marlow – landlord and J. Emberton – local historian.

**Lye, 16 Cross Walks Way** Apparently this was once a rough area where nail makers lived.

1873 listed as trading with William Smith as licensee; by 1892 Joseph Smith had become the landlord.

Between 1916 and 1921 the licensee was George Underwood.

c.1965 the house was pulled down and rebuilt further down the road, but by 2007 the pub had closed and the whole area had been redeveloped.

JH, Landlord and J.Emberton – local historian. 1828 P; 1912-1940 K.

# Netherton

**6 *Windmill End / Bumble Hole*** This is a suburb of Netherton.

1870 John Taylor was listed as licensee, he was also a nail factor and issued tokens from here.

1873 John Farmer, a shopkeeper and nailmaker was listed as licensee.

1874, 4th April – *The Dudley Herald* published a notice that the White Horse at Bumble Hole, Netherton was to be sold by auction with the adjoining building.

1881 the census lists the address as No. 12 Windmill End. The licensee, John Farmer, now aged 44 is described as a grocer and licensed victualler, his wife Charlotte is the same age. A domestic servant, Emily Taylor aged 22 also lived here.

1901, January – the premises were put up for sale.

1909 the address had changed to No. 6.

1912, 19th July – a licence renewal was refused.

The Compensation Authority claim was £855 17s. 6d., an offer of £550 was accepted.

1913, 22nd December – the pub closed, the licence being extinguished.

**62 *St. Thomas Street*** Recorded as trading before 1873, with a brewery attached, when John Cartwright was the owner, he also became the licensee.

1873, 17th May – the *Dudley Herald* published the following notice concerning the White Horse:

> 'Instructions from the trustees under the will of the late John Cartwright . . . To be sold by auction . . .
>
> Lot 2 – That capital freehold public house, the White Horse, with the brewhouse, outbuildings and appurtenances thereto, formerly occupied by the late Mr. John Cartwright, deceased.'

1938, 28th April – the licence renewal was refused and was extinguished on 31st May the following year. 1872 PO; 1900-1912 K. MWF and J. Emberton – local historian.

**Norton, South Road** 'The Old White Horse'. At one time there was a relief of a sad-looking white horse over the front door instead of the usual inn sign.

c.1930 the house was rebuilt and the name was changed to 'Old' White Horse.

1972 Newly formed Stourbridge and District Cricket Society held their meetings here until 1985.

2003, 28th July – The pub re-opened after alterations. A stained glass window depicting a white horse was destroyed while demolishing a wall. JH and Mrs. Carole Neale.

## Oldbury

***22 Church Street*** 1872 PO; 1900 3 listed – 1940 K.

***50 Newbury Lane*** 1850 listed as trading in the Post Office Directory with Benjamin Law as licensee.

The house was rebuilt during the 20th century and the name was changed to The Doll's House.

By 2008 it had been renamed the Valentino Restaurant but was demolished during the summer of this year, the site to become a nursing home.

J. Emberton – local historian. 1822-1842 P; 1851 K. 1886 S; 1872 PO.

**Quarry Bank, 166 New Street** Originally two cottages with outhouses at the back where animals were obviously kept as the water troughs and cobbled yard still exist. The old bowling green, which had been installed by 1892, is still in use.

This was a region of nail makers and colliers who had a reputation for their ale consumption.

1851 listed as trading with Joseph Parrock as licensee.

1864 William Tranter was listed as licensee, his family ran the house for the next 42 years. His wife Emily bought the premises, she had taken over the licence by 1870.

1916 Issacher Willetts was listed as landlord, he was also a chainmaker; his wife Lily took over the licence for eight years in 1920.

During the 1920s the pub ran a successful bowling club.

1996, July – Gordon Beasley, the licensee, was shot dead in Johannesburg, South Africa, while tackling intruders.

Between 1999 and 2006 bowling teams from here were the Lye and District Bowling League champions.

J. Emberton – local historian. 1860-1936 K.

**Sedgley, 12 Dudley Street** 1868 the landlord, E. Stanley, died and Moses Stanley took over as licencee.

1904 listed as trading with a beerhouse licence.

1908, 4th March – the house was granted a billiards licence.

1995, August – the pub reopened after being refurbished at a cost of £60,000. Landlady maintained "Definitely pre-war."

**Selly Oak, 29 Chapel Lane** 'White Horses'. 1884 a beer retailer was recorded as trading here.

1892 first listed as 'White Horse' with William Caesley as landlord. After this date it was always listed as 'White Horses'. 1973 K.

**Smethwick, Soho Street** By 2003 the public house had closed.

**Sutton Coldfield, Whitehouse Common Road** The White Horse, a coaching inn situated in a residential suburb of Sutton Coldfield, was listed as trading in Kelly's Directory of 1896 with Miss Ann James as licensee.

During the 20th century the stables and bowling green were converted into an hotel.

2005 the house was completely refurbished; it was redecorated again in 2006 and ceased to be known as 'Miller's'. J. Emberton – local historian. 1872-1912 K.

**Tipton** 1860 K.

**Wall Heath** 1850 listed as trading with William Worrall as licensee. The 1851 Census records him as a licensed victualler aged 45. He had a wife, Ellen aged 35 and a son James aged 11.

By 1854 John Legge had taken over as landlord. J. Emberton – local historian.

## Walsall

1822 Pigot's Directory lists a White Horse public house but gives no address in Walsall.

*Mount Pleasant* Listed as trading in 2003 and 2006.

*Park Street, 41 & 18* 1834 listed as a beerhouse owned by James Thomas.

1845 James Thomas was licensee, he was also a provision dealer.

1851 listed as trading with Henry Slater as owner.

By 1903 Butlers, the brewers, owned the house with Alfred Medlam as manager who was paid £3 10s. 0d. per week.

1864 Nicholas Roper, a maltster, was licensee; he issued tokens from here.

1902 G. Thomas, the licensee, was recorded as a committee member of the Walsall and district Licensed Victuallers' Association.

1912 the manager, George James was paid £4 per week.

By 1939 the house was owned by Atkinsons, brewers, who paid Frederick Day £3 per week which included the services of his wife.

1958, August – the house was closed. 1860 K; 1880 PO; 1900-1936 K.

**92 Wolverhampton Street** 1834 listed as a beerhouse in Wolverhampton Lane.

1845 listed as trading with William Oakley as licensee, he was also a tailor.

1851 listed as trading with Samuel Cartwright as owner.

1896 Job Sutton, a coal dealer, was listed as licensee.

By 1901 Edgar Evans of Wellhead Brewery at Perry Barr had become the owner with Emily Bliss as the tenant.

1933 Mr. G. S. Twist opened his White Horse Brewery adjacent to the house at No. 92, with Charles Pedley as manager of the pub. The famous White Horse minerals continued production until 1975.

1943, April – Twist Brewery Ltd, White Horse Brewery was registered as a private company.

1950 the brewery and the pub were taken over by Atkinsons.

1975 the house was closed. 1842P; 1960K; 1880 PO; 1900 K. J. Emberton – local historian, JC, and WLH.

**Wednesbury, 46 Bridge Street** At one time a boxing club met upstairs in this hotel.

c.1999 the building was demolished and rebuilt as the Purity Soft Drinks, which had previously been part of the hotel. 1822-1842 P; 1860 K; 1880 Darlaston PO; 1900-1936 K. Hotel.

**West Bromwich** 1860 K; 1880 PO.

**Willenhall, Short Heath** 1834 listed as a beerhouse with Joseph Jackson as licensee.

## Wolverhampton

**50 Worcester Street** 1818 listed as trading with Andrew Jones as licensee.

By 1851 the house was known as the Old White Horse. J. Emberton – local historian. 1822 P.

**Canal Street** This street was once the overflow for the pig markets held in the town centre. Every time there was a storm part of the street would flood and the inhabitants would have to move to their second storeys.

At one time there were 13 inns and beerhouses here which included The Clogg which by 1892 had been renamed the White Horse with George Green as the licensee.

*77 **Duke Street*** 1870 listed as a beerhouse with Miss Julia Meek as licensee.

1884 John Cecil; was listed as landlord, he was also a broker and china and general dealer.

*Powlett Street* 1892 James Dodds was listed as licensee – apparently the house became the White Rose. J. Emberton – local historian.

## Notes

1. An exclusive Society which never had more than fourteen members consisting of geologists, chemists, scientists, engineers and theorists. Amongst these members were:

> Dr. Erasmus Darwin (1731-1802) Physician, inventor, author and poet. Grandfather of Charles Darwin.
> Joseph Priestly (1733-1804) Chemist, political theorist and clergyman.
> James Watt (1736-1819) Engineer and inventor who did survey work for canals c.1760-74. His improvements to the steam engine led to the widespread use of steam power in industry. Col.
> Josiah Wedgwood (1730-1795) Potter and industrialist. The other grandfather of Charles Darwin.

2. 'The canals, by means of which Birmingham communicates with other parts of the island and with the sea, deserve some attention. The old canal opens a communication through the Severn with Shrewsbury, Gloucester, and Bristol, and through the Trent with Gainsborough, Hull, and London. This canal has also a junction with the grand line running through the potteries of Staffordshire, to Manchester and Liverpool, so that both the Irish sea and the German ocean are laid directly open to Birmingham traders.' JB.

3. The glass was held horizontally and cut at angles with a wheel, giving a crisp, sparkling line. ACr.

4. This pigeon fanciers' club no longer exists in Britain but in France is the Federation Colombophile Française and in Belgium the Royale Federation Colombophile Belge.

# WILTSHIRE

*Before the gods that made the gods*
*Had seen their sunrise pass,*
*The White Horse of the White Horse Vale*
*Was cut out of the grass.*

*Before the gods that made the gods*
*Had drunk at dawn their fill,*
*The White Horse of the White Horse Vale*
*Was hoary on the hill.*

*Age beyond age on British land,*
*Æons on æons gone,*
*Was peace and war in western hills,*
*And the White Horse looked on.*

The Ballad of the White Horse by
G.K. Chesterton (1874-1936), published 1911.

**Biddestone** Dating from the 15th century, the pub is a white painted stone building with a slate roof and small paned windows. It is situated next to the village green and was once a staging house for horse drawn coaches. Oliver Cromwell is reputed to have visited this inn before sacking Slaughterford Church nearby.

1789 Daniel Charmbury, one of 7 children and a wealthy builder, was left The White Horse Inn by his mother Betty Charmbury, a victualler.

2006 Inside there is an open fire and over the bar there is a fine collection of small Toby Jugs. There was a pets' corner in the large garden at the back.

MM and BHS. 1867-1939 K.

**Bratton** Situated below the Westbury white horse hill cutting – dating from c.1700.
1775, 20th November – the following two entries from the *Salisbury Journal* read:

> 'To be Lett, and entered upon immediately or at St. Thomas-day next
> – *29th December* – that ancient and well-known public-house called the
> White Horse, with a good garden and orchard adjoining, and a close
> of pasture it required, now and for many years past in the occupation
> of Mr. John Watts who is about to decline that business.'

9th December 1776 – 'To let: ancient and well known White Horse at Bratton Wiltshire formerly occupied by John Watts and now by George Grant.'

The house continued to trade until c.1880. WLS, MM. 1842 Westbury P; 1844P; 1851-1867 K.

**Calne, 1 Wood Street** 1758, 13th March – an advertisement in the *Bath Journal* reads:

'To be Lett or Sold immediately . . . The White-Horse Inn, and two small Tenements thereunto adjoining, situated in the Market-Place; with four Acres and a Half of Land, some Arable and some Pasture; together with Right of Commonage, and other Privileges.'

By 1981 there was a Co-op store on the site. The building then remained empty for a few years before Block Buster Ritz Videos took it over.
c.1996 Halifax Property Services moved into the modern, single storey building. 1939 K.

**Compton Bassett[1], The Road** This white-washed and rendered stone building dates back to the early 1700s. By 1848, it served as a beerhouse, bakery and grocer's shop but did not become an inn until 1855.
By 1859 the house was listed as White Horse; a traditional open fire was kept burning every day for the next 150 years, by which time alterations had taken place – the village shop becoming the kitchen and the old snooker room at the back being altered into an accommodation block.
2008 French boules could be played on the lawn and there was also a skittle alley. It retained its traditional character, with its low-beamed ceilings, open fire and mullioned windows. Mr and Mrs Adams, Landlords. 1867-1939 K.

**Cricklade, High Street** c. 1650 an undated ¼d. (farthing) token issued by Anthony Worme, a carrier of Kricklead (sic.) – *a horse trotting*. GCW No. 63. *This could have referred to the White Horse.*
According to W.C. Oulton writing in 1805, the White Hart had always been considered the principal inn, however the White Horse was known as the principal inn when it was kept by the Godbys during the 18th century.
c.1890 the name was changed to The Vale – perhaps influenced by local interest in the Vale of the White Horse Hunt. 1805 Principal Inn WCO; 1826 L; 1842-1844 P; 1851-1875 K.

**Devizes** 1839 Robertson's Directory lists an unnamed public house in New Park Street, whilst Pigot's Directory of this date lists the Plume of Feathers in this street with Daniel Cole as landlord.

Three years later in 1842 the White Horse is listed under Taverns and Public Houses and not under Retailers of Beer; also at this date Pigot's Directory lists Daniel Cole as landlord of the White Horse, but gives no address.

By 1848 the White Horse was no longer listed – but a William Cole is given as having a watch and clock business in New Park Street. Devizes Public Library. 1842-1844 P.

**Downton, The Borough** The original building is thought to date from 1420; the unnamed hostel mentioned in documents of the 13th and 14th centuries is believed to have been the White Horse.

1576 the justices were trying to control the number of lodgings in Downton; the landlady of the White Horse, Ann Maple, succeeded in being permitted to continue to take lodgers by making an injunction against the other innkeepers.

1599 first listed as an inn under the sign of the White Horse.

The borough cross, dating from c.1210, stands in front of the White Horse. All elections were held at the cross until 1832 and Parliamentary election results were announced from the balcony of the inn.

By the 17th century manorial courts were held in the school building behind the White Horse and fairs were held nearby. The stone cross has been badly damaged at least twice, once in 1642 by Cromwell's men; it was repaired during the 19th century, only to be damaged again in 1940 by a German landmine. In 1952 some of the original stone was found in a ditch nearby, an expert stonemason recorded that it had been 'cut across the lace', or the grain of the stone, which had not been done for hundreds of years. To commemorate the coronation of Queen *Elizabeth II* in 1953, Lord Radnor gave the local Chilmark stone needed to replace the broken shaft – which was surmounted by the original ancient cross.

1739, 15th May – an advertisement in the *Salisbury Journal* described the White Horse as 'a good new-built inn . . . with a good Brew-house, Stabling, and other Conveniencies; together with the Arable and Pasture-Land thereunto belonging.' Building materials from nearby Old Court – originally a Royal Saxon palace – were used, which included two carved stone busts purported to be of King John and Queen Isabel of Angoulême, which have been placed in niches on the front of the building at first floor level. Old Court was shown on a map of 1734 as 'the ruins'.

1745 the properties between the White Horse and Moulds Bridge were described as 'formerly tanshouses'. A tan yard was mentioned by James Bell in his Gazetteer of 1834.

1776, 23rd December – The *Salisbury Journal* published an advertisement for a Sale at the White Horse. Two years later on 19th October, they published the following notice:

'Stephen Rogers begs leave to acquaint the Public in general, that he has taken the White Horse Inn, in the town of Downton aforesaid, which will be immediately fitted up in the most commodious and genteel manner; and at the same time solicits a continuance of the favours of the usual customers, and others, where they may depend that his utmost endeavours shall be exerted to merit the same.

A stock of the best liquors of all kinds, and a good larder will be kept. There is good stabling, and plenty of ground for taking in of horses and cattle.'

1805 W.C. Oulton maintained in 'The Traveller's Guide' that King John had a castle behind the mill at one time called Old Court, 'some of the walls of which were still standing at this date'. The stables stood where the White Horse is now situated.

1981 The White Horse Distillers Ltd. presented a new sign to the inn – a reproduction of Lionel Hamilton Renwick's painting of a white horse.

At one time one of the two bars was known as the Downton Museum – displaying memorabilia of the village.

VCH Wilts. Vol. XI, WRO, and David Waymouth – local historian. 1842 P; 1867-1939 K.

**Edington, 43 Westbury Road** A 1782 map of the County of Wiltshire showing the late estates of Harry Duke of Bolton, gives a house on this site as number 30, and the field behind as number 33.

The Erlestoke Estate map of 1832 shows the house numbered 109 – later to be described as a hostelry.

About this time one of the landlords had a dog with retrieving skills, its master would throw a sovereign into the grounds and his clients would lay bets as to whether the dog would bring it back. Almost a century later, a former occupant of the house found a sovereign in the garden.

c.1900 the railway line opened with a station at Edington; excursion trains were popular as well as bicycling and the landlord built a tea room next to the public house for visitors who no doubt came to see the white horse carved on the hillside above Bratton – other visitors preferred the nearby Monastery Gardens Tea Room.

The Erlestoke Estate sale catalogue of 1910 describes Lot 92 as:

'The full-licensed Premises known as The White Horse Inn, with garden, stabling and allotment land. 3 acres, 1 rod and 26 perch.

The accommodation includes, on the ground floor, Porch, Entrance, Bar Parlour, Tap Room, Living Room, Wash House, Pump House, *with a 40 foot well*, and Cellar. On the upper floor, four bedrooms.

Adjoining is a large galvanised iron on brick Tea House, and on the opposite side of the road are brick and tiled Stable of three stalls and two loose boxes, timber and galvanised iron Trap House and Yard.'

The sale to brewers J.H. & H. Blake was completed on 11th August 1910; by 1928 the White Horse was owned by Ushers Brewery.

1927 about 22nd December there had been a fire at the inn, possibly caused by an exposed beam in the chimney which had continued to smoulder, this had been put out by Mr. Crook, the landlord. It is possible that this was the cause of the blaze which destroyed the house some two weeks later on Wednesday, 5th January. The following article was published on 7th January in *The Wiltshire Times*:

'Mr. Benjamin Crook, the licensee of the well known White Horse Inn, awoke shortly after 2 o'clock on Wednesday morning to find the house full of smoke and with flames making great headway. As access to the stairs was impossible, he hastily put on some clothes and then leapt through the window to safety. He hurried to the post Office, about 100 yards away and asked the operator to call the Fire Brigade, afterwards returning to the inn and doing his best to fight the flames with buckets of water. It was an hour later when the Westbury Fire Brigade under Chief Officer Fry, arrived and found the quaint old hostelry well ablaze.'

There was a difficulty in obtaining water, as the very adequate 40 foot house well was within the building. Frank Nicholas, as a five year old child, remembered being woken on the night of the fire to see flames coming from the White Horse next door – the Nicholas's well was likely to have been used – as the cattle drinking pond near the Priory Church had by then been completely drained of about 800 gallons.

'All efforts of the Brigade failed and by four o'clock the place was completely destroyed ... There now only remains the blackened skeleton of the building ... 'Everything I possess has gone', said Mr. Crook, 'and I have even had to borrow the clothes I stand up in ... '

The pavilion at the side of the house, which Mr. Crook had erected for summer trade, had already been rented by the County Council for cooking instruction classes – it escaped the fire, as did the stable.

1928, 25th August – the ruin, which had been left standing for seven months, was sold to William Colderick who rebuilt it as a private house on the old foundations, naming it Cynthia House. He sold it in October 1935 due to monetary problems.

Sometime later it was renamed Rosedale before becoming known as The Old White Horse – still a private house.

c.1986 the stable building was demolished.

c.1988 when the living room floor was repaired the old 40 foot well, which had been blocked up, was discovered; it was excavated and lined, to become a feature within the house.

In 2003 the tea room was still standing; until c. 1978 the word 'TEAS' could still be seen on the original match-boarded timber wall; sometime after the closure of the inn this building was used as a scout hut.

Fragments of many clay pipes have been found in the gardens – these with a fill of tobacco were probably sold in the public house.

Mr. Edwin F. George – owner of the house until 2003 and *The Wiltshire Times* of 7th January 1928. 1867-1875-1911 K.

**Fisherton-Anger** 'Farm of the Fishermen'. In 1242 Richard Filius Aucheri held the manor; this was later misread as Ancher and by 1295 a new spelling pronunciation of 'Anger' arose.

This village stands three quarters of a mile west of Salisbury and 'is separated from Salisbury by the River Avon with which it communicates by a very ancient stone bridge' – with six arches. The county gaol, bridewell[2], a chapel and two infirmaries were situated here. They were built in 1818 at a cost of £30,000. JB.

*New Salisbury / New Sarum was created in the 13th century when the cathedral and clergy were removed from Old Sarum to a new town in the river valley 1½ miles to the south. The clergy desired to move to a more commodious site from the inhospitable environment of Old Sarum on the hill, where water had to be brought from a distance at high cost. In particular they wished to leave their uncomfortable quarters in the castle, where their movements and those of pilgrims were subject to interference by the castellan[3].*

1713 listed as White Horse.

*The Salisbury Journal* published the following advertisements:

> 1748, 1st February – 'To Let. White Horse in Fisherton; Samuel Gambling, the late landlord being just dead. 'Tis an old accustomed inn and which had always a good share of business.'

> 8th August William Earll announced a move from the King's Head in Fisherton to the White Horse.

> 1752, 30th March – 'To Let. White Horse in Fisherton-Anger. A good accustomed house occupied by William Earle. Stock and furniture for sale.'

1764, 27th February – a Deed drawn up between John Thorpe of New Sarum and Thomas Williams, a shopkeeper, mentions that the property is 'bounded on the East with the river Avon, on the West by the Whitehorse Inn Cart Yard . . . ' and that John Thorpe and his heirs would keep ' . . . all the Elm Trees now standing and growing round the said barn and stable with free liberty . . . to fell cut down and carry away the same or to lopp and topp the same at his and their will and pleasure without any interruption or disturbance whatsoever . . . '

1767 the public house ceased to trade and an infirmary was built on the site.

WRO, PNW and RCH: City of Salisbury 1980.

**Highworth, Lechlade Road** This is a one and a half storeyed building with thick walls and a steeply pitched roof that is likely to have been thatched originally. The walls are substantially built of uncut stone. There was stabling for four horses here in the early 20th century but it is unlikely to have been a coaching inn. In one book it is said to be one of the oldest inns of the town but this is most unlikely as these would have been in the High Street or Market Place. The words 'West Country Ales 1760 Best in the West' are painted on this Cotswold stone building. MM. 1939 K.

**Hindon** c.1748 the public house was recorded as 'new'; it survived a disastrous fire in 1754.

1784, 12th April – *The Salisbury Journal* published a notice that Philip Green of Mere, miller, dealer and chapman[4] being declared a bankrupt, was required to 'surrender himself to the Commissioners . . . on twenty-ninth day of March and the twenty-sixth and twenty-seventh days of April at ten o'clock in the forenoon . . . at Mr. Harding's, the sign of the White Horse Inn, in Hindon . . . ' For many years the Assizes were held here. WLS.

**Kington St. Michael** An 18th century building originally trading under the sign of the White Horse Inn which was also a brewery; it used water taken from eight wells on the site.

c.1842 the name was changed to The White Hart. The sign was moved from the opposite side of the street when the inn of that name closed; the original White Hart at one time had been trading on the same side of the street as the White Horse but had moved to a house across the road to where, it is said, a tunnel ran under the street from the White Horse.

1880 the building was restored.

1972 the inn was renamed The Jolly Huntsman.

By 2007 seven of the wells had been sealed up but a red brick Victorian well is inside the building and still has clear water.

**Milford** 'White Horse Cellar'. 1739 listed as trading with Richard Dickmon as landlord. 1834 this village was described as being situated half a mile east of Salisbury. WLS and JB. 1844-1851 K.

**Minety, Lakeside / (Lower Moor) Station Road** It is thought that a building with stabling and coach houses had been on this site for many years; vaults under the present building extend under the road – there is also a good sized fishpond. Maps of 1773 and 1820 show the inn as The White Horse – the 18th century turnpike road between Malmesbury and Cricklade passed nearby.

After the railway came in 1837, several horse drawn conveyances were kept here for passengers alighting at Minety railway station. Because this was such a critical section of the railway line, due to almost two miles of the 20 feet high clay embankment which slipped during 1841, Isambard Kingdom Brunel (1806-1859) oversaw the rebuilding work himself.

By c.1860 a new impressive three-storey building, with at least twenty four chimneys had been built on this site, which in 1867 was listed as The Railway Hotel.

By 1880 it was named The Vale of the White Horse Inn: 'Vale' because passengers from London will have passed through the Vale on their way west, and 'White Horse' due to the many white horse chalk cuttings in this area.

The 1881 census lists John Godwin as Hotel Keeper, with a wife, two daughters and three sons who occupied the first and second floors – the first floor being level with the road. The lower floor, extending below the road, was divided into tenements where eighteen other people were listed as occupying five 'dwellings'. Their occupations included seamstresses, agricultural labourers, a railway porter, a general servant and scholars.

It is said that some people, who slept on sacks in the cellars, brewed cider with meat and ham bones in it, causing the deaths from diseased stomach linings in almost an entire generation of the village.

1999 the business closed, to reopen in 2003 after much refurbishment; the top floor remains unaltered as a private apartment.

2006 the main dining room still has oak beams and floors with doors leading out to a large terrace overlooking the lake.

MM and James Dedman – owner of the hotel – and Miss Felicity Ball of Steam Museum.

**Quidhampton, Moor Road / Lower Road** Named 'The White Horse Inn' but it was never a coaching inn; however there were loose-boxes built round the yard – three of which remain; these were used by racehorses which arrived by train at the local station the night before the August race meeting and were walked the following morning to the nearby Salisbury racecourse. Mr. Blain, Landlord. 1842-1844 P; 1851-1939 K.

## Salisbury

During the 18th and 19th centuries there were five White Horse public houses in Salisbury.

**38 Castle Street** *The Salisbury Journal* published the following:

1740, 10th June – 'Whereas the 28th May last, was left at Tho. Webb's, at the White-Horse in Castle-street, Sarum, a Sorrel Mare about 14 Hands high, a Whip Tail[5], a Blaze in her Face, (with some other Marks) the Owner by applying to the above-said Tho. Webb, and paying the Expense at keeping her, may have her again; otherwise she will be disposed of.'

1747, 28th September – 'Went away, on the 26th Instant, from her Mistress in Figon-street, Sarum, Anne Jones, Mantua[6] Maker: She had on (when she went away) a Blue Camblet[7] Gown, and a White Straw Hat; very tall of her Age, being but thirteen Years old, and is Pock-fretten.

Whoever can give Information (or brings her) to Mr. William Baden, in Figon-street, aforesaid; or to Mr. Phelps, at the White Horse, in Castle-street, shall be handsomely rewarded for their Trouble.'

1775, 25th September – 'To be Lett, a House . . . Enquire of John Footner at the White Horse, who will shew the premises'.

1779, 28th June – 'To be Sold by Auction, on Tuesday the 20th instant, at the sign of the White Horse, in Castle-street, Kept by John Footner.

Ninety-one Casks of Spirituous Liquors consisting of Foreign Brandy, Rum, and Holland's Geneva[8]; the same having been lawfully seized and condemned.'

1784, 13th September – 'John Footner *of the White Horse Inn, Castle-street, Salisbury* returns his most grateful thanks to his friends and customers for the favours he has received at the above Inn, and hopes to be honoured with a continuance of the same.

He also returns thanks for the favours he has received in the Coopering Business, which he has carried on upwards of twenty years in this city, and begs leave to solicit the future commands of his friends and customers, and of the public in general, assuring them it will be his constant care to serve them in the best and cheapest manner.

Casks and all sorts of Coopering Ware, new and second-hand, on reasonable terms, by their most obliged humble servant, John Footner.'

By 1981 Devenish Weymouth Brewery had sold the premises and it became a free house. By 2006 the White Horse Inn was advertised as an hotel.

***Fisherton Street*** Listed as trading between 1713 and 1767.

***London Road*** 'White Horse Cellar'. 1822 listed as trading, but closed c.1855.

***Milford Street*** 1739 Richard Dickmon was listed as landlord.

***New Street*** 1744 William Fry is listed as landlord of the White Horse.

   1774, 18th April – a house auction took place here; John Fry had taken over as landlord. MM. 1822-1844 P; 1851-1939 K. Hotel.

**Trowbridge, Frome Road** 1875 listed in Kelly's Directory as in Upper Studley[9]. The house continued to trade until c.1930. MM.

**Winterbourne Bassett** The White Horse Inn originally stood further to the west of the present site.

   1784, 2nd August – The Salisbury Journal advertises 'For Sale – The White Horse, Winterbourn'.

   1855 listed as trading.

   There was a hop field behind the pub which had its own brewhouse – later to become a garage.

   1913, 2nd November – the building was struck by lightening and was rebuilt on the new site in 1922. Mr C. Stone, landlord. 1867-1939 K.

## Notes

1. The edge of a stratum showing at the surface of the ground; an outcrop. SOE.

2. A house of correction; jail, especially for minor offences. Col.

3. A keeper or governor of a castle. Col.

4. A trader, especially an itinerant pedlar. Col.

5. A long slender tail like a whip-lash. SOE.

6. A loosegown of the 17th and 18th centuries, worn open in front to show the underskirt. Col.

7. Originally a name for a costly eastern fabric but substitutes were made of various combinations of wool, silk, hair and latterly cotton or linen. SOE.

8. A spirit made in Holland distilled from grain and flavoured with juniper berries and also called 'Holland'. SOE.

9. Probably from Old Norse meaning 'pasture for horses'. WAd.

# YORKSHIRE

The original Ridings of Yorkshire each formed a third part of the county. Dating from the Danish occupation – the original word 'thriding' came from the Old Norse 'thrithjung' – a third part, applying to the East, North and West Ridings. These were abolished in 1974 and became the new counties of North, West, and South Yorkshire – the East Riding was restored in 1996. B2.

*'Yorkshire has long been famed for its horses, and the north riding is particularly distinguished for its breed. The Cleveland horses being cleanly made, strong and active, are extremely well-adapted to the coach and to the plough; those of the northern part of the vale of York are, by the general introduction of the racing blood, rendered the most valuable breed for the saddle. The southern part of the vale, the Howardian hills, Ryedale, and the marshes, also produce a great number of horses both for the saddle and the coach. The dales of the eastern moorlands rear many horses, which, being of a smaller breed, are too low for the coach, but are a useful and hardy race. Horses also constitute a considerable part of the stock in the higher parts of the western moorlands. They are generally bred between the Scottish Galloways and the country breed, and are a hardy and very strong race, in proportion to their size; these are chiefly sold into the manufacturing parts of the west riding and Lancashire for the ordinary purposes . . . those employed in the western parts are small, but hardy, and capable of enduring great fatigue. In those parts there are scarcely any horses bred for sale. The farmers and manufacturers breed a few for their own use, and endeavour to get such as they think the most suitable to their business.'* JB.

'At one time it was said that if you called for the hostler in the yard behind the principal inn in any town in England a Yorkshireman would appear.' WAd.

# EAST
# YORKSHIRE

## Hull & Humberside

*The horses, in the middle and eastern districts are of a good size,*
*and sufficiently strong for all the labours of husbandry.*
JB.

**Bempton, 30 High Street** Situated on the corner with Cliff Lane. Originally built as a coaching inn with very low-beamed ceilings, the old men met here to smoke their pipes and tell their stories.

In 1938 the brewers built the new White Horse on the site of a market garden. It has a distinctive blue tiled roof which was painted black during the 1939 war.

The original inn – next to Maynard's Shop the general store, later the Post Office – became a private house named September Cottage. ASi. and Robert Traves, Landlord.
1823-1828 ERQ; 1857 PO; 1864 S; 1897-1913 K.

**Beverley, 22 Hengate** This town was named after the beavers which once inhabited this area, their bones have been found at nearby Wawne.

Although it is not known when the inn started to trade, records of 1585 show there were buildings on the site which were owned by nearby St. Mary's Church. The inn with its warren of rooms and substantial stabling thrived on the many different traders in Beverley. The front of the late mediaeval, timber-framed building is faced with brick, part of which originally projected over the street.

1666 Sir William Dugdale (1605-1686), an emissary of Charles II, stayed at the White Horse Inn, for the purpose of enrolling the pedigrees of families resident in Beverley and the East Riding, probably for the publication of his book, 'The Baronage of England'.

In 1675 the following dates refer to leases endorsed 'White Horse Inn' for three messuages, stables and two garths[1] on the south side of Hengate: 5th August 1732; 5th March 1753; 10th March 1787; 2nd January 1805 and 4th December 1862.

Beverley became involved in the final adventure of Dick Turpin (1705/6-1739), the notorious highwayman, when in 1738 he was detained in the Beverley House of Correction after shooting a game-cock in the street at Brough and attacking a local labourer who had objected. While here it was discovered that he was also involved in horse stealing. He was eventually moved to the prison at York Castle where he was hung the following year.

1841, 25th October – a sale bill at the White Horse refers to houses in Keld Gate.

In its hey day as a coaching inn, there was stabling for 70 horses; the majority of its trade being farmers and cattle men attending the local Norwood cattlemarket. They would stay at the inn and keep their horses and carriages in the stables and yard, this practice continued until the 1940s. Auctions and meetings were also held in the inn.

1927 the tenant of some years, Francis Collinson a saddler, bought the property from the Church. His large family continued to run the White Horse until 1976. It was known locally as 'Nellies' after the formidable Nellie Collinson who had been the licensee. The following quote comes from B. Pepper's 1960 publication, 'The Best Pubs in Yorkshire':

> 'She held court here for almost fifty years catering most of that time for men only and displaying her innate ability to cope with them. Nellie's was, and to some extent remains, an institution. Miss Collinson (she would never have allowed a Ms) ran it her way. There was no bar and drinks were served from a marble topped table. If the room filled up then customers overflowed into two tiny kitchens. She had a live-in lover known to all as Suitcase Johnny because of the number of times he and his possessions were slung out by Nellie after one of their many arguments. But her name is remembered with affection by her former regulars.'

1976 the house was bought by Samuel Smith brewery who discovered the pumps still on the wall and the beer being served on a big kitchen table; there were old candle stumps in the cellar near to leaking gas pipes which had been mended with chewing gum, also a wall in the back yard serving as a gent's loo! 'The general air of tattiness' was retained, along with the rocking horse sign over the front door, coal fires, huge mirrors, gas lights – including the original gas chandeliers, worn flagstone and quarry-tile floors with no carpets, tongued and grooved boarding, a hatch to the servery and a Victorian fireplace in the main bar; also a collection of ancient furniture with the old kitchen range and high-back settles in the 'Sliding Door Room'; the public rooms include the old Scullery and, from c.1990, a pool room.

Until British licensing laws permitted all-day trading, Beverley was the only town in this area to allow certain pubs to remain open after 2 pm on market days.

By 2006 the main coach entrance with a hay loft trap door although no longer in use can still be seen as can the old mounting block, together with the sloping floors, numerous dimly lit small rooms and the original gas chandeliers in the larger areas within the building. The inn becomes busy after race meetings – the racecourse has been at nearby Westwood for over three hundred years.

O.G.W. Smith – Samuel Smith Old Brewery, Tadcaster, WhB, EPN, HCA and the following lease references in catalogues held by East Riding of Yorkshire Archives: PE 1/383; PE 1/443; PE1/582; PE 1/684; PE 1/204 and DDPK 9/2. 1822-1842 P; 1849 S; 1857 PO; 1864 S; 1897-1913 K.

**Driffield / Great Driffield, Middle Street** 1822-1828 William Moody is recorded as landlord. 1842 recorded as trading. ERA. P; 1849 S; 1857 PO; 1864 S; 1897-1913 Great Driffield K.

**Easington, South Church Side** A hostelry is thought to have stood on this site from the 17th century. Early trade directories record this as a small thatched alehouse.

It has a frontage of 10 yards and a depth of 15 yards – about 3½ and 5 metres – it overlooks the market place, standing to the east of the church.

c.1910 a disastrous fire destroyed the interior; the owners, T. Linsley & Co., rebuilt the inn.

1952 the premises, being in a poor state of repair, were restored by the new owners Duncan Gilmour & Co. of Sheffield. The public rooms were of moderate size, but there was no proper bar and the cellar just a tiny place at the end of a passage.

The 1953 building, described as being constructed of thin red brick, has been concreted and painted white on three sides, the rear wall is harled[2] and dashed with small pebbles, the yard is enclosed partly by sea cobbled walls on which Victorian brick has been superimposed; similarly the outbuildings are a mixture of ancient and modern construction. The roof is of pan tiles.

1953, 31st January – during the disastrous floods which badly affected the east coast of England, the water flowed to the front door of the inn. Mr. Gill, the landlord, and his wife provided the homeless and volunteers with meals and refreshment for several days at all hours of the day and night; the White Horse with its ample parking space also became the headquarters of the RAF troop-carrying vehicles which were engaged on rescue work.

*Mr. Richard Hayton, author and historian, kindly supplied a copy of John Wilson Smith's manuscript dated c.1953 from which the above details have been taken.*

1864 S; 1879; 1913 K.

**Gilberdyke, Main Road** Situated on the main Howden to Hull road on land which was originally part of Bishopsoil Common – waste land of very little value being frequently under water in several places. The Ecclesiastical Commissioners were lords of the manor and received 6d. per acre as Bishop's rights, this land was part of the Bishop of Durham's manor of Howden.

1767 an Act was obtained to 'enclose, drain, divide and allot Bishopsoil to the ancient farms, having right of stray thereon'; in 1777 a parcel of land was allocated to the parish of

Laxton and it was here that the White Horse alehouse was built soon after this date.

1841, 3rd April – a Certificate of Justices records the highway to be in good repair from White Horse Inn to Skitam Lodge and in an east direction along the town street.

1851 William Healas was listed as landlord, he died in 1890 and two years later his wife Eliza, a victualler, is given by Bulmer's Directory as landlady of the White Horse, Gilberdike. The family lived here until c.1906.

1926 the landlord, Charles T. Foster, lived here with his wife Ada and two year old son Thorpe. Shortly after they arrived Ada opened a fish shop in the outside buildings, she also kept a cow and made butter.

1930 Charles Foster died and Ada closed the fish shop; she continued to run the house as a family pub until 1959; her regular customers were allowed to use 'the slate' during the week which would be paid every weekend, they created their own atmosphere and sorted out any trouble amongst themselves before it got out of hand.

Her son, Thorpe Foster, recalled that the Gilberdyke Feast used to be held in the White Horse field and that during the 1930s and 1940s a circus – which might have been Fossett's – with elephants and other animals, annually visited the village possibly having a rest period on their journey to the Hull Fair.

See Alford, Lincolnshire where we have recorded a description of an un-named circus which visited the White Horse during the Second World War – possibly the same circus!

During the 1939 War, the Fire Brigade stored their fire engine and equipment in the outbuildings.

There is a large room at the back of the house where the British Legion used to meet.

c.1982 extensive alterations were carried out, re-opening in July 1983.

Landlord and ERA. 1823-1828 Laxton P; 1827-1828 Eastrington ERQ; 1879 PO; 1913 Bishopsoil K.

**Hayton** There were two landlords recorded in 1824 – and another, William Walker, in 1825 who a year later changed the name from White Horse to the Blacksmith's Arms. ERQ.

**Howden, 10 Market Place** 1695, 4th/5th December – a Lease and Release of John Acklom, citizen and apothecary of London and Elizabeth his wife for £105 to Thomas Clarke, a yeoman, the 'House and three shops with the bulkers boxes, heirlooms and appurtenances in the Market Place in Howden'.

1702, 21st March – Mr. Clarke leased for one year to Thomas Peart, a blacksmith, the 'House and two shops in the Market Place'.

1720, 12th/13th October – a marriage settlement and a lease for £150 by Mr. Peart to Richard Worsop, gentleman, and William Dent, a saddler mentions the 'house in the Market Place with one close of ground, meadow or pasture lying at Knedlington Town End'.

1757, 15th July – a mortgage of £120 on the 'House in the Market Place, three acres of meadow ground in Howden called Yarmshaw and a croft at Knedlington'. This list of property remains unchanged in documents to 1764.

1773 the premises, which at one time had been within the Manor of Howden, were purchased by Mr. Thomas Carter who inherited a parcel of old title deeds dating from 1559 to 1764 referring to 'White Horse Inn and premises adjoining':

> 1559, 3rd April – a Surrender and Admission of 'one acre of land lying in the fields of Howden'.

> 1626, 2nd October – a Surrender and Admission of 'one acre of land lying in the Small Ings Bottoms'.

> 1634, 1st April – a Surrender and Admission of 'half acre of land in the fields of Howden'.

> 1678, 8th April – a Surrender and Admission of 'half acre of land arable, meadow or pasture lying in the Flatt Field in Howden'.

After this date the Manor of Howden is no longer mentioned and the inn is referred to as 'The House'.

Thomas Carter came from an influential family who owned breweries in Market Weighton and Knottingly. He came to Howden with his wife Sarah and 9 year old son John. They had 5 more children while at the White Horse, but John was the most successful: in 1823 he was maltster and brewer in Hailgate. By 1826 he owned the White Horse Inn and the Wheatsheaf Inn (then called the Howden Packet), together with the maltkiln on the sight of 52 Hailgate and the brewery behind the Board Inn. He had a number of other businesses too, all of which were based at the Board Inn. He died in 1850 aged 86.

c.1856 the coach house was burned down.

1955 mentioned as an hotel in a Whitbread Brewery Trust Deed.

2003, 16th October – HRH Prince Charles visited Howden and may have visited the White Horse!

ERA – The Catalogue for DDTR 219-231. 1834-1842 P; 1849 S; 1864 S; 1897-1913 K.

## Hull

In 1299 King Edward I acquired the land and properties of this growing township from whence it got its name 'King's Town upon Hull'.

c. 1390 Hull began its existence as a trading outlet for the Cistercian abbey of Meaux, under the name of Wyke of Myton. *See introduction to North Yorkshire.*

**51, Carr Lane** During the 15th century the inn was trading outside the town walls on land called White Horse Ings.

   1823 Billings Directory lists No. 45 Carr Lane.

   c.1830 listed as trading – by this time it was within the boundaries of the expanding town.

   1846 White's Directory lists No. 37 Carr Lane.

   c.1955 the house was demolished and replaced with a new building. 1834-1842 P; 1849 S; 1864 S; 1897-1913 Kingston upon Hull K.

**Fish Shambles** 1786 listed as trading, but is thought to be of an earlier date although there is no documentary proof.

   1806 the building was demolished.

**38-9 Humber Street** Originally trading as White Horse until 11th November 1809 when its name was changed to the Crown and Anchor, which continued to trade until c.1892. *The Hull Advertiser.*

**10 Market Place** The White Horse was said to have been trading since 1321; by 1387 there was still no record of it trading under this sign.

   1391, 16th November – a White Horse inn recorded as trading.

   1552 during the reign of King Edward VI, Hull was ordered to restrict its number of inns, or hostelries licensed to sell wine, to four premises, which included the White Horse; this was strictly adhered to until after the Restoration in 1660.

   1672 a Guild of Innkeepers was established – its charter stated that innkeepers should congregate in the Lord's Chamber of the White Horse; the title of this room may have applied to the De la Pole family, who owned the land prior to 1438.

   1727, 8th May – a letter referring to a Holderness watercourse was to be left at the White Horse in Hull.

   1746, Michaelmas – Joseph Waters of the White Horse, Market Place presented a Petition due to the loss of the use of his leg.

   1778, December. The inn closed. Up to this date it had been the terminus for the Hull to York diligence – or slow coach. ERA – CSR 19/97; QSF 154/D8; Hull City Records BRA 6/D – 169.

**Hutton Cranswick, Main Street** Originally a coaching inn situated on The Green by the duck pond.

   By October 2003 a restaurant and eight bedrooms had been added, along with conference and function facilities and a fully equipped gymnasium where there was a fitness and health club. Landlord.1857 PO.

**Kingston upon Hull** *See Hull.*

**Laxton Grange (near South Cave)** 1864 Slater's Directory listed William Healas as licensee.

It has been difficult to locate this inn although there was a White Horse public house at Gilberdyke which lies geographically midway between Laxton and South Cave – equally it might be the White Horse at North Cave.

## Nafferton

*6 Coppergate* 'White Horse Hotel', sometimes known as 'The White Inn'. This street was originally named Neatherds, becoming Neithergate or Nethergate.

By 1682 there was a two-hearth thatched house on the site of the future White Horse named either Tighill, Tighell or Tickhall House; there was also a thatched cottage to the rear which was used as a dissenting Protestant meeting house. Since 1719 all the waste ground in the parish belonged, for some reason, to the occupants of the White Horse – possibly because of the St. Quintin[3] family being Lord of the Manor.

1790 both the thatched house and cottage were taken over by a prosperous local blacksmith, William Hodgson the Elder, and his family; Tighill House was pulled down, tenements and two new shops occupied by a joiner and blacksmith were built on the site, together with a coalhouse and garth[4] or orchard. His third son, Benjamin, was listed as a 'Servant in Husbandry' at this date.

1794 William Hodgson died and by 1798 his two elder sons, William and Francis, had also died leaving young Benjamin – now listed as 'A Yeoman' – to inherit this property and another in nearby Westgate, a public house named White Horse – in 1800 this sign was to be moved to the Rose and Crown in Nethergate.

1799 a larger house was built on to the tenement buildings, this became The Rose and Crown; Benjamin Hodgson placed a stone plaque over the doorway with the initials BH 1799 which may still be seen. The first innkeeper was Benjamin's relative, William Pinder. This new, western end of the house was built of larger bricks – a brick tax had been introduced in 1784, so it was advantageous to build with fewer bricks. A six foot high cellar, which was later uncovered, may have been built at this date when all three properties were sold to a farmer, William Jefferson.

1800 re-named the White Horse, with Jefferson's son, Richard, as the innkeeper. Three years later, on the death of his father, he inherited the property, in trust for life. He employed William Storry, a smallholder, to take his place as innkeeper; tragically in 1808 William's son was killed by a heavily laden horse wagon.

1822 an Ale Recognizance shows that 'William Storey at the sign of the White Horse . . . acknowledges himself to be indebted to our Sovereign Lord the King in the sum of Thirty Pounds'.

c.1830 Nethergate was renamed Coppergate.

1840 William died and Marmaduke Storrey, took over.

1846 Reuben Levitt became the landlord, also running a successful blacksmith business which continued until 1874.

1856 the White Horse was enlarged by incorporating the two adjacent tenements, thus creating a large club room where the Court Leet[5] and Court Baron held their meetings and annual dinner until 1914.

The rateable value on the buildings during the Victorian era was £16 per annum.

1874, 31st August – John England Jefferson, aged only eleven years, inherited the property in trust. Francis Carter, a member of the local brass band, was the innkeeper from 1879 to 1896.

1891 the Ordnance Survey map shows both sites in 'West Gate' and 'Copper Gate'.

c.1891 a further extension was added to the western end of the building; 1894, October – the premises were sold to brewers, J. Russell & Son of Malton; they blocked the top window on the gable end and painted their advertisement for fine ales on the end wall, which could still be seen in 1973. On the ground floor level was the window through which the Church tithes were paid until c.1927.

1914 John – known as Jack – Slater was listed as landlord.

The following memories were provided by his daughter, Mrs. Dorothy Drape in 2003, with help from her brother and sisters who were all over ninety years of age:

> When Jack Slater and his wife, Hannah, moved into the White Horse Hotel they had six children; two more were born while they lived here. Mr. Slater played the cornet in the local band and every week he went to band practice proudly wearing his bowler hat and carrying his cornet in its case. He also helped to establish a football team in Nafferton.
>
> The large wooden barrels of beer stood on rails above a long trestle table down a passage from the bar. As there was no gas or electricity supply, all the rooms were lit by brass oil lamps, every day the lamps were cleaned and the wicks trimmed. The rooms were heated by coal fires.
>
> There was a separate room where young men could entertain their young ladies, and a large room at the far end of the hotel was used for meetings – each Friday agents from the local Mill would come for a meeting, followed by high tea. There were four large bedrooms upstairs and one very large club room where men and youths were coached by an elderly man in the art of boxing.
>
> The kitchen fire heated the oven for cooking and supplied hot water for the copper at the side; on Friday nights a large tin bath was brought into the kitchen and filled with hot water for bath night.
>
> There were three steps down from the kitchen to the outside where three very large tubs stood, one on each step; for swill, bran for the pigs, and sawdust – used for sprinkling over the floors. My friends and I used to play in the sawdust with my mother's old pan lids, weighing and measuring in our play shops. One day my elder brother upended my friend head first into one of these tubs of sawdust – she was very distressed! All the outside buildings were white-washed, and my friends and I used to 'powder' our faces with the powdery lime from the walls.

As children we were never allowed into the bar or associated rooms – we could play in the yard behind the hotel, but never in the street. The earth closets were in the yard, which was also used by farmers who brought their Shire horse mares to be put to the stallion – *which at that date would have been walked from village to village.*

We had two pigs, two pet rabbits and a goat who was milked every day – she was tethered nearby on the grass surrounding the Church; we had to move her peg each day to make sure the grass was grazed regularly.

We continued to live in the house after the closure in 1927 until we found another house in Nafferton.

1925 Mrs. J. Slater was listed as innkeeper.

1927, 21st March – Russell & Wrangham of Malton closed the business and the buildings were sold for £375 to Walter Horner, a blacksmith, who renamed the private house 'Springhead'.

1935 the fee farm rent[6] of one shilling per annum paid to the St. Quintin family of the Manor of Nafferton ceased, due to an Act of Parliament. The Lord of the Manor had to be paid compensation for twenty years purchase, plus solicitors fees which amounted to £1.1s.0d.

1973 the buildings, having become derelict were sold by the blacksmith's son, Michael, for redevelopment. The following year the original buildings had become totally derelict and vandalised which was when Mr. and Mrs. William Pugh bought the property. Some original features were saved and incorporated into the new building, including a gable of 1799 which can still be seen in the roof space; the remnant of a 6 inch x 6 inch beam built into the wall of the western upstairs room, which would have protruded out over the street and from where the inn sign would have been suspended, had been cut off square with the outside wall can still be seen. The blacksmith's shop (a large coalhouse) and a very steep ladder staircase to the first floor joiner's shop were still in evidence. A six foot high cellar which extended under the footpath with two steel up-opening hatch doors had been covered by tarmac; in 1975 this was filled in and sealed. The chimneys in the east section of the building are thought to be part of the original thatched 17th century Tighill House; the other chimneys date from the two later extensions. During the refurbishment, Mr. and Mrs. Pugh used as much of the salvaged materials as possible.

By 2003 new owners had built an extension to the back of the building where they manufactured plastic finger nails.

*The Driffield Times* of 24th September 2003 published an article which appealed for information regarding the White Horse Hotel at Nafferton and showing two photographs – one of a brass band outside the hotel – a member of this band was Francis Carter, innkeeper of the White Horse from 1879 to 1896; the other photograph dated 1918 shows Mr. and Mrs. John Slater in the doorway with one of their children. The descendants of the Slater Family, Mr. William Pugh and Mr. Clifford Horsman, Local Historians. 1842 P; 1864 S; 1897-1913 K.

**13 *Westgate*** From the middle of the 18th century The Hodgson family were listed as alehouse licensees in Westgate; at this time the building would have been built of chalk and thatch.

c.1800 the house was rebuilt in brick and pantile and the sign was moved to The Rose and Crown in nearby Coppergate.

According to the detailed list mentioned in a mortgage dated 1800, which was provided to Benjamin Hodgson by Robert Boddy, this property consisted of a 'newly erected messuage, cottage tenement or dwelling house used as and for an inn and known and distinguished by the sign of The White Horse'. Here there was a fold yard, outbuildings, grass garth and orchard 'now in the tenancy or occupation of Ann Hodgson, widow' - which would have been his mother.

By 2003 this house had once again become an inn called the Blue Bell.

Mr. William Pugh and Mr. Clifford Horsman, Local Historians.

**North Cave** 1778 The Christmas sessions record an Indictment[7] of inhabitants of North Cave for non repair of the highway at the west end of the town to the White Horse Inn. 1822-1828 ERQ; 1834 [Cave] – 1842 P; 1849 S; 1857 PO; 1897 Brough K.

**Ottringham, Keyingham Road / Patrington Road** Keyingham Road is a continuation of the modern Patrington Road.

There was probably a medieval alehouse on this site.

1823 trading as White Horse; once a small posting house ideally situated at the cross-roads, with substantial yards and stabling at the back. The building, with two adjacent cottages on the west side, is believed to date from the late 17th century and are thought to have once formed part of the original inn. The present inn is haunted by the ghost of a General!

1905, July – a photograph shows Heavy Battery Gunners of the 2nd East Riding of Yorkshire Royal Garrison Artillery (Volunteers) 'breaking step' at the White Horse to rest and water the horses, while they too took liquid refreshment. After coupling up the teams of horses, usually on loan from traders in Hull, they marched on to summer camp at Kilnsea on the East Coast. The photograph shows that the inn sold John Smith's Tadcaster ales and there is an advertisement on the wall for 'Good Stabling and Trap Hire'.

1918 the cavalry horses used the pinfold[8] paddock opposite the inn.

c.1953 the building, described as 'T' shaped, was built of narrow red and yellow brick with a red pantiled roof, the frontage being plastered and painted white. The inside had massive half round beams and a red tiled floor in the bar; a table by the window has a large statuette of a white horse standing amongst flowers.

c.1997 the original pinfold was taken over for building.

Mr Richard Hayton, author and historian, kindly supplied a copy of John Wilson

Smith's manuscript dated c.1953 from which the above details have been taken.
1822-1828 ERQ; 1834 Patrington – 1842 P; 1849 S; 1864 A; 1857 PO; 1897-1937 K.

**Skirpenbeck** East Riding Quarter Sessions record William Dickins as licensee between 1823 and 1828. 1864 S.

**Stamford Bridge** 1842 P and 1849 S.

**Swanland** 1857 PO; 1864 S.

**Thornton** 1823 listed as trading with John Gray as innkeeper.
1825 the East Riding Quarter Sessions listed the house under Grey Horse.
1826 listed under White Horse. ERA.

**Tibthorpe** 1857 PO.

## Notes

1. A courtyard surrounded by a cloister; a yard or garden. Col.
2. To cover a building with a mixture of lime and gravel. Col.
3. The St. Quintin family was Lord of the Manor of Nafferton from 1719 until the line expired in 1933; Sir William St. Quintin was MP for Kingston upon Hull from 1695 to 1723. The baronetcy ceased when the family was forced to sell much property and the tithe lease in Nafferton.
4. See note 1.
5. A special kind of manorial court that some lords were entitled to hold. Col.
6. This rent was peculiar to certain freehold holdings and was payable to the Crown, but like the Tithes they were owned by laypeople. C. Horsman.
7. A formal accusation of crime.
8. A fold or pen for sheep or cattle. Col.

# NORTH YORKSHIRE

*'It is significant that most of the great northern abbeys bred white horses. The Cistercian[1] abbeys of Jervaulx, Fountains and Rievaulx, and the Praemonstratensians – or White Canons – of Easby bred white horses and cattle, and the colour of their stock served as a trade mark . . . At the Reformation, in the 16th century, this white stock was dispersed, at first causing a general 'wave' of whiteness, which then gradually disappeared under the dominant darker colours.'*

There were also the flourishing Cistercian nunneries of Byland, Rosedale and Wykeham in this area. AAD.

*At one time ten White Horse Inns traded along the forty mile stretch of the A170, which runs eastward from the Great North Road, through Thirsk to Scarborough; these inns were spaced some ten miles apart, as during the coaching era the need to bait, change horses and keep to time was vital.*

Could there be a connection between the many Cistercian monasteries and nunneries in this area, and these inn signs?

**Amotherby** This had been a large public house but by 2004 had become a Chinese restaurant. 1837-1840 WDG; 1842P; 1849 S.

**Ampleforth, West End** The building, with a stone front and pantile roof, dates from the seventeenth century and has been an inn since the middle of the eighteenth century. The bar is furnished with wood settles and has an open fire; pool and darts can be played. 1822 BDR; 1842 P; 1849 S; 1857 PO; 1864 S; 1897-1913 K.

**Beadlam Nawton, Main Road** Situated on the village boundary – the village of *Bodlam* in the Domesday Book. 1856, 5th April – the Malton Messenger recorded that Mr. Robert Barker was living at the White Horse. Deeds dated 1892 refer to William Younger – brewer. The celebrated Mally Clash Well, now filled in, was situated at the top end of the present car park;

it was perpetuated in the following jingle displayed in the pub sometime before 1971:

*"If you want good, clear water, there's plenty to sell,*
*At a town called Beadlam at Mally Clash Well."*

c.1981 the new signboard, depicted the White Horse at Kilburn.

By December 2003 the inn had ceased to trade. ML and P.A. James – Local Historian. 1822 BDR; 1834 Helmsley P; 1857 Nawton PO; 1897 Nawton – 1937 K.

**Boroughbridge, 4 Fishergate** An important coaching inn – the High Flyer left the inn yard for York at nine every morning except Sunday and left for Ripon at half past three in the afternoon on its return journey.

Early in the 19th century The Duke of Newcastle imposed a high rent of nearly £70 a year, this was thought to be due to it being a busy inn and to the inland revenue and excise office which operated here, rather than the amount of land connected to it, which was worked by the landlord, Samuel Morrell – his father ran the Three Horseshoes.

The revenue office continued to operate here after the inn closed in c.1860.

2004 the inn yard was occupied by Rostlea Upholstery and Northern Insurance Financial Services; the archway to the yard still led on to Fishergate.

WmB. 1822 BDR; 1834-1842 P; 1849 S.

**Borrowby** 1857 PO. 1897 Northallerton K.

**Brearton** The inn held a license up to 1938. 1897 K.

**Church Fenton, Main Street** Until 1981 the Poulter family had held the tenancy continuously for 100 years. Robert ran the inn from 1880 to 1905 when there was a joiner's shop at one end of the premises, two cottages stood at the other end. 1822 BDR; 1837-1840 WDG; 1889-1913 K.

**Easingwold, White Houses** 1840 Easingwold Directory of Trades and Professions lists both a White Horse and a Grey Horse.

1837-1840 WDG; 1857 PO; 1864 S; 1897 K. Some of these dates may have referred to the 'Grey Horse'.

**Falsgrave, 94 Falsgrave Road** An 18th century coaching inn which has been modernised.

1994 the inn was renamed the Tap and Spile. The authors were unable to contact the pub – there was no landline telephone. JWC. 1822 BDR; 1834-1842 P; 1849 S; 1857 PO; 1864 S; 1897-1913 K.

**Great Edstone** 1872 the inn was still recorded as White Horse but by 1897 it was listed under Grey Horse.

1991 'The Grey Horse' was still trading.

2004 it had ceased to trade.

P. A. James, Local Historian and NYR. 1822 BDR; 1837 WDG; 1849 S; 1857 Great Edstone PO.

**Huntington** Listed in Street Directories from 1822; it was demolished in c.1950, the site becoming the end of Pear Tree Close. Its name survives in White Horse Close some two hundred yards away.

Huntington Library. The following Trade Directory dates were listed under East Huntington – indicating that the inn was situated on the east side of the River Foss.

1822 BDR; 1837-1840 WDG; 1913-1937 K.

**Kirkbymoorside, Market Place** The building developed from an early purpose-built inn with its own brewhouse; two wings were then added – each with a staircase giving access to a separate block of rooms. It is thought that the main letting rooms were those facing the market place and served by the main staircase which was accessible from the inn yard – an important feature for a Posting House. W.C. Oulton visited this inn on his travels in c.1800.

Substantial stables and coach houses were added in the late 19th century when horse-drawn transport became an important link between the new local railway stations.

In 1905 the inn was still brewing its own ale.

By 2000 the signboard depicted a prancing carthorse; the original stable yard and buildings were still in use.

RCH 'Houses of the North York Moors'. 1987. 1822 BDR; 1826 L; 1834-1842 P; 1849 S; 1857 PO; 1864 S; 1897-1913 R.S.O. Hotel K.

### Knaresborough

An 1891 edition of the Ordnance Survey map of Knaresborough shows an hotel where Finkle Street joins the High Street, this might explain why the address is sometimes given as Finkle Street and at other times as the High Street.

***High Street*** 1842 listed as an hotel in Pigot's Directory with John Layfield as licensee.

***Finkle Street*** 1889 Henry Johnson is listed as licensee.

***Corner Street*** By 1951 the house had become a restaurant named after a local inhabitant, Eugene Aram.

LH and NYR. 1822 BDR; 1823 P; 1849 S; 1864 Finkle Street S;1871 PO; 1912 K.

**Malton, 45 Yorkersgate** The following ditty mentions 'Javel' which in c.1606 would have described a low fellow, rascal, or a very decrepit nag:

> *'To Malton come I, praising th'saile, Sir,*
> *Of an horse without a tail, Sir.*
> *Be he maim'd, lam'd, blind, diseased,*
> *If I sell him, I'm well pleased;*
> *Should this Javell dye next morrow,*
> *I partake not in his sorrow.'*

c.1650 a ½d. token issued by Will Snary in New Malton – a horse trotting. *This may have referred to the White Horse.*

Dating from the 17th century, this very large coaching and posting house was recorded by W.C. Oulton in his Travellers' Guide of 1805.

The White Horse Commercial Hotel was mentioned in a number of issues of the *Malton Messenger* during 1855 and 1856, including the following dates:

1855, 2nd June and again on 20th October when an article was published on Thomas Weatherhead who died on 18th July 1786 aged 68 years, his epitaph is in St. Leonard's churchyard. He presumably lived at The White Horse.

1856, 12th April Mr. William Stewardson, who was taking over the White Horse, published an advertisement for the sale of household furniture and on 19th April the following advertisement appeared:

'William Stewardson begs most respectfully to inform his Commercial and Agricultural friends and the public in general, that he has entered upon the above Hotel, and hopes, by strict attention to the comfort of his guests, to be favoured with their patronage.

Well-aired beds.
Good stabling and coach houses.

*N.B. A Cab is in attendance at the Railway Station on the arrival of every train.'*

1856, 6th September – Ann Stewardson was now running the inn; on 22nd November a Grand Ball was held here.

1859 the inn closed to become a boarding school for boys.

By 1982 the building had become the Malton surgery.

ML, Barnabee's Journal, GCW No. 228 and SOE. 1822-1842 P; 1849 S; 1857 PO.

**Marrick** The White Horse was built in c.1730 as a farm and inn; it was a longhouse, consisting of an attached barn – now the west wing – which has a date stone of 1738. 'Monks hood' mouldings are above the windows and a sundial above the front door, the metal gnomon[2] is original but the markings have been eroded. Many pieces of clay pipes are still found in the flower beds and the original privy still stands in the garden.

1774 listed as trading by the Licensed Victuallers.

1782 Jackson's Map of the Manor of Marrick shows the house with two barns flanking the building, giving it a U-shape.

1790 William Whaley was listed as licensee, in 1794 he was joined by his son Matthew; William died in 1821 aged 73.

1823 Jane Whaley, daughter of William, was listed as licensee.

1950 the Food and National Registration Office visited the inn – the owners did not wish to install a flush toilet in the house. Chickens were found to be quite at home in the bar – which would perhaps account for the building remaining empty until 1976; it is thought the two flanking barns were demolished at about this time.

1998 the property was advertised for sale as a private house with three reception rooms, a breakfast room, kitchen and five bedrooms. It is a Grade II listed building.

Mr and Mrs Howard Young – owners. 1822/23 BDR; 1837-1840 WDG; 1857 PO; 1913-1937 K.

**Mickley** Originally a terraced private house thought to have been built during the eighteenth century by William Goodyear.

Dates and information are not clear as the deeds were lost in a fire during the nineteenth century.

Sometime after the fire the building was converted to a public house known as 'The Ham and Firkin'.

By 1871 the Post Office recorded it as trading under White Horse.

1940 Theakston Breweries closed the pub and sold it as a private dwelling to the Hitchen family on condition that 'no intoxicating liquor be sold'. The family used it until 2003 as a holiday home, changing its name to The White House.

1889-1936 K. Mrs Margaret Parking – née Hitchin.

**Northallerton, 227 High Street** 'White Horse Cafe' was named c.1994 after the Kilburn White Horse hill cutting.

**Pickering, Burgate** 1856, 7th June – Robert Appleby was listed as licensee.
   The pub once stood next to a popular sweet shop. By 2004 the premises had become a private house named 'White Horse Cottage'.
   ML. 1834 P; 1849 S; 1857 PO; 1897-1937 K.

**Richmond (North side of Pottergate)** Richmond Library records the White Horse as trading from 1708-1729; 1897 K.

**Ripley** 1871 PO; 1889 K.

**Ripon, 61 North Street** Conveniently situated for most of Yorkshire's racecourses and is the home of England's 'Garden Racecourse'.
   In The Square at 9 pm every day the Ripon Hornblower sets the watch at the four corners of the obelisk.
   A coaching inn with eight original stables standing around a yard – four on one side have been converted into bedrooms; the other four with grooms' accommodation above, have become a joiner's shop.
   Landlord. 1834-1842 P; 1849 S; 1864 S; 1871 PO: 1889-1936 K.

**Rosedale Abbey, West Side** The name is not derived from the flower but from the Old Norse 'hrossa dalr' – horse valley / dale – the inn stands on the west side. 'Abbey' refers to a Cistercian nunnery founded in the reign of Richard I (1157-1199).
   c. 1690 originally the Lane Head Farm, illustrated by the sign which shows a white working horse. It is believed that the farmer had a 'taproom' at the end of the farmhouse to offer refreshment to the iron ore mining community. At the time the beer house was known as the Pick and Shovel.
   By 1702 it was listed as White Horse Farm Hotel.
   c. 1960 the hotel was modernised.
   By 1991 the owner had built an extension to the restaurant using material from a redundant church in Wakefield where he used to be a choir boy; an ancient door, a window and an archway are in the restaurant, miserechords support new beams and a font lid and carved Tudor and Stuart pew ends are also used. This owner kept two white donkeys which were still there in 2000. *See the introduction to North Yorkshire*!
   WAd. 1872 West Side PO; 1913-1937 Westside K. John Rushton: 'The Rydale story'.

## Scarborough

**14 St. Thomas Street** 1857 PO; 1864 – 9 St. Thomas Street; 1897-1913 K.

**8 Tanner Street** 1837-40 WDG.

**Seamer, 15 Main Street** The building is over 300 years old, and was originally a farmhouse. It once had a large acreage of farm land to the rear, which was sold for housing.

By 1968 the name had been changed to The Copper Horse. 1857 PO; 1897-1937 K.

**Settle, Market Place** Situated on the corner with Church Street. 1671 is recorded on a datestone with the initials 'RLP' – found when the rendering was removed in the early 20th century – suggesting that Richard Preston had inherited the property from the owner, Robert Preston, and was living here at this date with his wife Lettis. At the back of the property there was stabling and a paddock.

1822 listed as White Horse in Baine's Directories of North and West Ridings.

It is thought that during the late 19th century a new storey and frontage was added, to be used by a grocer Charles Frankland until early in the 20th century. At one time the premises had been a barber's shop.

On the top storey at one side of the building, there is a blocked doorway with a hoist and gantry giving access to a loft.

c.1940 the inn closed and the licence transferred to the Settle Social Club.

By 1970 the old stone mounting block at the side of the building had been removed – probably to make more parking space.

By 2000 the building was occupied by a newsagent and Garnet's electrical shop.

PH and RPH. 1834-1842 P; 1864 S; 1871 PO; 1889-1936 K.

**Snainton** There is no record since c.1960 of this inn being named White Horse.

1837-1840 WDG; 1849 S; 1857 PO.

**Spofforth** 'The Old White Horse' is situated in the centre of the village next to the church.

There is a wall plaque on the inn with the date and initials '1748 TWC'.

1848 Slater lists Elizabeth Groves as licensee.

1889 listed as trading in Kelly's Directory with Richard Wells as licensee.

1919 the inn was advertised for sale and a year later it had become a private house.

1979, 5th May – it was advertised for sale in the *Yorkshire Post* – described as being built of stone, with a large dairy off the kitchen and having a 'compact range of stone outbuildings, garages, court house and stone walled garden of good size'. 1822 BDR; 1837-1840 WDG; 1849 S; 1864 S; 1871 PO; 1897 K.

**Staithes, The High Street** Staith / staithe from Middle English means land bordering on water-especially a waterside landing-stage.
Thought to have started as an inn during the 1836-44 depression. Comment by a local: "Not been a White Horse here for past forty years"! SOE and Mrs. J. Eccleston, Postmistress. 1837-1840 WDG; 1849 S; 1857 PO; 1879-1913 K.

**Stokesley, Beckside** Situated on the banks of the River Leven.
1823 an un-named beerhouse – probably The White Horse – is listed in Beckside with the landlord, George Robinson, brewing ale; by 1834 he is recorded as licensee. Mrs. Irene Ridley, local historian. 1842 P.

**Tadcaster, Bridge Street** Originally the inn was named The Angel, but by 1803 the house was listed under The White Horse. During the early 20th century the name was changed to 'The Angel and White Horse'.

The stables in the coach yard housed the Sam Smith Brewery team of dappled grey shire horses, one named Hercules was reputed to have been the largest horse in the world and his shoe is on show in the bar as proof, he is depicted in an oil painting over the fireplace.

2009 there were still four working grey shire horses, two of which were housed in the stables at the inn while the other two were at the farm. GAC, GPG. 1805 WCO; 1822 Excise Office P; 1826 L; 1842 P; 1871 PO; 1889 K.

## Thirsk

***Ingramgate*** 'The White Mare'. 1879, 2nd August – the following description of the inn sign appeared in a letter published in *Notes and Queries*:

' . . . a vilely painted modern signboard attached to a public-house in the outskirts of Thirsk, where a jockey is going up in the air like a balloon, and a grey mare descends to the Inferno . . . '

1912 the 25" Ordnance Survey map shows the White Mare Hotel situated at the Helmsley and Easingwold crossroads at the north east corner of the town.

Later, this house had a new sign showing a white mare in the act of leaping over the cliffs carrying its frightened rider into the waters of Gormire Lake. Both these signs illustrate a legend which tells of a trick played between the Abbot of Rievaux[3] and a knight, Sir Henry de Scriven, who rode a race at night after a carousal[4]. Sir Henry, riding the abbot's white Arab mare disappeared over the cliff, whilst the abbot, on the knight's black horse, turned into the Devil calling out to him:

> *"If you must play a trick,*
> *Try it not on Old Nick;*
> *I'll see you below,*
> *When I visit the sick."*

**Long Street** Christopher Forster listed as landlord. 1864 S.

**Market Place** White Horse Café. 2009 the café was still trading.
NYR and MF. 1822 P; 1857 PO; 1897-1913 K.

**Upper Poppleton, The Green** The village is on the north western outskirts of York; the White Horse Inn overlooks the village green and has always played a leading role in village life, including being a meeting place for the Overseers of the Poor of Poppleton.

1793, 18th October – an Indenture concerning a property sale of £100 by two yeomen, Thomas Swailes and Joseph Hall – with his wife Ann – to the occupier, Charles Knapton. The property was described as a 'messuage, dwelling house or tenement . . . together with outbuildings and orchard or Garth . . . '

*It is thought that the present White Horse is probably built on the orchard but there is no documentary evidence.*

1802, 27th May – an Indenture records a lease for one year to William Knapton, however a further Indenture dated one day later shows that William bought the property for £140 – it was about this date that the first ale house was recorded. He and John Fearby, a 'gentlemen', sold it in 1805 for £220 to Thomas Eden – parish clerk, shoemaker and mender.

During these years in the early 19th century – the population was 210, to rise to 319 by 1831 – a room was set aside as a temporary mortuary; deceased persons were conveyed to York for burial, the local parish church at this time being merely 'a chapel of ease'.

1841 Richard Tindall, the landlord, also carried on a butcher's business.

c1844 – c.1865 Thomas Tindall is listed as landlord.

The inn had stills which were used for making peppermint, villagers took bottles to the inn to be filled with the sediment when these mint stills were cleaned, which they used for coughs and colds. There is also the suggestion that the poppies grown by the monks of St. Mary's Abbey were also distilled here to make laudanum for medicinal purposes.

c.1890 John Illingworth was both landlord and a market gardener; he provided dinners for the Poppleton branch of the United Order of Free Gardeners. On Easter Monday of 1897 they processed with a band through the villages of Upper and Nether Poppleton, assembled for a service at All Saints Church and had a dinner where 'loyal and other toasts were given, and a very pleasant evening was spent'.

1903 the new inn building opened and the former White Horse became a private residence named 'Old White Horse'; this was destroyed in a fire and a commercial building – Sycamore Stores - was erected on the site during the 1960s.

c.1920 – c.1940 the premises became the White Horse Hotel. By this time Tadcaster Tower Brewery was the owner – it was known as 'the Snob's brewery' because it was founded by a group of baronets' sons.

During the 1970s the inn was patronised by bikers.

Until 1994 local residents worked allotments on the brewery's land at the back of the inn – they were rent free but there was an unwritten rule that each tenant would provide the landlord with garden produce every seventh week.

1988 the inn came under new management and the building was refurbished.

The inn, with its function room, is a venue for the many village clubs; it also has two pool teams, and holds quiz nights and musical events for charities.

Mr. and Mrs. G. Watson – Proprietors and the Poppleton History Society. 1822 BDR; 1837-1840 WDG; 1897-1936 K.

**West Rounton** 1841 the only public house listed in White's Directory under West Rounton was the Blacksmith's Arms.

1849 there was still no White Horse listed here in Slater's Directory.

1876 listed under White Horse in Slater's Directory.

1881 the census lists John Fortune as Innkeeper and Blacksmith at the White Horse; he had a wife, Mary, and two grandsons – Denton, a scholar, aged five and James aged one.

1891 the census records Mary as a widow, publican and cowkeeper. She lived with her unmarried daughter, Mary, and her grandson James now aged 10 years and a scholar.

## Whitby

Shortly after the dissolution of Whitby Abbey in 1538, the entire precinct was first leased, and then bought (in 1545) by Richard Cholmley. The Cholmley family coat of arms depicts a white horse and a griffin. It is then that the two pubs, The White Horse and Griffin, and The White Horse, took their names. There is evidence that their origin was as a pilgrimage centre, and so it is possible that the two pubs were once a single establishment. They were certainly separate, however, by the 18th Century. Whitby has been the centre of many trades throughout the past centuries, meaning that the two pubs have been frequented by seamen, fishermen, whaling men, wool merchants, traders, jet carvers, pilgrims and tourists. The result of the town's industrial success in the 18th century was immense wealth. Whitby could afford to have municipal paving and street lighting, the lighting being of particular interest: in 1790, three great lanterns were installed on Church Street, two of which were placed outside the two White Horse Pubs, which were both operating as termini for stage coaches at the time.

Though the two pubs were separate by this time, they still shared a common coach yard. They used this to operate an ingenious scam:

'Before setting out on the long journey to Whitby (10 days from London) it was normal practice for a gentleman to send a letter containing a bank note, as payment for a room and horse, to be made available to him upon his arrival.

From time to time, a less careful individual might simply address the letter to The White Horse, Church Street, Whitby and would be greeted upon arrival with the news that his letter had not been received. The infuriated traveller would have little choice but to pay again for his night's lodging with the intention of taking up the matter with the local magistrate in the morning.'

P.J. MacFarlane, D.H. Kelly & A.J. Slater 'The Black Horse Inn'

The gentleman would be taken to a lawyer who was also involved in the scam. The Magistrate would have to point out that the man had probably gone to the wrong White Horse, and he would then be presented with a lawyer's fee on top of his two bills for the night. This was all put to an end in 1828 after a Magistrate ran out of patience. Either the pubs change their names or they would get no licence. As a result, the White Horse became the Black Horse.

***87 Church Street*** 'The White Horse and Griffin'. Built in 1681 and situated in the centre of the old town.

1761 Captain James Cook used the inn to sign on his crews and also to pay them on their return.

1788 the first stage coach from Whitby to York began to operate from the large stable yard – the inn was a post house. A sign on the wall reads 'The first stagecoach from Whitby in 1788 operated form this inn'. The journey to York took 2 days.

Charles Dickens (1812-1870) stayed here and noted that oyster-shell grottoes were a feature; in recent years BBC Television used the inn to film scenes for his novel 'David Copperfield'.

1939 the hostelry was abandoned until c.1989 when it was bought by Stewart Perkins who took ten years to restore the building.

1993 it was reopened as an hotel.

2000 the inn appeared in the film of A.S. Byatt's book 'Possession'.

2003 after ten years it was still a successful hotel and restaurant.

By 2006 the White Horse also owned several self-catering cottages in Whitby.

Landlord and www.whitehorseandgriffin.co.uk. 1805 WCO; 1822 listed as The White Horse BDR; 1823 P 1826 L; 1834-1842 P; 1849 S; 1864 S; 1857 PO; 1897-1937 K.

**91 Church Street** 'The Black Horse' There are indications there has been a hostelry on the site since the 12th Century. The current building has been built on its foundations, and is thought to be of the same date as the White Horse and Griffin. A refit in the 1880s gave it a Victorian frontage and a public serving bar (believed to be one of the first in Europe, having been installed shortly after its invention in America in the 1870s).

1973 The Black Horse is declared a building of Special Architectural and Historical Interest.

2005 The most recent refit: amongst other things, the guest rooms were all converted to be 'en-suite'.

PJ MacFarlane, DH Kelly and AJ Slater: The Black Horse Inn. 1805 WCO; 1822 listed as The White Horse BDR; 1823 P 1826 L; 1834-1842 P; 1849 S; 1864 S; 1857 PO; 1897-1937 K.

## York

**6 Bootham** There has been some confusion between this house and the White Horse in Fossgate which was trading in 1502. 23rd January, 1770 is the earliest recording of the house in Bootham.

1894, 18th December – the inn was rebuilt and opened the following year, the date being recorded on an oriel window. It was built hard against the walls of St. Mary's Abbey with no back yard. On the ground floor there were two smoke rooms and a dram shop while upstairs there were four

bedrooms – used by the family – and the kitchen from where food was supplied, with difficulty, to the customers on the ground floor. In 1902 the cellars and rooms were recorded as being damp. HM. 1842 P; 1849-1864 S; 1897 K.

**Coney Street** 1810, 4th June – opened in the old Judges' Lodgings by Mr T. Crow.

1814, 15th August – the inn was advertised to let.

This house was also known as 'Crow's Coffee House'

**Coppergate** 1733 common carriers could be found at this inn.

1818 William Hargrove wrote in his 'History and Description of the Ancient City of York: 'There is also in Coppergate, a good inn, called The White Horse; where will be found comfortable accommodations.'

1823 Mary Sowerby was listed as innkeeper.

1901 owned by J.W. Craven, a confectioner, who rebuilt the inn on a smaller scale.

1909-11 William Sowerby was listed as innkeeper.

1968, October – the inn was closed; by this date it was owned by York Corporation who demolished it for the Coppergate Development.

HM, PEB, HM, HTD. c.1803 GAC; 1805 WCO; 1842 P; 1849 S; 1864 S; 1897 K.

**Fossgate** Recorded as trading from c.1502.

1520/1, 6th February – the innholder, John Butterfield bequeathed his house, trading under the sign of the White Horse, to the Master of Trinity Hospital'.

1548 listed as trading. HM.

**Goodramgate** 1834 P.

**Grape Lane** 1733 common carriers could be found at this inn. HM.

**Marygate** 1855 listed by H. Murray in his 'A Directory of York Pubs'.

**St. Sampson's Square (Thursday Market)** Prior to 1711 The White Horse was renamed The Nag's Head, but was also previously known as The Cutt-a-Feather and the White Swan, which might have been confused with White Horse.

By 1789 the inn had closed. HM.

**23 Skeldergate** 1688 in his diary, John Webster records the White Horse as being near to St. Mary Bishophill Senior.

1745, 26th February – mentioned in the *Yorkshire Gazetteer*.

All six bedrooms were occupied by the family. The ground floor was comprised of a smoke room, dram shop, bar parlour and a kitchen from where food could be supplied, but it was scarcely ever asked for. There was also a cellar. The WC was shared between the family, customers and the occupants of an adjoining shop.

1931, 7th October – the premises were closed. HM. 1822-1842 P; 1849 S; 1864 S; 1897 K.

**Walmgate** 1822 BDR.

**West End**

## Notes

1. The Cistercian monasteries are, as a rule, found placed in deep well-watered valleys. They always stand on the border of a stream. The valleys, now so rich and productive, wore a very different aspect when the brethren first chose them as the place of their retirement. Wide swamps, deep morasses, tangled thickets, wild impassable forests, were their prevailing features. EB.

2. The stationary arm that projects the shadow on a sundial. Col.

3. This must have been prior to the dissolution of the monasteries in 1539.

4. A merry drinking party. Col.

# SOUTH YORKSHIRE

*About a mile from Doncaster 'is a celebrated race-ground, on which the St. Leger stakes are run, and which has for many years been increasing in its attractions . . . being a thoroughfare on the great road from London to Edinburgh. The High-street – for length, width, and beauty, is generally allowed to be the best on the road betwixt the capitals of South and North Britain.'*
JB.

**Barnsley, 7-9 Shambles Street** 1880 recorded as trading with George Davis as landlord.

1889 Kelly's Directory lists Thomas Alsopp as licensee, while Paul Mitchell is listed as publican and an omnibus cab proprietor.

1897 John Shaw is recorded as publican and a fried fish dealer.

1901 Stuart Verity is listed as landlord until 1903 – there is no trace of the White Horse after this date.

By 2006 the site had become a Chinese take away, the New Jade Palace. ET and PO.

**Chapeltown, Market Place** 'White Horse Hotel'.

1822 listed as trading in Baines Directory with Charles Hoyland as publican and victualler.

1825 Charles Hoyland was listed in Gell's Directory as the publican and a butcher.

1851 White's Directory recorded that Charles Hoyland was still licensee.

1861 the census listed the White Horse.

It does not appear to have been listed from the late 1840s to 1856 probably due to the fact the directory compilers sometimes did not include the surrounding areas of Sheffield.

1861 William Hoyland is listed as publican.

1871, 1879 and 1881 listed as trading.

1891 listed in the census.

1893 Robert Unwin, victualler, is recorded as publican; and by 1902, William Kilner, victualler, was at the White Horse, Market Place.

It was listed as trading up to the First World War in 1914 and is thought to have continued for some years although it does not appear in the Sheffield directories.

c.1970 the site was redeveloped, the building becoming a private house with shops attached. The Chapeltown and High Green Archive, Sylvia Pybus – Sheffield Local Studies Librarian and W. 1842 P; 1864 S.

**Doncaster, Market Place (on the corner of High Fisher Gate)** In 1663 the Turnpike Road Bill was passed and toll-bars were established – the almost impassable state of the roads rendered such a measure necessary. A journey of two hundred miles at this time would normally have taken more than a week.

Tokens were issued at Doncaster and Stilton, which are the only known tokens of tollmen. GCW No. 68. If a race was run for a wager through toll gates, they were opened without charge as was illustrated in the poem 'John Gilpin¹' by William Cowper (1731-1800), when the two wine bottles on his belt were taken to be the wager bags of gold;

> *'Away went Gilpin – who but he?*
> *His fame soon spread around;*
> *"He carries weight, he rides a race!*
> *'Tis for a thousand pound!"*
>
> *And still, as fast as he drew near,*
> *'Twas wonderful to view,*
> *How in a trice the turnpike men*
> *Their gates wide open threw'*

1770, June – An entry in The Doncaster Courtier – the title of the volumes recording Borough Council proceedings – reads:

> 'Ordered that the several shops and tenements in the Market Place in the possession of William Jackson and other tenants of the corporation extending from the Parsonage Gate to the White Horse be pulled down and a new wall, if thought necessary erected.'

Colin Walton in his publication of 1980 'The Changing Face of Doncaster' described the White Horse as a famous hostelry; the forecourt of the inn on market days was always full of articles for sale, specially chosen and displayed to attract the many visiting country buyers.

The vats of the White Horse were not always used for brewing ale, sometimes they were used for storing eels – on one occasion an eel measuring about 18 inches escaped into the gutter, it was captured and replaced in the vat.

1864 listed as trading by Slater.

1866, 1ˢᵗ August – an agreement for the tenancy of the inn between the Mayor of Doncaster and the innkeeper, Thomas Bullas, recorded that a rent of £19 be paid every six months. A year later on 2ⁿᵈ September, a rent of £3 3s. 4d. was required to be paid every month.

1868 the inn was demolished to make way for an extension of the cattle market. On 3ʳᵈ June at the Quarterly Meeting of the Town Council the minute book recorded that ' . . . the main or principal entrance to the Cattle Market be placed at the South West corner, on the ground formerly occupied by the White Horse Inn.'

RBD, CWa, PT and DA AB7/4/979, AB7/4/1000, AB2/2/5. 1822-1842 P; 1849 S.

## Sheffield

In the past Sheffield had a reputation for having an exceptionally large number of public houses and beer houses per head of population. In 1843 the ratio was 1 to every 105 of the inhabitants, probably due to the nature of work in the steel and cutlery trades, which was extremely hot and exhausting and workers would flock to the public houses in their free time.

***Blast Lane*** 1849 listed as trading.

***34 Copper Street*** Listed as trading at the following dates: 1822, 1825, 1833, 1842 – listed at No. 22 with George Shepherd as licensee – 1849 and 1871. In the 1881 census there was a White 'House' listed here – presumably a misreading as it was listed as White Horse in 1891. The final listing was in 1942.

***87 Creswick Street / Grammar Street (situated on the corner)*** At one time this was an unnamed beerhouse. Trading as White Horse between 1937 and 1965 – it is not listed in 1968.

***18 Effingham Street / Cattle Market*** 1841 first listed as trading under White Horse.
1842 Edward Shaw is listed as licensee.
1849-1863 listed as trading.

***Henry Street*** 1901 the only White Horse to be listed in the census.

***65 / 57 Malinda Street*** At one time this house was a work information place for carters.
1937-1974 listed as trading but by 1975 it was not listed in the telephone directory. J. P. Turley in his book on Sheffield of 1998 described it as a sing-a-long pub which ended by collapsing into rubble.

***76 Matilda Street*** Mentioned in two local books with no information.

***Norfolk Road*** Listed in the 1871 census.

***275 / 22 Solly Street*** Listed as trading at the following dates: 1822 P, 1825, 1833, 1842,1849, 1861 census and 1871. It continued to trade as White Horse until 1940.

GCH, DgL, Sylvia Pybus – Sheffield Local Studies Librarian and SLS.

## Tickhill

**Sunderland Street** 1803 the building was licensed at this date but it has not been possible to identify the name.

1867 the sign of the White Horse was possibly transferred to these premises from the White Horse in Northgate.

1877 Kelly's Directory referred to the Gleadall family at these premises who became tenants of the inn.

1907, 8th March – the *Doncaster Gazette* recorded that a licence had been refused.

**Northgate** There are no trading dates after 1867. It is thought that the sign name may have been transferred at this date to the licensed premises in Sunderland Street as the licence had been withdrawn, yet a new licence was granted in this same year. PT 1986 and 1990; 1838W; 1842 P; 1849 S.

**Wadsley Bridge, 104 Halifax Road** 1940, Saturday 6th April – an article in the *Telegraph and Independent* advertised the opening at 11.30 a.m. of the 'new White Horse Inn' an hotel built by the brewers, Thomas Rawson and Co., Ltd. It was described as a fine example of the modern public house with no expense spared – a handsome building and one of the most imposing hostelries in the city. It was built of brick with a steeply pitched roof of green glazed tiles; there were 'modern furnishings for the comfort of all patrons', a spacious concert room with a large stage, a lounge – and rubber floors throughout. There was central heating and 'artificial ventilation of the most modern type to prevent smoke-laden or over-heated rooms'. Below the hotel there were large, cool cellars installed with the latest beer engines and patent stainless steel pipes. There was ample car parking on all sides of the building.

The inn would be worthy of the firm and the neighbourhood it was to serve.

By 2006 the building was boarded up; however in January 2007 apparently it was doing some trade.

## Notes

1. The story of John Gilpin's ride was related to William Cowper by Lady Austen, who had heard it as a child; she married Sir Francis Austen (1774-1865) who was brother to Jane Austen (1775-1817) the author. Cowper published it anonymously in the November 1782 edition of the Public Advertiser. It is believed that the original John Gilpin was a linen-draper, Mr. Beyer, who lived at the Cheapside corner of Paternoster Row in London; he died in 1791 aged 99 – so it would appear that the original escapade would have taken place soon after the Turnpike Road Bill was passed.

*West Yorkshire clothiers carrying undyed cloth on pack-horses, 1814.*
*The white or grey pony was then still common in the Dales.*

The Foals of Epona

A.A. Dent and Daphne Machin Goodall, 1962

# WEST
# YORKSHIRE

*'. . . the demand for pack-horses was greater than it had ever been or was ever to be again in England. The textile trade had expanded enormously in the last generation but it was in the hands of a multitude of small masters whose stock was not counted in the usual way by the wagon-load. It will have been carded, spun, woven, waulked[1] and bleached in different places: some of the processes were put out as piece-work in the worker's own homes distributed in small parcels, and collected again, by pack-horses. Every clothier must keep a horse, perhaps two, to fetch and carry for the use of his manufacture.'*

*There was still no local network of carriage roads to carry this great traffic of raw materials, part-finished cloth and piece-goods. Even if some wholesalers sent it to London or to the ports by wagon, many of them supplied pedlars whose stock was retailed from panniers carried on a pony.'*

Report on Cloth Towns of West Yorkshire 1724-1726 by Daniel Defoe (1660-1731). AAD and DD.

**Adwalton** 1834 Drighlington P; 1864 C&An; 1897-1912 K.

## Ardsley

***East – 58 Main Street*** 2006, November, the hotel became bankrupt, ceased to trade and was demolished to be converted into residential property. By this date it was listed at Fall Lane which is the continuation of Main Street.

***West*** 1822 and 1853 listed as trading in both Baines and Newsome's Directory and White's Directory.
 1889 Kelly's directory listed Thomas Tempest as licensee.
 1912 listed as trading in Kelly's Directory.

**Armley, 87 Town Street** 1834 James Bell gives Armley as 2miles W by S. of Leeds, standing on the Leeds and Liverpool Canal and on the River Aire – where there were several fulling[2], cotton and corn mills.
 1856 listed by Gillbanks Directory in Town Gate with John W. Charnock as licensee.

1888 listed as an hotel.

1950 listed as an hotel, but in Wellington Road.

LLS. 1826 WP; 1834-1842 P; 1849 S; 1864 C&An; 1870 W.

**Bingley, 80 Old Main Street** 'The Old White Horse'. 1086 The Domesday Book lists an alehouse on this site. It is thought that the De Gant family were early owners of the tavern. In 1212 King John granted William de Gant the first market.

The present white washed building with a stone slate roof dates from the mid 17th century, but parts still date from the 13th century. Being a Knight Templars' Tavern, their sign of stone lanterns can still be seen as finials on the gables.

During the 16th century when it was already known as 'Old Whitehorse Inn', the town Magistrates' Court was held here; there was a police cell within the building which became a cellar, stocks stood at the front of the house and a gibbet in the courtyard at the back, where over the years between five hundred and six hundred people were hung! At one time The Tax Office was also in this building.

The doorway to the old Court Room has a stone carved lintel, there is an oak-panelled dado with an overmantel and pulvinated frieze dating from c.1720; there are two painted heraldic coats of arms, the right hand is of the Ferrand family showing two crosses of the Knights Hospitallers and the crest being a cubit arm wielding a battle axe, the cinquefoil below is taken from the arms of the Paganel family – 1120. The motto reads 'Justus Proposti Tenax', meaning 'Just and firm of purpose'. The arms on the left are those of the Currer family who acceded to Lords of the Manor in 1600, the motto reads 'Merere' – 'Merit it' or 'Be worth it'. In 1600 Queen *Elizabeth I* took ownership of properties and rents originally owned by the Knights of St. John of Jerusalem - William Currer is recorded as paying rent for his lands of 18 pennies to Her Majesty Elizabeth I 'as due to St. John of Jerusalem'.

John Wesley (1703-1791) preached in the doorway to the right of the main door – now blocked up.

Bingley was on main coaching routes to Leeds, Bradford, Skipton and Kendal. The White Horse had large livery stables and in 1869 carriers' carts were advertised as travelling to Harden for 6d. and to Cullingworth for 1s.

A post box was built into the wall during the 19th century and is still used in the 21st century.

During the 20th century some alterations took place when the old Court Room became a pool room.

The stable block, coach house and prison cell in the main bar area with its two stone slabs can still be seen. Kevin Roberts, Landlord; Christopher Walbank, local resident and HT. 1834 P; 1842 P; 1849 S; 1864 S; 1871 New Road Side PO; 1889 K.

**Birstall, 274 Leeds Road** *See Howden Clough.*

# Bradford

The following dates refer to a White Horse in Bradford, there is no way of telling to which they refer to: 1823-1842 P; 1871 PO; 1889 K.

*Adolphus Street* This beerhouse was situated in a very poor slum district and according to police records opened in 1853.

1869 the landlord, Isaac Broxup was refused a licence. The newspaper account of the licensing sessions described the house as a haunt of dog-fighters, dog-runners, cock-fighters, prostitutes and thieves, especially juveniles.

The whole of this area has been demolished long ago.

*221 Bowling Back Lane* This is the main thoroughfare of what was the district of Bowling – mediæval Bolling[3]. For more than two hundred years the large Bowling Iron Works were on the opposite side of the lane, it had its own quarries in the locality. The pub and neighbouring terraced houses served the foundry.

1860 it was opened as a beerhouse. During the 1860s and 1870s rum and coffee were served to workers from the iron works from 5.30 a.m. at a cost of 3d. Everyone could have as much soup as they liked for 3d. or a full meal for a shilling. Managers and senior staff from the works used the dining room and had a waitress service.

The landlord kept his own pigs which were killed on the premises.

There was piano entertainment, games of darts and 'diddlum[4]' and the daughter of the house read the newspaper to customers in the bars.

1884 John Naylor of Warley springs Brewery was listed as owner. *A John Naylor was listed as landlord of the White Horse in Queensbury in 1895.*

1906 the Ordnance Survey shows the building on the corner of Lake Street, which is now known as Bowling Back Lane.

1912 Mrs. Alice Baines is listed in Kelly's Directory as a beer retailer in Lake Street.

At one time the house was known as the New Image, but up to c.1995 this Freehouse was trading as 'The Shamrock', it was then sold to T&D Industries and became Brook House.

2005 by this date the building had become the offices of Francis Ward Ltd.

Dr Paul Jennings – Author and Local Historian, P.W. Robinson, – Local Historian and J.S. Kliene, chairman of Francis Ward Ltd.

*Kirkgate* At the end of the 18th century there were fifty public houses for a population of 6,000; the streets were lit by dingy oil lamps and there were only six constables.

c.1750 the White Horse Inn opened.

1779 it is recorded as having only three rooms and a back kitchen.

1802 the site is shown on a map giving the principal inns of Bradford.

1825 The Improvement Commissioners – who were responsible for paving, lighting and policing the town from 1803 onwards – met here and confirmed a rate of 1s. 6d. in the pound – 7½ p. Appeals against this sum were also held here.

From time to time other administrative bodies also used the inn, such as the overseers of the poor; the turnpike trustees who financed and maintained new roads and from 1834 the highway surveyors.

1849 listed as trading in Slater's Directory.

c.1850 the building was demolished, and by c.1877 most of the old town with its inns had been pulled down and redeveloped. PJ and Sc.

**Bramham, Tenter Hill** 1849 S; "Old White Horse"; 1864 S; 1871 Tadcaster PO; 1897. Boston Spa-1936 K.

### Burley-in-Wharfedale, The Grove / Main Street

Over 100 years old, the small building is set in a terrace, white fronted and neat with railed steps up to the front door. A corridor leads to the bar which contains an amazing collection of bric-a-brac, including photographs of all the pubs which have traded in Burley – but not of the White Horse! The dog grate in the stone fireplace weighs two hundredweight. Two windows are etched with a bowing courtier, the trademark of the former owners, the brewers Melbourne of Leeds. Dominoes and darts are played here.

1996, 15th March – in her article published in the *Ilkley Gazette*, Louise Auty described the building as compact with an open plan bar and a small darts room. Involved in the community; the pub supported two local football teams, darts and dominoes team and a sea angling club. The Best Pubs in Yorkshire, JTe, Landlord, BCL.

**Castleford, Albion Street** 1857 Aaron Sidwell is recorded as landlord, however a licence was not granted until 1862.

1869 the house closed; the renewal of the licence was refused by the Brewster Sessions of 1870 because of offences committed here. Rod Kaye – Local Historian.

**Cullingworth** 1849 S; 1871 Bingley PO.

**Deighton, 761 Leeds Road** 1848 listed as trading in Slater's Directory with George Berry as licensee.

1925, 27th May – an inventory and valuation amounting to £331 14s. 4d. records a commercial room, a snug, and vaults and bar; there were three cellars – 2 for ale and one for spirits; a kitchen and a porch; outside there was a pig trough, 2 stables – one with a hen cote adjoining, a wash house, a coal place and a shed.

Amongst the many items listed there were ten spittoons – 3 of earthen ware in the commercial room, 2 copper and one brass in the snug and 4 cast iron in the vaults and bar; this bar also had 2 stuffed pigeons in a glazed mahogany case and 4 sets of dominoes. In the kitchen there was a salt box, and a bread creel[5] with hooks.

1942, 21st October – another inventory and valuation amounting to £351 13s. 8d. recorded that the license had expired 20 days previously. There were still the same number of rooms and outbuildings, but the snug had been renamed the tap room.

At this date there were black-out blinds[6] in every room and one to the back door. Only two of the cast iron spittoons were still there in the tap room and one set of dominoes had disappeared. In the commercial room there was a pianoforte on a deal platform. The kitchen still contained the bread creel and hook, but the salt box had been moved to the porch. The stone pig trough still stood outside at the back of the property.

The stock-in-trade included 7,330 Woodbines, 1,260 Players, 90 Wills Whiffs, 113 Britannia cigars, 3 Rajah cigars and 26 Castella cigars.

1953, 18th and 19th August – the valuation amounted to £544 0s.1d. The wash house contained a meat safe with a gauze panelled door – not mentioned previously.

There was a further inventory in 1956, this time a garage is mentioned.

1967 Richard Whitaker Ltd. described the house as the only one of their inns with twin front doors.

c.1997 the name was changed to 'DB's'.

2005 the name reverted to The White Horse and is listed under Huddersfield.
1849 Huddersfield S; 1871 Huddersfield PO; 1889 -1897 K; 1936 K.

**Denholme, Well Heads Road** By 2005 the building had been turned into flats, the outside remaining unaltered. PO. 1936 K.

**Emley, 2 Chapel Lane** Situated on the old Huddersfield Road, this was an hotel.
1889 Kelly's Directory lists George Smith as licensee.

By 2005 it was no longer an hotel, but had a restaurant. 1842P; 1864 S; 1897-1936 K.

**Friendly, Burnley Road** Originally a beer house at the end of a row of cottages on the Spring Gardens estate.

According to the landlady it is a modernised 19th century building. By 2008 it was known for fundraising events and prize winning floral displays. P.W. Robinson – Local Historian.

**Gomersal, 298 Oxford Road** The White Horse Inn was built on the site of an old coaching inn named The Fox and Grapes; it was recorded as trading under White Horse in the late 18th century with Mr. Lang as landlord.

1822 listed in Pigot's Directory with Joseph Lang as landlord.

1838 Nancy and Mary Lang were the landladies.

1843 Sir Charles Ibbetson of Denton ceased to be the owner. The heirs of Mr. William Sykes became owners of the inn and the attached brewhouse, butcher's shop, mistal[7], barn, shed, stable, garden and yard.

1850 Nancy Lang married John Sykes.

The following celebrations took place at The White Horse:

> 1872 the anniversary of the Free Trade Lodge No. 15 of the Birstal District of the Yorkshire United Order of Oddfellows.

> 1875, May – Whit Tuesday – the 50th Jubilee of the Gomersal Friendly Society.

> 1886, Saturday 30th October – the annual supper of the Gomersal Pig and Poultry Society.

1887 John Sykes, who had married Nancy Lang, became landlord. He prepared an inventory of the inn. Upstairs it consisted of a lodge room, lumber room and five bedrooms. Included in the contents of the taproom was an oil painting entitled 'Black Horse' and eight iron spittoons. Contents at the back of the property included a large water butt, a manger, a wood pig cote, pig trough, piggeries, hen house and stable with pigeon cote above; there was also a washhouse containing swill tubs, sawdust box, firewood and a mangle. The total value of furniture, fixtures and stock in trade was £384 6s. 5d.

1888, Friday 22nd June at 7 pm – the inn was put up for auction at The Commercial Inn, Heckmondwike. It was described as a freehold property and 'all that fully licensed house or inn known by the name of The White Horse Hotel, situate at Hill Top in Gomersal. Together with the yards, stables, greenhouses, gardens and outbuildings . . . also two dwellinghouses and plumber's shop adjoining the inn or near to with the gardens and the outbuildings.' The property was withdrawn from the sale at £1,850.

1890, Saturday 20th January – the annual supper of St. Mary's Church Cricket Club was held here. Landlord and Rod Kaye – Local Historian. 1834 P; 1864 Hill Top S; 1897-1936 K.

**Great Horton, 731 Great Horton Road** The building, which has been extended to include three adjacent old cottages, is thought to date from c.1790. It is set back from the main Bradford to Halifax road and is recorded as trading from c.1833.

Being a substantial inn, it was a venue for auctions and political meetings and was at one time the Bradford City Transport terminus for the trams.

1847 a Liberal election dinner for 80 people was held here.

1849 a dinner was held to commemorate the Repeal of the Corn Laws[8].

1965 the inn was sold to Whitbread by Richard Whitaker Ltd., brewers of Halifax.

1984, 5th September – an article in the *Telegraph and Argus* described the stained glass window in the porch in which a white horse is depicted which was lit up at night. The bar front was of carved oak and there was an old fashioned tall weighing machine for customers' use. There is a large function room upstairs.

A top Bradford Sunday League football team based at the inn has taken the name 'White Horse' and play in blue jerseys with a white horse on the front.

Ten years later on 14th June 1994, the *Telegraph and Argus* published another article on the White Horse describing it has having an 'ultra-modern appearance' and a varied clientele, from briefcase-bearing business men  to young people. It is thought there is a ghost – there is sometimes a sudden strong smell of perfume, although nobody can be seen, which disappears as quickly as it came; the fire door opens and closes although there is no wind. RW and BCL. 1849 S; 1864S; 1897 K.

**Halifax, 33 / 39 Southgate** c. 1648 an undated ½d. token issued by John Gersed in Stainland (Parish of Halifax) – *a horse prancing* – this may have referred to the White Horse.

c.1745 listed as trading.

1819 listed as trading.

1842 Pigot lists James Brier as licensee.

1864 Slater lists the address as 12 Southgate.

1882, December – Hebden Bridge Council Building Plans record that Rev. W.H. Patchett applied for an additional sitting room at the White Horse Hotel.

1871 the Post Office lists George Yeates as licensee.

c. 1898 Demolished.

1899 Rebuilt and opened on 24th June.

1981 the pub was completely refurbished; it stands  next to the covered market.

CCL and CDA Ref: BIP/HB175. 1822; 1834 2 listed–1842 P; 1849S; 1871 2 listed PO; 1889-1936 K.

**Hebden Bridge, St. George's Square** Trading from the late 18th century, it was one of the two principal inns in the town, the other being the White Lion.

The following early history of the inn has been gleaned from a Description of Machpelah & Lee's Yard in Commercial Street dated 1816:

> 1796 the White Horse was built by Mr. William Patchet of Bankfoot House, who founded the White Horse estate; he bought Bannisters Farm, rebuilding part of it as an inn. The winter of this year was very severe, and the frost so keen as to freeze the water held in a glass outside the house of Mr Patchet. His house, afterwards let as a private residence to Mr. Lee, consisted of extensive grounds with beautifully laid out gardens. A high wall surmounted by smooth coping stones ran around the grounds, up to the walls of the White Horse Hotel, forming a fence on one side to the small plot garden under the window of the hotel's billiard room.
>
> A considerable trade in this district was the dying of fustians[9]. The founder of one firm of dyers, John Whitely, married a niece of Captain Patchet of Bankfoot House.
>
> Important meetings and political rallies were held here in the large upstairs room.

1834 Pigot's Trade Directory listed George Jackson as landlord – as did the Local Almanac of 1842.

1845-c.1855 William Jackson, a victualler, was listed in Walker's Directory as landlord of the White Horse Posting and Commercial House.

1849, 31st October – a meeting was held at the White Horse Inn concerning a scheme to seek Parliament's permission to build a Todmorden Turnpike Road *to become the A646*. There were also meetings advertised for schemes concerning the Rochdale Canal, *which came to fruition,* and the railway which only reached as far as Oxenhope – some seven miles north of Hebden Bridge – to become the Keighley and Worth Valley Railway.

1858-1868 Archdeacon Musgrave received various bills for suppers from David and Maria Brier of the White Horse Inn.

1864 Slater lists the White Horse Hotel as a Posting and Commercial House.

The General District Rate Book for Hebden Bridge listed the White Horse Hotel under the following dates:

1874, 24th June. *See HA.* HEB577.

1876 the Revd. James Winterbottom Patchett of the White Horse Hotel & additions St. Georges Square.

1879, 4th June – William Jackson, late of the White Horse Hotel died aged 64. Halifax Archives HEB 577 and 578.

*The Todmorden and Hebden-Bridge Local Almanac* of 1867 and 1869 recorded the following meetings at the White Horse:

> 1855, 17th July – details were given of the winding up of the Hebden-bridge and Keighley Railway Company.

> 1859, 17th June – the first meeting to discuss the raising of a company of Rifle Volunteers and on 27th October, a meeting was held resolving to light the streets with gas.

> 1860, 16th May – a public meeting resolved to lay down footpaths through the village, the cost to be defrayed by voluntary subscriptions.

A gravestone in nearby Heptonstall Church Graveyard bears the following inscription:

> 'In Loving Memory of Simpson Bowes, White Horse Hotel, Hebden Bridge, Born August 2nd 1869, died July 1st 1909 also of Minnie Dutton, wife of the above, who died Feby 5th 1940 aged 67 years.'

1887, 8th June – *The Todmorden and Hebden-Bridge Local Almanac* of 1888 recorded the marriage of the owner of the White Horse, Mr. Winterbottom to Miss Margaret Hey at St. James's Church, Mytholm. She was the youngest daughter of Mr. Thomas K. Hey, chemist and druggist of Bridge-gate.

The following sales by auction took place at the White Horse Hotel:

> 1888, 25th January – the fancy dealer's and hairdresser's business lately carried on by Mr Job Lindley. The stock in trade fixtures, etc. were said to be worth £306 and were knocked down to Mr. James Fielding for £115. From the Almanac of 1889.

> 1895, 9th May – a great property sale by order of the trustees of the will of the late Mr. Wm. Barker of Wood-top consisting of 20 lots. Only one lot was withdrawn. From the Almanac of 1896.

1879 and 1894 the hotel was advertised in two local guide books. HA GS3/5 and GS3/3.

Towards the end of the 19th century the following article appeared in the local newspaper:

> 'December the 28th. Fancy Dress Ball at Hebden Bridge. On Thursday last, the White Horse Hotel was a point of considerable interest to the people of the village, from the fact that Ganalkiel Sutcliffe Esq. of Stoneshey Gate was entertaining a number of invited friends at this hotel. The form which was given to it was a novel one – that of a fancy dress

ball and a good deal of curiosity was manifested and hopes entertained that perchance glimpses might be caught of the richness, variety and perhaps eccentric fashion in which some of the ladies and gentlemen were apparelled, but we believe the privacy with which matters were conducted disappointed this curiosity. The invitations were issued for seven o'clock and by carriage and cab the privileged parties arrived at the hotel with their differing gear for the evening's wear. The company numbered seventy-two persons brought from diverse parts of the country, Halifax way and beyond Halifax, neighbours and quasi-neighbours, elderly persons and youthful, as various in their ages as they were in their habiliments – law, physic, divinity and trade were respectively represented. Of course Mr. Winterbottom put his best foot forward to provide and arrange things for so exceptional event at the White Horse Hotel.

He drew upon the resources of Manchester in the departments of cuisine, attention and attendance, fitly to meet the wishes of the entertainer, in which, we believe, he was entirely successful. The upper rooms of the hotel were placed wholly at the service of the invites as attiring and retiring rooms . . . Supper was set out in the billiard room, the table of which was taken away to make room for other tables. It was served to the strains of music *from a pianoforte and a string band* . . . The party was certainly a picturesque and rather a grotesque sight in some respects, either in the ballroom or at supper, the mixture of medieval and modern, oriental and other which it represented . . .

Guests came from families of leading social status in the neighbourhood and the Hotel was kept open until three o'clock in the morning . . . for the gratification of the party, such as, we may now venture to affirm, had never before been seen in Hebden Bridge nor in these parts and probably, for some time, we shall not see the like again'.

1904 an advertisement gave the telephone number for the Proprietor, Thomas Wadsworth, and the fact that the tram terminus was only 100 yards away from the hotel.

1912 the will was proved of Mr. Thomas Wadsworth formerly of the White Horse Hotel.

1919, Tuesday, 23rd September, at 6 p.m. – Dan Crossley and Crosland at the White Horse Hotel offering at a Sale by Auction 'Plans and Particulars of Valuable Freehold Shops and Dwellinghouses, Fully Licensed Public House, Business Premises, Building Land and Ground Rents, in the centre of Hebden Bridge.

1960 the hotel was closed.

1962, Monday 26th November – during the afternoon demolition workers moved in. In the evening of the same day, the Light Opera Society gave a performance of the operetta 'White Horse Inn[10]' at the Carlton Ballroom across the street.

The site with its two holly trees was bought by the council for £3,500 and by 2004 had become a car park. HGH, CDA Ref: FW: 39 287, Mrs. Diana Monahan and P.W. Robinson – Local Historians. 1842 P; 1871 PO; 1889-1936 K.

**Holme** 1849 S.

**Howden Clough, 274 Leeds Road** c.1687 the Gott family are thought to have owned the inn with a number of surrounding cottages which were shown on an 18th century map of Yorkshire.

1788, Christmas Eve – the publican, John Gott, died, his family had lived in Howden Clough for many years.

1791 Jeremiah Carter was listed as landlord.

From c. 1820 to c.1840 the pub was owned by John Gott's grandson, also named John, he lived at Clough House and was the local Squire; he was recorded as a distiller and gentleman. He also owned the houses – including the historic ABC row – and the fields on both sides of Leeds Road, one of which became the Birstall cricket field. Some of his domestic tenants also occupied the pub as landlords while continuing their own trades, their shop being incorporated into the pub – in 1826 Arthur Lobley, a victualler and bobbin and shuttlemaker, was landlord; there was also a butcher, Benjamin Rhodes and John Brooke, a cordwainer[11].

The outbuildings included stables and a cart shed which was on the opposite side of the road.

During the 19th century Samuel Jackson owned the inn and his tenants living on his estate would hold their Rent Dinners here.

c.1845 Eliza Jackson, the granddaughter of the last John Gott, took over the ownership.

1865 the house was partially rebuilt to have a split-level adjustment with both newer and older features; the frontage was extended up to the nearest cottage and it is thought the ground floor was lowered by several feet to a new street level which could be attributed to the coming of the railway when a shallow cutting was dug in Leeds Road from one side of the pub, lowering the road to accommodate the building of a railway bridge.

1872 a celery show for local amateur gardeners was held here, this became an annual event which developed into the Howden Clough Floral and Horticultural Society. In 1880 one exhibitor cheated by placing two pieces of lead in a brace of celery – he had to forfeit his prizes and was barred from future shows.

1882 Mrs. Eliza Jackson entertained her tenants to Christmas dinner at the inn, her oldest tenant William Clegg aged 74, presided for the evening.

1883 a bowling green of competition standard was made on the premises for the Howden Clough Bowling Club. The green became a car park during the 20th century.

1891 Eliza Jackson died and as she had no children her property, which had formed the basis of the hamlet, became a trust and was sold in separate parcels over the next ten years. Elizabeth Child was landlady at this time and paid £1 rent per week. The pub, with its outbuildings, adjoining cottages, the old cart shed and some land was bought by Kirkstall Brewery.

1902 Sam Fozard was licensee until 1913 when Joshua Knowles took over briefly to be followed by Fred Fozard in 1914.

1912, 4th June – listed in a Trust Deed of Dutton's Blackburn Brewery Ltd.

c.1927 the landlord was said to have banned the playing of darts, dominoes and cards.

1930 Samuel and Elizabeth Walker ran the pub for twenty two years.

1970 Whitbread Brewery became the owner.

*The Author was informed by the barmaid "We're haunted: a nun – several seen it luv";* *when asked if the building was old, she said Yes, **really** old luv".* The nun was allegedly executed during the 17th century and her body thrown down the well at the back of the pub. One winter a mysterious female voice was heard singing 'The Holly and the Ivy' for about 10 seconds by Brian and Valerie Watt – landlord and landlady from 1989.

2007, Thursday, 20th December – the *Batley News and Birstall News* published an article by Malcolm Clegg whose great-great grandfathers were John Brooke, a publican of the White Horse and William Clegg, both former tenants of the Gott family. It recorded that Howden Clough would face its first pub-less Christmas in at least 230 years with the closure and demolition of the White Horse – it had been one of the oldest pubs in the district. Rod Kaye, Local Historian and WP. 1834-1842 Birstal P; 1849 Birstal S; 1864 S; 1889 Batley – 1936 Birstal K. An hotel.

## Huddersfield

*2 Cornmarket* The detailed survey maps of the Ramsden Estate dated 1778, 1780 and 1797, show the substantial Plot No. 470 as the White Horse inn. It appears that the fold, barn, stables and shops standing nearby the cornmarket may have belonged to the owner of the White Horse Inn, as these are bracketed together for the valuation of properties in the 1797 survey book.

1842 Pigot lists William Booth as licensee and George Clay as proprietor.

1857 the White Horse inn was pulled down in 1857 – according to a pencilled note added to the 1797 survey.

*761 Leeds Road* 1864 listed as trading in Slater's Directory.

There is an undated reference to this pub in the Huddersfield Police records giving the owner as J. Ainley, brewer of Lindley Moor.

WYA and P.W. Robinson – Local Historian. 1822 -1842 P; 1849 and 1871 PO.

**Hunslet, Low Road** 1849 the address was listed as 87 Low Road.

1856 listed in Gillbanks Directory with Thomas Wood as landlord.

1864 the address was listed as 102 Low Road.

1888 listed in Kelly's directory as 76 and 78 Low Road.

C&A. 1826 WP; 1842 Hounslet P; 1834 B&N.

**Ingrow, Wesley Place / New Road Side**
1794 the new Keighley to Halifax road was opened and the population of Ingrow expanded, thus new suburbs such as Wesley Place and New Road Side were opened up. By 1854 three beerhouses had opened.

1863 Timothy Earnshaw was recorded as proprietor of the beerhouse and brewery which was to become the White Horse.

By September 1865 it is thought that Mr. Earnshaw had either leased or sold the brewery to Henry Hargreaves Thompson – known as Harry Tap – who was to become the owner of a number of public houses.

c.1868 Mr. Earnshaw secured a full licence for the White Horse.

1871 the premises were sold to Henry Thompson who expanded the brewery, moving it over the road opposite the pub.

1877, 2nd December – Henry Thompson died aged 54. His executers continued to manage his large business for many years until it was taken over by Frederick Binns, a former manager of the White Horse brewery, which by this time had a number of tied houses.

1884 the White Horse Hotel is recorded A. Craven's Directory as one of the principal hostelries.

1897, March – the brewery was sold to Messrs Scott and Co. Ltd. of Skipton Brewery.

1898 the White Horse Brewery plant was offered for sale by auction.

1969, Easter Tuesday – the Brewery and the inn closed as the licence renewal had been refused due to the condition of the building, which backed on to a cliff and was very damp.

By 1988 all the public houses which had been owned by the White Horse Brewery had closed. BLS and E. Kelly – local historian. 1889 Keighley K; 1897 'White Horse Brewery' BHS; 1912-1936 K.

**Jackson Bridge, Scholes Road** An impressive stone-built inn dating from 1830. Described as 'a very old, characterful pub' which, from 1982 was the main location of the BBC's comedy, *'Last of the Summer Wine'* and where much of the action was filmed; the car park being used to park the film crew's generator. The publican at the time, Mr. Ron Backhouse, was the only

publican in the country to appear in a TV show behind his own bar.

2006, January – a new licensee took over and refurbished the inn – wooden beams were restored; original stonework was exposed in one bedroom with back-lit alcoves; an oak wall with an inset wardrobe was uncovered in another, and original floorboards and wooden beams were exposed along with original stonework in a third bedroom. Framed photographs of the series are displayed on two of the bar walls – the inn owns some two hundred.

Excellent accommodation is provided and well-behaved dogs are welcome.

L&M 6th November 1996. 1834-1842 Holmfirth P; 1849 Holmfirth S; 1864 S;1889-1897 K; 1936 New Mill K.

**Kirkheaton** 1834 P.

**Ledston, 30 Main Street** Built in c.1774, with sturdy limestone walls. A farm was attached to the inn until 1948. The family's living room was also the taproom which had a huge Yorkshire fireplace with an enormous chimney and two capacious ovens. Stone steps were found inside the chimney when alterations were made for the installation of a modern range. Dick Turpin and other similar characters are reputed to have sheltered here. DGJ. 1912-1936 K.

## Leeds

In 1871 the Post Office listed nine White Horse inns under Leeds and in 1889 there were eight listed. With the advent of the railway many inns lost their trade.

*1 Armley Road* 1888 Charles Edwin Martin was listed as licensee.

*14 Boar Lane / White Horse Street* In 1822 Pigot's Directory recorded that coaches ran from the White Horse Inn, Boar Lane – George Newlove was licensee:

'Royal Union – to London, through Ferrybridge, Doncaster, Retford, Newark, Grantham, Stamford, etc. daily at 12 noon, arrives in Leeds (on return) at ½ p. 9 morn.

True Briton Post Coach (6 insides) daily, at 12 at noon, by Millbridge, Huddersfield, Junction & Oldham, to the Swan Inn, Manchester, returns to Leeds at ½ past 8 night.

Royal Union (4 insides) arrives from Ripon at ½ past 9 morning, and returns through Harrogate at ½ before 2 afternoon, daily.

The Hope (4 insides) every day (except Sunday) at 6 morn. To Selby where it meets the Steam Package for Hull, whose arrival is waited for to convey passengers to Leeds.'

*As seven of the above towns have a White Horse inn – the Royal Union to London having three on its route and the True Briton Post Coach with two – could the coaches have stopped at these other White Horses?*

1834 listed as the White Horse Hotel.

1856 listed in Gillbanks Directory as the White Horse Hotel – Commercial and Posting.

1864 listed as a Commercial and Family Hotel, but it was not listed in the West Riding directory of 1867.

1869 a photograph of the inn was included in a booklet of photographs entitled 'Lantern Slide Leeds'.

1904, 31st December – the Yorkshire Evening Post mentioned 'the old White Horse Inn' in an article on Boar Lane. YHD, B&N and DL. 1826L; 1842 P; 1849 C&A; 1898 hotel K.

**21 Carr Square** 1849 C&A.

**243 Meadow Road / Court 5 Meadow Lane** 1842 W; 1888 K.

**183 Meanwood Road** 1864 C&An.

**129 Wellington Road / Wellington Bridge** This bridge spans the Leeds and Liverpool Canal and the River Aire.

1864 listed as an hotel at Wellington Bridge.

1888 Charles Edwin Martin was listed as licensee of the White Horse Hotel at 22 Wellington Road. 1870 W.

**202 Woodhouse Lane** 1864 Slater lists Thomas Greenwood as licensee.

**360 York Road** Situated about a mile from Leeds city centre on the A64; the house is thought to have replaced an earlier building.

1888 Kelly's Directory gives the White Horse Hotel as at Osmondthorpe Terrace, York Road.

2008 the pub is listed in the Leeds Beer Drinkers' Companion.

1822 DL; 1826 L; 1834 B&N; 1842 W; 1849 C&A; 1864 C&An 1870 W.

**Lepton, 315 Wakefield Road** The original inn was a typical two storey 18th century building with beer and keeping cellars; the tap room, bar, snug, kitchen

and a 'best room' were on the ground floor, the first floor comprised the innkeeper's accommodation and a club room. The outbuildings consisted of a mistal[12] for twelve beasts, a three stall stable, a loose box, three sheds, two pig sties, two earth closets, an ash place and a wash house'.

It is thought to have been a coaching inn as it was situated near to the toll gate on the Wakefield to Austerlands Turnpike – there was also a smithy on the opposite side of the road.

1780 the inn is shown on the Lepton Township Map set diagonally from the road, surrounded by a semi-circle of land.

1798 The Lepton Field Book records George Wood, innkeeper, as a tenant who held the fields near to the inn – namely Rugh Royd, Square Royd and New Close – amounting to 17 acres.

1800 the building was described as 'moderate good' but the barn needed repairing. The rental records George Wood as publican, his family ran the inn for at least the next eighty years.

1820 a new turnpike road was built, the old one having been built in 1759.

1928 the buildings and land were sold by Mr. H.R. Beaumont.

c.1930 the buildings were demolished and the present public house was built.

HLS and GEM. 1822 B&N; 1842 P; 1849 S; 1864 S; 1871 PO; 1897-1936 K.

**Lightcliffe, Leeds Road** Originally this building was three cottages situated on a highway known as the Leeds and Whitehall Road.

1832, July – Mr. Samuel Butterworth, a currier[13], applied for a licence to sell excisable liquors for consumption on the premises, which consisted of four rooms on the ground floor, three chambers, a cellar, a brew house and sundry outbuildings – the licence was granted on 28th August, at least 61 respectable inhabitants had signed a petition supporting the application. It was the only inn on this road between Leeds and White Hall, a distance of twelve miles; built next to a chapel it was a convenient venue for funeral teas.

1881 the address was recorded by Kelly as Bramley Lane – the end of the road between a shop and the White Horse Inn led into this lane.

1889 Crossley Wolmersley is listed as licensee. K.

1903, 5th January – a Deed of Exchange between Joseph Stocks & Co. Ltd with William Shillito of Halifax, Clerk records the exchange of 'a triangular plot of land near the White Horse Inn in Lightcliffe for a piece of land part of a close called the Goff Close near the White Horse Inn in Lightcliffe'.

1847 Whites Directory lists the White Horse under Hipperholm, Bramley Lane, this

was a misprint for the 'Whitehall' at Hipperholm which was still trading in 2008.

1982, Thursday 7th January – the *Evening Courier* published an article about the area around the old Lightcliffe toll bar, this included a picture showing the old cast iron bollards, two large ones to support the gate and small ones at the side to stop carts and animals going through without a toll being paid. The White Horse is situated just beyond the toll bar and apart from the addition of a porch entrance at the front and a car park to the rear it has hardly been altered.

CLH, CDA Ref: FW: 113 21, P.W Robinson – Local Historian and 'Bar to Progress' by Jack Redfearn. 1842 P; 1849 S; 1853W; 1864 S; 1870W; 1922-1936 K – these dates were listed under Hipperholm. 1897-1936 K.

**Lofthouse** 1849-1864 S.

**Luddenden Foot** 1898, June – Halifax Brewery Co. applied to build a urinal at the White Horse Inn.

1924 James Alderson and Company Ltd. applied to build WCs.

CDA Ref: BIP/L 128 and 237.

**Morley, Town End** This small public house was still trading in March 2008.

1830 B&N; 1834-1842 P; 1849 S; 1864 C&An; 1888-1936 K.

**Normanton** 1889 Listed as trading. K.

**Northowram, Lands Head** A beerhouse which was also a farm. The main customers were workers from nearby quarries which gradually declined and from 1913 no more licence applications were made. P.W. Robinson – Local Historian.

**Old Sharlston / Sharlston, 49 West Lane** The original licensed house, which lay well hidden from the main road, was named The Bay Horse until the mid 19th century.

The date of the building is unknown, although it is said that the highwayman Dick Turpin (1706-1739) visited this inn.

1797 the kitchen was extended to accommodate both the brewing of the ale and the cooking.

1842 listed in Pigot's Directory as White Horse.

1845 Abraham Atack is listed as landlord; his family continued to run the inn until 1955.

1870, Saturday 5th February – The New Sharlston Coal Company Limited held their annual supper for their workmen here.

1892, Thursday 25th February – The Old Sharlston Social Club held their annual dinner here.

1897 listed in Kelly's Directory.

1930, 22nd March – an article was published in *The Yorkshire Observer* together with a photograph showing builders at work as alterations were being made. The author records that his ale was served in a pewter tankard which had belonged to four generations of the landlord's family – the landlord being the great great grandson of Abraham Atack. Seating in the taproom consisted of long, high backed seats, one dating from c.1830 and another had once been in the Earl of Westmorland's private chapel – built in 1591 at Sharlston Hall[14] – along with the ancient church pew upon which the author was sitting. The double-faced fireplace with its huge chimney dated prior to 1780 – it had a hiding place big enough to conceal a man. Above the fireplace was a large stone slab divided into twelve numbered squares, this is where the customers' bills were written in chalk. Seating consisted of long, high backed seats – Other curiosities included an old oil-painting of King Charles, the handle of a stage coach door and the spit which once roasted the beef.

1998 the licence and the landlord, Edwin Bell was transferred to West End House, Sharlston which was renamed The White Horse.

Wakefield Local Studies Library and Rod Kaye – Local Historian. 1842 P; 1897-1936 K. Hotel.

**Osmond-Thorpe** Osmondthorpe or Osmanthorpe. *See Leeds.* 1842 P; 1849-1864 S.

**Otley, Manor Square** 'Royal White Horse Hotel'. 1718 is the earliest record of an inn on this site.

1776 it is thought that the inn had been rebuilt by this date when it was first recorded as trading under the sign of White Horse.

1796 Yorkshire's oldest agricultural show – the Otley Show – took place for the first time and provided many customers for the inn.

From c.1805 the White Horse was mainly associated with the Wharfedale Agricultural Society who organised the Otley Show.

It became a prominent coaching inn and during the first half of the 19th century the Leeds to Kendal Union Coach was horsed here.

1865 the late 18th century building was encased in a re-built block.

1876 the twenty six year old Duke of Connaught, who had been promoted to Lieutenant Colonel in the Rifle Brigade, visited Otley with his regiment en route from Liverpool to Edinburgh – and stayed with his staff officers at 'The White Horse'. As a result the inn secured 'letters patent' to use the Royal Coat of Arms and prefix the word 'Royal'.

1905 Mr. John Mudd, the proprietor, had a natural gift in the art of catering, he was assisted by his wife and their food was described as 'perfection'. The inn was a remarkably convenient and comfortable commercial hotel with every modern appliance, a tall building standing in a prominent and commanding position at the corner of Manor Square and Westgate, less than a minute's walk from the Jubilee Clock tower in the Market Place. On Market Day farmers congregated here to transact their business; property sales frequently took place in the upper room. At the rear of the building there were lock-up coach-houses and loose boxes.

At this date the hotel was the headquarters of the Cyclists' Touring Club.

1973, 28th September – the inn was closed. There had only been seven landlord families here since 1876; the final family being the Churchmans who came in 1946 and stayed until the end.

1976, April – the premises became Barclay's Bank.

Paul Wood – Museum Consultant and Leeds Central Library, AR and HW. 1826 L; 1834-1842 Posting House P; 1849 S; 1864 S; 1871 PO.

**Pudsey, 2 Hough Side Road, Lowtown** 1849 and 1864 listed as trading in Slater's Directory.

By1889 Kelly's Directory lists it as an hotel.

1936 K.

**Queensbury, Thornton Street** 1895 listed as a beerhouse with John Naylor as landlord and owned by Major Stocks, an important land owner in Queensbury who had several coal mines in the area; he also owned a large brewery at nearby Ambler Thorn.

1925 the pub was closed with compensation.

*A John Naylor of Warely Springs Brewery was owner of the White Horse, Bowling Back Lane, Bradford in 1884.*
P.W. Robinson – Local Historian.

**Rastrick, 46 Rastrick Common** 1837 listed as trading in White's Directory and again in Slater's Directory of 1848.

1871 listed as trading with James Cardwell, he was succeeded by Sarah Cardwell.

1882, October – Sarah, who had kept the pub for 14 years, appeared before the Brighouse Magistrates court. One Sunday morning at 10.20am she had refused to open her door to the police when they were doing a routine inspection on local pubs because it was before opening time. She was fined 2s. 6d. with 7s. 6d. costs.

By 1892 the brewer, Richard Whitaker & Sons Ltd. had bought the house.

1967 it was described as 'a small popular pub which becomes rather crowded . . . '

1968 Whitbread & Co. Ltd. took over Richard Whitaker.

Owned by Richard Whitaker Ltd., brewers of Halifax before Whitbread took over in 1965. RW and P.W. Robinson – local historian. 1922-1827 K.

**Rishworth, Turner Bottom** Standing in a terrace of stone built cottages, the White Horse was known to have been trading before 1820 but had closed by c.1860.

2008 the cottages still stand beside the A672. P.W. Robinson – Local Historian.

**Sharlston, 49 West Lane** *See Old Sharlston.*

**Thornton, 33 / 34 Well Heads** Situated high on the Pennines above Thornton, it is thought that the White Horse Inn – a free house – was once one of the principal public houses in the area.

1842 listed as trading at Well Heads.

1999, 9th July – advertised in the *Telegraph and Argus* as a traditional country inn with large fireplaces and a beer garden, having opened again after refurbishment.

2005 the inn was bought by the Timothy Taylor Tied Estate.

2006, 24th November – the house re-opened after alterations. The bar had more than doubled in size and a new kitchen had been installed, along with new lavatories. The owner, Timothy Taylor, is the last independent family run brewery in West Yorkshire. BCL. P.

**Triangle, 2 Stile Bank** Originally two cottages, The White Horse Inn was situated on the mules' route – the Rochdale / Sowerby Bridge Road at the edge of the village; it was a typical small country inn of the middle nineteenth century and was first licensed in 1865.

An elderly resident remembers he first visited the inn as a child having been invited to the birthday party of the daughter of the licensee. He remembered there were two entrances into the building, one to the tap room and the other to the best room. The inn, which only sold ale direct from the barrel, was always popular and thrived throughout the 1920s and 1930s. One of the last licensees was Jim Harrison, a retired Metropolitan Police officer.

1955 the inn – a free house – was closed as the owners decided that it would not be practicable to modernise the premises; the renewal of the licence was refused and the inn referred for compensation.

A bungalow has been built on the site. *Halifax Evening Courier* of 16th July 1955 and 3rd October 1996.

**Wakefield, 76 Westgate** A coaching inn and hotel.

1558 a 'Whyte Horse' is recorded as trading.

1793 the White Horse was listed as trading.

1842 Pigot's Directory lists George Lyle as licensee and in 1848 Slater lists Benjamin Potter.

The following article was supplied by Mr. Rod Kaye, a Local Historian:

1889, 25th June – Police Constable Pearce went to the White Horse at about 9.30 pm where there was a great commotion due to a boxing competition which was about to take place in the singing-room, he asked if he might go in; he found two people taking tickets of admission from those coming to watch the fight. There was a ring formed with a rope and seats were set around it; the room was full of people. After a few minutes the landlord made an announcement that there must be no betting, or he would have to stop the fight – *probably due to the presence of the policeman.* Shortly afterwards two young men accompanied by four others came into the room, two pulled off their coats to reveal fighting dress – tights, vests, trousers, shoes and stockings. Another man held up a gold medal and announced that the men would fight under the Marquis of Queensbury's rules for the medal valued at five guineas. The two men put on boxing gloves and examined them. Shortly before 10 o'clock the ring was cleared of people and the combatants exchanged blows and fought until they were exhausted, one of them being unable to rise from the floor without assistance. The room was cleared at the end of the fight.

1889, Wednesday, 17th July – Samuel Duke, the licensee appeared in court charged with unlawfully permitting violent, quarrelsome and riotous conduct in his public house.

The Magistrates only imposed the mitigated penalty of £2 and court fees of 16s.0d. because it was customary for these sparring matches to take place on licensed premises, although in this case it was considered that there had been an infringement of the Act of Parliament.

1897 listed as trading in Kelly' Directory.

1900 the public house was rebuilt.

1996 it was renamed the Forehorse[15] and Firkin, having been the Frog and Firkin – it later became Bing Bada Boom.

The stable yard, which became known as White Horse Yard, is one of the oldest yards in Wakefield; it was renovated in 2000.

BC, D&W and Wakefield Local Studies Library.

1822 P; 1848-1864 S; 1871 PO; 1936 K.

**Warley, Spring Gardens** 1842, April. The White Horse was sold by the Stansfeld family of Sowerby. It was still trading under this sign in 1975. CDA Ref: STN 189.

**Wetherby, High Street** There were a large number of hotels and inns, originally being situated on the Great North Road – half way between London and Edinburgh.

Twelve of these were sold by auction at the 'Great Sale of Wetherby' in 1824.

This inn was known as the Blue Boar and was renamed White Horse; by 1871 George Dreusis is listed as a publican and blacksmith at the White Horse.

WDH and PO. 1864 S; 1889-1912 K.

**Whitkirk** 1834 P.

**Woodhouse Carr** Between 1826 and 1834 Benjamin Rhodes was listed as licensee and a carrier; by 1842 Henry Kennedy was licensee and listed as a carrier.

WP. 1856 listed in Gillbanks Directory; B&N; 1842 W.

**Wortley** 1842 P; 1864 New Wortley S; 1871 New Wortley PO.

## Notes

1. Possibly 'sorted'.

2. The process of cleansing and thickening cloth by beating and washing. SOE.

3. A pollard (tree). 1691. SOE.

4. A slang term for a dice game called 'Crown and Anchor'. Also a word used in the Victorian period to describe a savings plan.

5. A large wicker basket. SOE.

6. The extinguishing or hiding of all artificial light from an enemy attack from the air, especially during the 1939-45 War. Col.

7. A cow shed or byre. Col.

8. Introduced in Britain in 1804 to protect domestic farmers against foreign competition by the imposition of a heavy duty on foreign corn. Repealed in 1846.

9. A hard-wearing fabric of cotton mixed with flax or wool with a slight nap. Col.

10. See Note 1 for Gort, County Galway in Southern Ireland.

11. A shoemaker or worker in Cordovan leather. Col.

12. See note 7.

13. A person who grooms horses – hence currycomb; a person who cleans, dresses and finishes leather after it has been tanned. Col.

14. The residence of Nicholas de Fleming, the Earl of Westmorland and the Stringers' family.

15. The foremost horse in a team, leader. SOE.

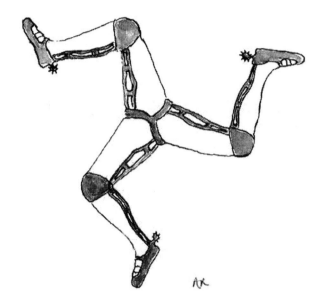

# ISLE OF MAN

*According to Manx mythology, the Celtic sea god, Manannán mac Lir, ruled the island and protected it from invaders by drawing his misty cloak around it.*

*He defeated intruders by rolling down a hill in the form of a triskelion, three bent legs joined at the thigh with a spur at each heel. This gives the island its symbol and motto "Withersoever you throw it, it will stand".*

# ISLE OF MAN

*The world's oldest parliament, dating from at least 979AD,
was formed on the Isle and continues to this day. In 1079,
Godred Crovan created the Norse Kingdom of 'Mann and
the Isles', which by the 14ᵗʰ Century was ruled by the British.
Since 1765 it has been a self governing Crown dependency
whose head of state is the Queen.*

## Douglas

***King Street*** 1837 listed as trading in Pigot's Directory with John Clague as innkeeper.

***St. Martin's Lane / Crooked Lane, North Quay*** The Wood's and Taggart's town plans of 1833 and 1834 show the street as St. Martin's Lane, by 1866 the Ordnance Survey map shows it as Crooked Lane.

    1846 listed as trading in Slater's Directory with Eleanor Clague as licensee.

    1857 listed as trading by Slater.

    By c.1890 the whole area was being demolished.

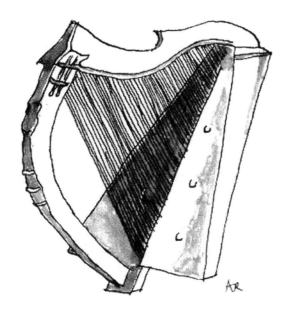

# NORTHERN IRELAND

*Traditionally King David, an early Irish king, took the harp of the Psalmist as his badge. Henry VIII (1509-1547) was the first to adopt it as the Irish device, and James I (1603-1625) placed it in the third quarter of the royal achievement of Great Britain.*

# County ANTRIM

**Aldergrove, 20 Dungonnell Road** The building dates from c.1880 and might have been built by John Moore (c.1858-1948) who was the first landlord when the pub opened in c.1940 during the Second World War. Aldergrove Airport was a busy flying base – 'Langford Base', both the RAF and American camps patronised the White Horse.

The isolated house was situated at the centre of the Kilultagh, Old Rock and Chichester Hunt, whose kennels close by at Dundrod, had good sport chasing the airport hares – followers were entertained generously at the Officers' Mess before moving off – this was sometimes destabilising for them, as was the wire mesh concealed in the mud of the runways which often removed the horses' shoes.

1948 John Moore died. He was blind, but always recognised people correctly by their voices – he spent his days making fishing nets. The pub was taken over by his son, David Moore.

1974 the building was bombed.

The business was eventually taken over by Bobby Barrow which he ran from 1978 to 1994.

The oldest part of the building is the snooker room; the inn was gradually extended over the years. The friendly and attractive bar and restaurant were busy at weekends.

By 2006 the licence had been withdrawn, the house uninhabited and up for sale – though the door was unlocked and the author was able to enter and look around!

### Belfast

**Corporation Street** Situated in the dockland of Belfast and catered mainly for sailors and dockers. By 2003 the inn had closed. MG.

**25-29 Garmoyle Street** 1891 listed as trading by the Federation of the Retail Licensed Trade, Northern Ireland.

By 2003 the inn was closed but by 2007 it was listed as trading under The Clarendon.

# County ARMAGH

**Armagh, Scotch Street** A 17th century inn. By 2003 the inn had closed. ACM.

**Lurgan, 2 Church Place** By 2003 the inn had closed.

# County DOWN

**Ballynahinch, 17 High Street** A coaching inn on the route Downpatrick – Ballynahinch – Lisburn. At one time named 'Rogan's Bar'.

The cellars used to be used for an illicit trade in corpses which were stolen from the Maghadroll cemetery and sold to medical students for dissection. The corpse was no use if allowed to remain in the ground more than three days, and anxious relatives used to keep watch to circumvent the thieves.

1946 the bar was bought by Kevin Poland. Soon afterwards the neighbouring house was burnt down, probably in a bid to get insurance money. The Polands bought the burnt out shell and built the White Horse Hotel.

There is an outstanding collection of 'Snaffles[1]' prints and sporting pictures; also elegant glass engravings of horses on doors and windows which was the reason the house was renamed The White Horse because of the engraving on the bar window depicting a grey hunter mounted by a member of the Co. Down staghounds, with whom Mrs. Kevin Poland used to hunt. Ballynahinch is the centre of the Co. Down staghound country and the hounds meet every St. Stephen's Day in The Square.

It is run primarily as a commercial hotel and has a large function room.

2006, Sunday 23rd April – A helpful resident buying his newspaper maintained that he had stayed in the White Horse Hotel in 2004 – since when this large site has been converted into a car park and a branch of the Ballynahinch Credit Union Bank due to start trading during 2006. MG.

**Carrowdore, 39-41 Main Street** A quaint building. There is an oil painting of the inn with a meet of the North Down Harriers taking place in the foreground.

By 2003 the house had closed. MG.

**Saintfield, 49-53 Main Street** The building dates from the 17th century and is likely to have been a coaching staging post. Although not listed, this is a fine long four-storied building, including the cellar; it is situated next to the church and is surrounded by old buildings, which are listed.

Horses are of particular interest locally. A former owner of the inn, Sam Spratt, had been a farmer and had worked with horses all his life, this is shown in the large collection of paintings, horse brasses and harness which adorn the walls of the many rooms and Bars.

The ancient small stone building in the yard was at one time a workshop for building carts and other horse-drawn vehicles.

The inn has three bars – the Snug Bar, the Paddock Bar and the Cellar Bar, which has its own entrance from the street via an iron gate and stone steps. By 2006 the cellar had been converted into the restaurant, known as The Oast. JJT.

# County **LONDONDERRY**

**Dungiven, 160 Main Street** An old building, originally the stables attached to a hotel.

1947 by this date the hotel had been demolished and the bar was bought by Hugh O'Hagan when he retired to his home town after serving in the American army. He named the house 'O'Hagan's White Horse Whisky' after the popular brand of Scottish whisky.

2003 the illuminated sign is a landmark in the town and the third generation of O'Hagans are proud of their family's long association with the bar. MG.

**Eglinton, 68 Clooney Road** Built on the site of an 18th century inn named The White Horse.

c.1910 the house was owned by the Cole family and for a time was known as 'Cole's of Campsie' – Campsie being the name of the district.

A photograph dated c.1950 shows the original small wayside inn with the local hunt meeting outside, this would probably be the Route harriers, Northern Derry being their country.

1983, 20th November – an article in The Sentinel records that 'In the fanlight above the entrance door a little figure [of a white horse] stood for many years. It may well have been issued by a well-known firm of distillers . . . '

By 1991 the house had become a large and luxurious hotel bearing little resemblance to the original roadside inn, but it still retains the stone-flagged floor, fireplace, snug, porch and front door of its predecessor.

Some local people still believe in the legend of the phantom coach drawn by four elegant white horses – it was seen by the Cole family and their servants on several occasions.

1965, 4th December – James McElwee gave this report of his experience in 1959 for the *Coleraine Chronicle*:

> "I saw four beautiful white horses and a coach without wheels, encircled by a glow of light. It rumbled along the left hand side of the road, and drove round to the yard of the Inn, where the old mail coach once watered and awaited its passengers while they had a meal. The driver was dressed in old time garments and was a tall, stout, friendly sort of man. I enquired inside the inn, but no-one else had seen the driver of the coach, and when I looked out again, the coach and horses had gone."

Later the same year, Henry Doherty also saw the glittering coach passing his house:

> "Beheaven, the bite I was eatin' almost choked me. It was floatin' in the air, and although I couldn't see the horses, I could still hear the rat-tat of their hooves fading away in the distance."

This apparition is said to appear every seven years and often called for celebrations in the Campsie district as it was never associated with a crime or weird haunting, but in honour of the food and wine for which the inn was famed.

c.1976 the hotel was bombed by the IRA.

2002, 26th April – the hotel, had been rebuilt and extended as a conference and leisure centre to serve the large commercial community in the surrounding area; there were fifty six 'en suite' bedrooms, swimming pool, sauna, steam room, and gymnasium. The new complex was opened by Sir Reg Empey, MLA, the Minister for Enterprise, Trade and Investment; the project had been funded by the International Fund for Ireland and the Northern Ireland Tourist Board.

MG, JJT & Mr. & Mrs. Ross Mowbray.

**Knockcloghrim, Nr. Maghera** By 2003 the inn had closed.

# County **TYRONE**

**Dungannon, 69-70 Scotch Street** Listed as trading by the Federation of the Retail Licensed Trade, Northern Ireland, but by 2003 the inn had closed.

**Notes**

1. The military and sporting artist Charles Johnson Payne (1884-1967).

# SOUTHERN IRELAND

## EIRE

*In the 5th Century, St. Patrick introduced Christianity to Ireland; he used the three leafed shamrock to illustrate the Holy Trinity.*

# County CAVAN

**Cootehill, Market Square / Market Street** As there was stabling, the inn was probably a coaching inn on the Dublin – Enniskillen – Ballyshannon route. Cootehill became a railway terminus, horse-drawn coaches being provided for passengers wishing to travel further.

There appears to have been a White Horse Inn early in the 19th century as Mrs. Theresa McBeen, whose grandfather was waiter at the inn, remembers the story of her great grandmother, as a child, being asked to go up to the White Horse for a glass of white whisky for her mother, the cost being 2d.

A modern mirror behind the bar is engraved to portray riders at the gates of the local castle. A horse's head is depicted on the doors leading into the lounge bar.

2007 The White Horse Hotel was still flourishing. MG.

# County CORK

**Ballincollig** 1840 established as a public house. It was popular with the army when soldiers were stationed in Ballincollig Barracks, but later veered towards republicanism and by 1901was known as Walter's Pub. Walter Murphy had two sons; one, Leo, became a commandant in the IRA and his movements were monitored by the Royal Irish Constabulary stationed in the square opposite the pub. They were alerted to his being at his parent's place by seeing his white horse tied up outside. He was shot at Ballyonore on 29ᵗʰ June 1921 and his body brought back to the Royal Artillery barracks in Ballincollig.

1970 the last of the Murphy family died and the premises were sold to Jonathon O'Donoghue. It was he that named the White Horse Inn after Leo Murphy's white Horse. MG and Beamish & Crawford Ltd.

**Cork, 26 Parnell Place** c.1940-1965. Beamish & Crawford Ltd.

**Mitchelstown** 1989 renamed 'An Capall Bàn' – literally, 'Inn Horse White'. MG.

# County DONEGAL

**Ballyshannon, Riversdale, Westport** Sweeny's 'White Horse Bar and The Cellar'.

White sea horses painted by Barry Sweeny abound in and around this thriving inn and restaurant run by a fourth generation of Sweenys. The inn is on a tourist route which links the lovely Fermanagh Lakeland area of Upper and Lower Lough Erne with the sea-side holiday town of Bundoran.

A landmark in Ballyshannon's history was the new hydro-electric scheme entailing the building of the great barrage and a necessary realignment of water-ways and streets. Salmon and Sweenys were equally inconvenienced. The inn found itself out on a limb and it was necessary to move it a short distance. Easy enough to secure a new building but the transfer of the publican's license seemed a legal impossibility. By law the old building had to be demolished – or at least have its roof removed – before the publican's license, with the white horses, could be moved a few yards down the road.

In 1978 the present house was built a short distance away from the previous White Horse Bar next door to a filling station where motorists from the North could avail themselves of cheaper petrol prices than those to be found in Northern Ireland.

2007 the house was still listed as trading. MG & DCL.

# DUBLIN

### Dublin

*1-9 Benbulbin Road Drimnagh* 1950-1960 listed under P. Timmons, a Wine and Spirit Merchant.

By 1977 the house had been enlarged and was known as 'White Horse Inn, Vintners'.
1982 listed as trading.
1992 listed as White Horse Inn, publicans and vintners. The following year the house was renamed Fitzgeralds, Marble Arch. T.

*Burgh Quay* 2007 P. McCormacks, or The White Horse Inn is situated in the city centre; looking big from the outside, it is compact inside with an exotic-looking staircase leading to a higher floor.

*1 George's Quay* Originally named The Anchor. Admiral Bligh (1754-1817) – of The Bounty – made this inn his headquarters while re-designing the port of Dublin.

Situated beside the River Liffey on the corner with Corn Exchange Place and close to the offices of the Irish Press, this Georgian style inn was therefore popular with newspapermen, printers and journalists.

1935 the house was listed as White Horse Inn, when it had the all Ireland agency for White Horse whisky.

1946 listed in Thom's Street Directory as trading under White Horse Bar.

1950 the house was trading as 'White Horse Bar, Grocers, Wine and Spirit Merchants'.

1963 listed under M. O'Connell & Sons, White Horse Bar.

By 1995 the name had returned to the White Horse Inn and various bands would perform in an upstairs room called The Attic.

1997 the site was scheduled for redevelopment and the building was demolished the following year, to be replaced by a new building with a metal staircase.

2006 the house was still trading. MG, DUB, KE and IAA.

***Winetavern Street*** 1568 known as The Chamberlain's Inns.

1619-1646 listed as trading under White Horse House Inn and Cellar.

c.1862 a fire station was built in the yard. There is still a White Horse Yard off 12 Winetavern Street, presumably on the original site. RSA & ED.

# County **GALWAY**

**Gort, Crowe Street** 1972 the house was bought by Mr. and Mrs. Connelly who re-named the inn The White Horse. Mrs. Connelly told the Author they had chosen this name because "we love the music from 'The White Horse Inn'" – anything rather than Connelly's Bar."

By 2003 the inn had been renamed The Blackthorne.

**Headford, Main Street** Originally a coaching inn on the Galway – Castlebar – Westport run.

1975 purchased and renovated – black timbers and white stone walls. Food and accommodation were provided for both sexes, as previously there had only been stand-up bars for men in this town.

1982 the landlord, Mr. P.D. Woodgate, designed the inn sign showing the head of a white horse – the sign 'Whitehorse' (sic.)

# County KILDARE

**Kilcullen, Main Street** Situated on the coaching route from Dublin to Kilkenny. c.1910 the pub was known as 'Darby's'.

c.1940 Joseph O'Connell, the proprietor, was interested in light opera and named his establishment 'The White Horse Inn' after the musical of that name. *(See Note 1).*

# County LAOIS

**Mountrath, The Main Street** Named after the White Horse River. At the time of publication it has not been possible to discover any historical information.

In c.1995 there were three things of interest – its sign 'The Whitehorse Inn' (sic.); a trotting white horse engraved in the glass of the window; the advertisement "Guinness for Strength" displayed on the front wall, which showed a strong wagoner balancing on one hand his horse and cart. This was a white horse sitting in the cart playing a violin!

By 2003 the premises had closed and were up for sale.

**White Horse Lodge (on the approach road to Mountrath from Dublin)** Named after the White Horse River. Trading as a small restaurant with bed and breakfast since 1978, it was originally a cottage which has been extended over the years. White horse pictures and finely worked embroidery adorned the restaurant walls:

"The White Horse River flows from the Slieve Bloom Mountains, through the town and joins the River Nore. The White Horse Falls in flood, cascading down the mountainside are a fine sight." MG.

# County LOUTH

**Drogheda, 31-32 West Street** In 1690 the Battle of the Boyne was fought below Slane a few miles to the west of Drogheda.

Established in 1758, this was an important coaching inn on the Dublin to Belfast route.

1987, 29th January – the house was put up for auction, by this date the ballroom had become The Pegasus Nite Club.

By c.1990 the inn had been renamed The West Court Hotel.

2007 the hotel was still flourishing. MG. 1898 K.

**Dundalk, Linenhall Street** 'White Horse Salon Ltd. Lounge Bar'.

# County MAYO

**Ballyhaunis** Ceased to be a public house many years ago. The building became a supermarket. MCL.

# County MEATH

**Athboy, O'Growney Street** Originally 'Murphy's Bar' and an undertaker.

c.1980 the whole corner site was redeveloped as 'A Prestigious Commercial/Residential Development' and named White Horse Inn. ML2.

# County SLIGO

**Dunkineely, Main Street** VFI.

# County WICKLOW

**Newtownmountkennedy** 1982 listed as trading, but later burned down.

## Notes

1. 'Im Weissen Rössl' an operetta composed by Ralph Benatzky (1884-1957). It was used as the base of a musical comedy written by Oskar Blumenthal and Gustav Kadelberg during their stay at the White Horse Inn at Wolfgansee in Austria, it was first performed in Berlin in 1930 and later in Paris and New York. Araminta was taken to her first play at the Coliseum in London – The White Horse Inn – where she counted 21 live animals on the stage – since when most plays have been a great disappointment!

The inn at Wolfgansee dates from 1440 and has had many different names. In 1878 it was enlarged by incorporating the adjacent boarding houses and named Gasthof Pension Weisses Rössl, or White Steed or Charger.

Helmut Peter – owner in 1982 whose grandfather acquired the house in 1912.

# SCOTLAND

*"There is nothing yet which has been contrived by man,*
*by which so much happiness is produced as by a good tavern or inn."*

The Life of Samuel Johnson by
James Boswell, 21st March 1776

# GRAMPIAN

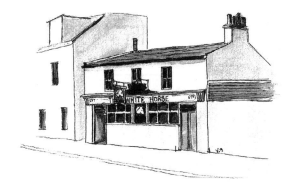

**Aberdeen, 695-697 Great Northern Road** 'White Horse Bar'.

**Balmedie, Ellon Road** A small hotel providing hospitality for golfers, as it is situated between some of the links in the north east of Scotland.

**Craigellachie** White Horse Distilleries Ltd.

**Elgin, 160 High Street**

**Strichen, 65 High Street** The building dates from 1794.

1985, June – the sign was changed from The Freemasons Hotel to White Horse Hotel, named after the white horse cutting on the Hill of Mormond nearby.

2002 the hotel won the UK Pub Food Awards.

Owner.

# LOTHIAN

## Edinburgh

*266 Canongate* 2007 the White Horse Cellar Bar with its narrow black frontage was a traditional local working person's pub and the smallest on the Royal Mile[1]. There is a single room on the ground floor with a bar in the corner and seating along the side wall.

*White Horse Close* Site of the famous ancient White Horse Inn in the Canongate dating from c.1603, the main hostelry in Edinburgh. It is thought that the Royal Mews of the Palace of Holyroodhouse once occupied the site.

1623 the inn and coaching stables were built by Laurence Ord, merchant and burgess of Edinburgh which he is reputed to have named after Queen Mary's white palfrey, 'Rosabelle', who carried her up and down the Royal Mile. He built houses and hay-lofts round the courtyard, presumably intending to keep it as a stablers' inn. It probably provided shelter for travellers to both the Abbey and the Palace, as described by Sir Walter Scott in 'The Abbot' using the synonym St. Michael's. A back entrance was for a time known as Lawrence Ord's Close and next to it a row of great arched doorways led to the stables. It is believed that the smithy attached to The White Horse was situated in nearby Duncan's Close – coaches and horses, including those for funerals could be hired nearby.

Marie Stuart in her 'Old Edinburgh Taverns' published in 1952, described the courtyard of The White Horse as being the scene of a gathering in 1639 when the leading Covenanting nobles 'Stoppit Stravaig' prepared to set off to meet Charles I at Berwick where he had 'invited them for a parley'. The Edinburgh ministers were suspicious of the King's intentions and organised a mob to forcibly prevent the nobles from leaving. Montrose alone escaped and reached Berwick:

'It must have been a dramatic moment when he galloped off from the White Horse, a tragic contrast . . . when as a wounded prisoner he . . . passed the White Horse Close . . . to trial and death'.

In 1745 the Young Pretender's[2] officers 'were quartered here while their leader was in residence at the Palace. One can picture the gay and gallant company carousing in the old rooms, the stable-lads busy in the great vaulted building at the back and the coming and going between council chamber and inn.'

1745 while Prince Charlie's army was occupying Edinburgh, the Prince was in residence at Holyrood and Jacobite officers used the inn as their headquarters. 1825 Chambers, in his 'Traditions of Edinburgh' stated that:

'The White Horse has ceased to be an inn from a time which no oldest inhabitant of my era could pretend to have any recollection of. The building still delights artists with . . . its 'cockit-hat' gables, dormer windows, outside stairway, overhanging balconies and flagged courtyard.'

The inn was patronised by literary and theatrical people, such as Dr. Samuel Johnson and his biographer James Boswell who had been educated in Edinburgh. In 1773 they stayed here on their way to The Hebrides; the actor and dramatist Samuel Foote was also a frequent visitor – *all three were painted by Sir Joshua Reynolds*. At one time the inn was well known to runaway couples - impromptu marriages could be arranged here.

The Inn was important for many years but trade diminished with the popularity of travelling by stagecoach and the increased use of the east coast route. The stagecoach left for Newcastle and London from the back entrance:

'The White Horse Cellar (was) the starting place for the London stagecoach, which departed on Mondays and Fridays at five in the morning. The journey was advertised as taking eight days (if God permits) and the intrepid travellers were allowed 14 lbs of luggage with a charge of 6d. a pound for excess luggage.'

Sir Peter Mackie chose the name 'White Horse' for his whisky having been inspired by the legendary background and the rich past of the White Horse Cellar. The White Horse label showed this historic link with the old White Horse Cellar with the original announcement, dated 1754, and the stagecoach pulled by a team of horses.

1845, an artist Horatio MacCulloch painted a view of the close, showing it in a dilapidated state.

William Dick (1793-1866), founder of the Edinburgh Veterinary College was born here. 1965 the Close was restored.

More recently the area has undergone several redevelopments, meaning that few of the buildings on the site are original. LH, GRM., WHD, B, and MS.

# STRATHCLYDE

**Dalrymple, 22 Barbieson Road** Originally a former church and manse which was converted into a public house and flat.

**Glasgow 4, Borron Street** White Horse Distillers Ltd.

# TAYSIDE

**Dundee, 8 Harefield Road**

**Kirriemuir, Bellies Brae**

**Perth, 5 North William Street** Hotel.

**Notes**

1. The road running from Holyrood House and the Abbey to the Castle.
2. Charles Edward Stuart (1720-1788), popularly known as Bonnie Prince Charlie. He was the son of the Old Pretender, James Francis Edward Stuart (1688-1766).

# WALES

*Allwedd Calon, Cwrw da*
*The way to the heart is through good ale.*

Old Welsh Proverb

*Tafarn y Ceffyl Gwyn*
*The White Horse Tavern*

*Both the leek and the daffodil have been adopted as national emblems,*
*as the welsh word 'Cenhinen' applies to both.*

# ANGLESEY

**Beaumaris, Church Street** 1835 listed as trading in Pigot's Directory; by 1844 it is no longer listed under White Horse.

**Bryndu** Between 5th July 1857 and 5th March 1859 Thomas Williams records his address as 'White Horse' in his Anglesey Savings' Bank book. ACR.

**Llanerchymedd** 1832, 10th September – Hugh Williams is recorded as having a Publican's Licence for the White Horse. ACR.

**Llangefni, High Street** 1828 listed as trading in Pigot's Directory.
   1850-1856 Frances Williams is listed as licensee; by 1868 Slater does not list the White Horse.

**Menai Bridge** 1895 listed as trading in Slater's Directory.

**Pentraeth** The author, George Borrow (1803-1881) lived here, the building features in his work 'Wild Wales'.
   1862-1869 Owen Jones, a blacksmith from Liverpool, paid five shillings Land Tax when he became landlord.
   1870 John Jones is listed as landlord – not Owen's son, who was only 13 years old at this date.
   1871 Owen is not listed in the census, but his wife is listed under Penrhos in Pentraeth as a blacksmith's wife with a seven month old daughter.
The business was closed sometime after 1900 and by c.1920 had become a shop. By 1999 the building was empty. 1844 P; 1868 S; 1895 Llanddyfnan S.

# BLAENAU GWENT

**Beaufort** 1868 S; 1895-1937 K.

**Bryn Mawr** 1868 S; 1895-1906 K.

# CAERPHILLY

**Egwlysilan / Abertridwr** James Bell in his Gazetteer of England and Wales notes that in 1801 the population of this village was 865, but thirty years later it had risen to 2,818. He also points out that 'the Taff Well in this parish is greatly esteemed.' This was a very beautiful area where people would have come to take the waters – until coal started to be mined and many colliers' houses built – by the 1870s it had been transformed into an industrial area, swallowing up the old villages and settlements. This would explain the rapid growth in the population.

1827, 26th September – an alehouse recognizance lists Thomas Evans as innkeeper. Gla.

**Pentwyn, Davies Row** Pentwyn – 'top of the hill or slope of the hill' – is a small hamlet at a crossroads outside Fochriw on the road to Deri. It is an old mining village situated above the Darran Valley on a Roman road now known as the Mountain Road. 'Tafarn y Ceffyl Gwyn' (White Horse Tavern) is possibly one the highest public house in Wales at 1,300 ft. The back of the building dates from 1700 and the front from 1820. Landlady. 1895-1926 K.

**Tafarnaubach** 1868 S.

# CARDIFF

### Cardiff

**Vicarage Street** 1838, 28th July – *The Cambrian* reported the death of Mr. Philip Hedges, landlord of the White Horse Inn. On 30th March 1839 the newspaper advertised the building to be sold by auction on 3rd April at three o'clock precisely:

' . . . that well-frequented free public-house . . . now in the occupation of Mrs. Ann Hedges; together with the coal yard adjoining thereto.
The above premises are held by lease for the term of 60 years, from the 1st August, 1834, at the yearly rent of £25. The House has been recently built, is well finished, and in good condition.
The Coal Yard is well walled, and extends to the Glamorganshire Canal, to which it has an extensive frontage; it is now underlet at the low yearly rent of £12, to Mr. Thomas Evans, as tenant at will.' 1822-1844 P.

**Working Street** 1849 listed as trading with T. Miles as licensee.
By 1859 no White Horse is listed in Cardiff. H&C.

# CARMARTHENSHIRE

**Abergwili** 1813, 16th October – *The Cambrian* reported the death of Richard Allen, at the age of 105, of the White Horse.

1844 listed as trading under Carmarthen.

1868 listed as trading by Slater's Directory under White Mill with David Morris as licensee. 1895 Abergwili – 1906 K.

## Carmarthen

*7 Chapel Street* 1829 the White Horse was trading with William Evans as licensee. This information comes from George Eyre Evans' article 'Carmarthen Inns and Taverns' dated 1929. Mr. Evans maintained that the word 'hotel' was not then in use – only 'taverns' and 'inns'.

1862, 25th April – *The Cambrian* reported the death of James Jones at the White Horse. He had been an overseer at the local journal office.

By 1978 the inn had closed and was derelict, it was demolished together with the public convenience next door.

*104 / 112 Priory Street* 1829 trading with William Woozley as licensee. From George Eyre Evans' article 'Carmarthen Inns and Taverns' dated 1929.

1843, 7th October – *The Cambrian* reported the death of Mrs. Woozley aged sixty, wife of the landlord. On this same day *The Cambrian* also reported the death of the landlady at the White Horse Inn at Llandeilo, some fifteen miles away!

Two years later, on 15th February 1845, the death was reported of the sixty three year old landlord, William Woozley.

1846, 3rd April – the death was announced of another William Woozley of the White Horse – aged 94.

*The Cambrian* reported the following deaths concerning the White Horse:

> 14th July 1848 – Mary Bowen on 6th July, wife of David Bowen, the landlord, who died aged fifty on 5th January 1855.
> 5th August 1853, – John Lawrence aged fifty three on 25th July; his daughter Anne died on 25th November 1858.
> 4th May 1866 – James Thomas aged twenty six on 3rd May.

1868 listed as 104 Priory Street with Margaret Bowen as licensee.

1871 listed with David Morgan as licensee.

1896 the assignment of a lease for Buckley's Brewery is recorded.

Ref. BB 400. PO. 1844 P; 1895-1926 2 listed and 1906 K.

**Llandeilo, 125 Rhosmaen Street** *The Cambrian* reported the following concerning the White Horse:

17th January 1829 – the death of Mr. William Rydderch.

7th October 1843 – the death of Mrs. Jane Rhydderch, aged 64. On this same day the newspaper also reported the death of a woman at the White Horse in Carmarthen, some fifteen miles away!

28th June 1845 – the marriage of Ebenezer Rhydderch to Eliza Parry, who also lived in Llandeilo.

27th October 1854 – the death of Mr. Hendry Rhyddercyh aged 42, presumably another son of William and Jane.

27th April 1855 – the death of Henry Stephen who had been the ostler at the White Horse, which may imply that this was a coaching inn.

2nd October 1857 – the death of Mr. John Davies aged 61 on 26th September.

25th June 1858 – the wife of Mr. John Evans gave birth to a son on 18th June – seven years later on 8th December 1865, John Evans aged thirty five died on 1st December at the White Horse.

1875, 16th July – a visit of the Swansea Working Men's Club to Dynevor Park:

'The excursionists, numbering about 300, including several of the gentle sex, left Swansea by special train from the London and North Western Railway Company's Station shortly before nine on Monday morning accompanied by the Swansea band, and on arriving at Llandilo proceeded at once to the Park, some to roam and inspect its varied beauties, including the remains of the old castle, some joining in cricket, others in football, races, etc. A tent with refreshments, provided by the hostess of the White Horse, was liberally patronised; whilst the band played throughout the day a selection of music which seemed to afford much satisfaction to the townsfolk, many of whom appeared on the scene. At one o'clock the bugle sounded for dinner, which was served in the large room at the back of the White Horse Inn, at the moderate price of 1 shilling and 6 pence . . . Several toasts were proposed . . . The health of Lord Dynevor[1] was then received with much enthusiasm, when his kindness was dilated upon in affording the

Working Men's Club the use of his magnificent Park, and throwing open to them those delightful grounds which overlook the Towy in one of its most beautiful reaches.

The company afterwards dispersed, some again to the Park others to explore the town and visit friends, and then finally left by special train for Swansea, which place was reached without any accident to mar the day's pleasure. The weather was very fine, a cool breeze and an unclouded sun combining to make the lovely and picturesque scenery more than usually attractive. The catering of Mrs. Evans of the White Horse, gave unqualified satisfaction . . . ' *She was probably the widow of John Evans.*

1844 Llandilo-fawr P; 1849 H&C; 1868 S; 1895-1926 K.

**Llangadog** 1895 K.

**Sandy, Pembrey Road** At one time the White Horse Inn was situated on the opposite side of the road to the Stag's Head, it was demolished to widen the road. The Stag's Head was then modernised and renamed the White Horse Inn.

1871 listed under Llanelli with John Richards as innkeeper.

A lease dated 1911 concerning Buckley's Brewery is held by the Carmarthenshire Archives. Ref. BB 664. 1868 (Llanelli) S; 1895-1926 (Llanelli) K.

**White Mill** *See Abergwili.*

# CEREDIGION

## Aberystwyth

*Great Darkgate Street* The inn was owned by Mrs. Drew, as mentioned by W.J. Lewis in his 'Born on a Perilous Rock: Aberystwyth Past and Present' published in 1980.

She eventually also owned the White Horse in Terrace Road.

*Terrace Road* This three storied Victorian / Edwardian hotel is situated on the corner with Portland Street.

1809 by this date part of the area had been developed and the inn may have been built at this time.

1834 the building was shown on a town map. Additions were made to the existing building during this year.

*The Cambrian* reported the following concerning the White Horse:

8th December 1838 – the death of Mr. Edward Jones, landlord, who had served in the 1st Royal Horse Guards.

17th February 1860 – the birth of a daughter on 3rd February to the wife of G.T. Smith.

28th October 1864 – the birth of a daughter on 14th October to the wife of Mr. John Rae.

1871 listed as trading with John Rees as innkeeper.

c.1900 the owner, Col. John Rae of Worcester, commissioned an elaborate Art Nouveau exterior of glazed tiling, showing grapes and vine leaves around a circular panel depicting a white horse. Ieuan Gwynedd Jones, in his 'Aberystwyth 1277-1977', published in 1977, describes this as 'a particularly good example of Edwardian engraved glass and ceramic work . . . The broad entrance has fluted pilasters and a cast iron parapet. A shop front wraps around the corner the whole length of Terrace Road with splayed entrances and tessellated doormats.'

1999 the name was changed to The Varsity and was no longer a hotel.

RCW, CC, and NMW. 1895, 1906, 1914 Hotel, and 1926 K.

**Cardigan** 1844 P; 1868 S.

# CONWAY

**Capel Garmon** The building dates from c.1540 – a coaching inn with beamed ceilings.

Two legends are connected with the inn:- The daughter of a one-time owner of the inn fell in love with an unsuitable local farm hand. One night he rode to the inn on his white horse, helped the girl to climb from her bedroom window and galloped off with her, never to be seen again.

It is said that the inn is haunted by the ghost of a friendly, blond-haired lady who watches over the activities of the inn. It is not known if the two stories are connected.

It is situated near to an Iron Age fort. Inn Brochure and GPG. 1874 Wo.

**Conway, Upper Gate Street** 1844 Aberconwy P; 1868 S; 1874 Wo; 1892 GRe; 1895 S.

**Mochdre** 1868 S; 1874 Wo; 1886 S; 1891 RPD; 1895 Llandrillo-yn-Rhos S; 1918 RPD.

**Penmachno** The following documents concerning the White Horse Inn are held by Gwynedd Archives and Caernarfon Record Office:

1841 an agreement concerning the White Horse.

A conveyance dated 2nd February 1885.

A mortgage document dated 3rd March 1886.

A indenture dated 4th October 1899 mentions Cadwaladr Andrew Vaughan, innkeeper of the White Horse Inn and Thomas Elias, a brewer and malster. The inn was described as a Messuage Tenement or fully-licensed Public-House called or known by the sign of The White Horse with stables and buildings, lands and gardens in all measuring approximately 400 square yards.

1916 listed as trading.

The inn became known locally as 'The Ring' – from the Welsh 'Yr Inn', or The Inn. Local people called it 'Rinn' and finally 'Ring'. Later the building became Victoria House which, by 2005 had been demolished.

There is an old packhorse bridge at Penmachno, possibly used by horses transporting the slate which was manufactured here in large quantities in the early 19th century. The population here in 1831 was 984. JB. 1868 S.

**Towyn** 1868 Isaac Roberts was listed as licensee.

1871 the census listed Mary Roberts, a publican, living at 2 Railway Terrace, but did not mention the White Horse; by 1874 Sarah Roberts held the licence. S. and Wo.

# DENBIGHSHIRE

**Bettws Gwerfil Goch** 1847 listed as trading in Worrall's Directory.

1916 purchased by Marston, Thompson & Evershed Ltd. 1874 Wo; 1895 S.

**Denbigh, Swine Market** 1828 listed as trading in Pigot's Directory.

By 1835 the address had been changed to Pig Market.

1895 listed as trading but with no address in Slater's Directory.

**Hendrerwydd** Situated in the beautiful Vale of Clwyd. The building, thought to date from the 17th century, is of a fine oak beam structure with a whitewashed exterior, is thought to date from c.1600; the inn has been trading from about this date.

1838 the building is shown on the tithe map.

1886 listed under Llangynhafal when Margaret Roberts is described as a licensed victualler and shopkeeper.

By 1968 the building had been extended to take in the village shop which was run by the same family.

2006 the house was still listed as trading. PDD and FP. 1874 Llangynhafal Wo; 1895 S.

**Llandyrnog** 'Y Ceffyl Gwyn' – 'The White Horse'. According to the landlady, the building was said to date from 1600. 1822 P; 1874 Wo; 1895 S.

**Llanelidan** 1883 S; 1891 and 1918 RPD.

**Llanfair-Dyffryn-Clwyd** White Horse – 'Ceffyl Gwyn'.

From 1868 to1874 Samuel Owen is listed as licensee, by 1886 his wife had taken over the licence.

c.1991 renamed 'The Haymakers'. CCR and PO. 1844 Ruthin P; 1868-1895 S.

**Llangynhafal** 1886 Margaret Roberts, the licensee, was also listed as a shopkeeper. 1874 Llanganhafal Wo; 1895 S.

**Rhuddlan, Market Street / High Street** Peter Campbell was listed as licensee from 1856 to 1883. 1856 Rhyll S; 1874 Wo; 1883 – 1886 S.

**Rhyl, Bedford Street**

**Ruthin, Market Place** This inn was run by women – 1828 listed as trading with Eleanor Roberts as licensee; by 1835 the licensees were Margaret and Grace Lloyd, and by 1856 Harriet Jones. By 1868 it was no longer listed under White Horse. P. and S.

**St. Asaph, Lower Street** 1858 Joseph Lloyd, a miller, raised a mortgage of £300 on the White Horse – he was still the owner forty five years later in 1903 with William Jones as licensee. FRO. 1856-1895 S.

# FLINTSHIRE

**Bagillt, High Street** Once the lead smelting centre for North Wales. Originally a building with a small adjacent cottage constructed of 'undressed stone'; predating the church next door, as the public house was built to provide refreshment for the workmen who built the church. It was owned by the Hughes family of nearby Sea View House.

1840 listed as a Principal inn.

1844 listed by Pigot as a Post Inn.

1892 the Police Return gives the owner as William Pierce & Co. with George Edwards as licensee; there was no accommodation.

1903 by this date the house had been sold to Greenall, Whitley & Co. of Warrington.

1916 the property was sold to Marston's, the brewers.

1920s-1930s The Bagillt Homing Pigeon Club met here and let their pigeons fly from the front room in the pub through a small pigeon hole – about 200 pigeons would be let go each time.

The original front door was between the two windows; the 'undressed stone' could be seen inside. Landlord, MTE, FRO and CCR. 1822 P; 1826 L; 1846 EP; 1874 Wo; 1883-1895 S; 1868-1883 S.

**Cilcaen / Cilcain, The Square** The building is thought to date from the 15th century.

1856–1895 listed as trading; the firm Robert Davies of Ruthin were recorded as the owners in 1892.

By 1903 the house had been sold to T.J. Roberts and Robert Davies of Mold.

2006 the landlord would organised visits to the theatre and other outings by coach for the village.

FRO, S, Wo and PO.

**Holywell** The town took its name from the celebrated St Winifred's well. On her feast day, 3rd November 1629 between 1,400 and 1,500 people from Lancashire and Cheshire visited the well, this caused the Justices of the Peace seven years later to use all means to hinder pilgrimages to Holywell[2].

During the 18th century large sections of the town were rebuilt – in 1699 there was no market and only one hundred dwellings, mainly thatched – one hundred years later there were 1,000 including twenty inns and sixty shops due to industrial growth. Lead ore auctions, or 'ticketings' were held fortnightly at The White Horse during the peak of the trade – Bagillt, part of the parish of Holywell, was the lead smelting centre for North Wales.

1747, June – an advertisement in *Adam's Weekly Courant* announced:

'There will be a main[3] of cocks fought at the sign of the 'Golden Lion and White Horse' at Holywell twixt the Gentlemen of Flintshire and the Gentlemen of Cheshire for five guineas each battle and fifty pounds the main.'

During Easter there were large county cockfights which often lasted for over a week.

1775, 27th December – a hotel bill with a beautifully illustrated heading shows Mary Hughes as landlady of the 'Golden Lion and White Horse'; items include 'Servants eating and ale – 3s. 2d. (about 30p)' and 'Horses hay and corn – 4s. 6d. (about 40p)'.

1785 the first mail coach ran from Chester to Holyhead via Holywell. A coach called the 'Harkforward' stopped every day at The White Horse for a change of horses on its way from Bangor to Chester, a considerable distance of some fifty to sixty miles.

Between 1790 and 1810 'Old Williams', a blind man, played the harp to the customers in the inn.

1805, 28th September – *The Cambrian* newspaper recorded the death of Mr Price of the White Horse Inn.

c.1820 the building is described as an attractive three storey, high pitched flat roofed building with a plastered front.

1832, 14th October – Princess Victoria, aged thirteen, left Plasnewydd on the island of Anglesey with her mother, the Duchess of Kent where they had been visiting the Paget[4] family. The children of Lord and Lady Anglesey were of contemporary age. The Royal party attended the National Eisteddfod where the Princess invested the Chaired-Bard, Gwilym Caledfryn for his successful ode 'The Wreck of the Rothsay Castle[5]'.

Having left Anglesey, they stayed for a night with Lord Dinorben at Kinmel Hall. The following day 'she had the greatest honour of her life in that she was escorted to Holywell by an angel in the flesh[6]'. She and her mother had lunch at the White Horse Hotel. They visited the holy well – a scarlet cloth had been laid from the porch of the hotel to the well and thousands of people lined the route.

She wrote in her Journal en route in a carriage to Eaton Hall in Cheshire:

> "10 minutes past 3. We have just changed horses at Holywell where we were met by the Flintshire Yeomanry, commanded by Major Jones, and the officer riding on my side is mounted on a lovely little horse. Just here it was a lovely view . . . "

Unfortunately the Princess does not mention the inn in her Journal, however the Curator at the Royal Library, Windsor writes 'As I assume that the change of horses would have taken place at the local pub, then the story that she had visited the White Horse Inn might well be true, but unfortunately I can't find any more detail to positively confirm it.'

They continued their journey to stay with the Grosvenors at Eaton Hall where the children were rather younger than the Princess. Their arrival at 'this exquisitely beautiful mansion of Gothic style' which had been completed twenty years earlier by the architect William Porden (1755-1822), was built on the site of the old Palladian house, erected by Sir Thomas Grosvenor during the 18th century.

1833 the inn was recorded as a posting house, by which time it had been renamed 'The Royal and White Horse'.

Excerpts from a lecture given on 4th December 1912 by Mr. Lloyd Price gives a glimpse of life in 'Holywell in the Early part of the Nineteenth Century':

> 'The annual Races commenced the first Sunday after the 14th October; coursing with greyhounds took place on the Monday and the week ended on the Friday with 'a grand and colossal dinner at the Royal and White Horse Hotel'. On one occasion after this dinner, the guests found the carcase of a dead donkey which they 'placed in a vertical position upon its hind legs against the front door of one of the principal residences in the town. When the housemaid opened the door in the morning to clean the steps the ghastly apparition fell into her arms and rolled to the floor with a thud whilst the domestic screamed and fainted to the great alarm of the household.'

The Duke of Westminster took part in the races and always hired the hotel for his own use, until he was angered by a commercial traveller who refused to move out, claiming he had a legal right to the accommodation. After this, the Duke bought a house nearby and converted it into Halkyn Castle.

The Royal Ball of the Flint and Denbigh Hunt was also held in the hotel as there was a large ballroom extending the width of the building; an awning was erected across the pavement to the middle of the street. It was usually attended by the Marquis of Westminster, Lord Mostyn, Sir Watkin Williams Wynn, Sir Edward Mostyn, Sir Richard Bulkeley, Mr. Price Brynypys and members. On one occasion Sir Richard Bulkeley challenged Sir Watkin that he would carry him without a halt up and down the High Street – a wager was laid but at the close of the Ball, Sir Richard said "I engaged to carry you, but not an ounce of clothes" – Sir Watkin paid the stake rather than remove his clothes!

During one of his visits, Mr. Price Brynypys asked his coachman to bring water from the holy well for his morning bath, the coachman refused as his duty was to drive and not run errands – whereupon every morning Mr. Price ordered his coach for 8.30 am, attired in his dressing gown and carrying a pitcher in each hand, he travelled to the well to collect water for his bath – this journey had to be repeated several times.

1850, 30th September - a survey of this date for the White Horse Hotel, with its farm gave the rent as £50 p.a. for the hotel, outbuildings and garden. A letter sent to the landlord, Mr. Marsden, by the agent of Sir Pyers Mostyn of Talacre, asked if he:

> 'intended remaining at the White Horse after May, his offer to you is; to remain at the same Rent, his returning your 7½ per cent upon the land as to other tenants but not upon the Hotel; to farm the land by a proper rotation of cropping not to have two white crops in succession etc. and to continue tenant for not less than five years, if I do not receive your acceptance of the above terms on Thursday next, Sir Pyers will consider himself at liberty to let the hotel and farm to another applicant on Friday morning.'

1868-1871 the building was empty, demolition was proposed as it was in need of repair.

1871 the building was altered to become the North and South Wales Bank – a plain 18th century building with a formal five-window façade – this may be when the ornate side door was added.

An alleyway between the White Horse and the Boar's Head led to a courtyard behind the hotel where there were at least eight small cottages of 'two up and two down'. The 1871 census records a family with six children between the ages of one and fifteen living in one of them.

1908 the bank became the Midland Bank.

1912 part of the next door building, behind the Boar's Head was demolished to make a new road to the new Holywell Town Station.

1923 further alterations took place; the old porch in the centre of the hotel was demolished and replaced by another built off centre towards the street corner.

1999 the bank became the HSBC.

The building is Grade 2 listed. The Royal Archives, Windsor; FRO – Ref: D/MT/1039; RCW, KD&DW, LP and JB. 1822 P.

**Mold, High Street** 1807, Friday 2nd October – a number of houses in Mold were sold by auction from three o'clock in the afternoon at the Black Lion – one being the White Horse Inn in the occupation of David Morris, with its good brew-house, cellars and other conveniences. He also owned a pig sty adjoining a house in Church Street – this house and the sty were also for sale in the auction; it had the right of water from the pump immediately behind the White Horse, along with the inn and a messuage and shop which were also for sale.

1822 Thomas Williams was the licensee, in 1835 William Baskell took over.

1844 – 1856 listed as trading.

1868 Slater does not list the White Horse.

1892 no longer listed in the Police Return. FRO – Ref: S/LE/671 and P.

**New Brighton** 1886 listed as trading with Isaac Edwards as licensee.

The returns of 1892 list a White Horse at nearby Soughton. PO, FRO and DRO.

**Northop** 1850 listed by Slater, but by 1856 is not listed under White Horse.

**Rhes-y-cae** Sometimes alias Hendrefigill. Previously in the parish of Halkyn.

1868 listed as trading by Slater.

1892 no longer listed. FRO.

**Soughton** 1892 listed as trading in the police return with Edward Chambers as the licensee.

By 1903 Maria Chambers had taken over the business.

There is some confusion with this White Horse and the one previously listed at nearby Rhes-y-cae. FRO.

**Trelawnydd / Newmarket** 1868 listed as trading in Slater's Directory.
By 1892 the White Horse is not listed in the police return. FRO. 1874 Trelogan Wo.

# GWYNEDD

**Bala** 1835 listed by Pigot, but by 1844 it is not listed under White Horse.

**Bangor** 1835 listed by Pigot, but by 1844 it is not listed under White Horse.

**Caernarfon, Bangor Street** 1818 Daniel Williams is recorded in the Quarter Sessions as obtaining a licence. 1828 P; 1868 S; 1895 S.

**Harlech** The name is thought to have been derived from Hardd-lech meaning 'beautiful slope'.
The inn was situated just off the centre of the town and has become a private house called The Bronallt next to the Lion Hotel. It is a double fronted building with a wall surmounted with the original railing protecting the old doors and wooden steps leading down into the cellar which still exist. D.A. Maidment and 'The Harlech Town History Trail'. 1844 P; 1850 S–1895 S.
*Apart from the trade directories we have yet to find any archive material referring to this inn.*

**Llanengan** 1892-1895 listed as trading.
1913 a survey of the property was drawn up for a compensation claim; it recorded that the inn, with a frontage of 48 feet, stood in a most commanding position in the village on the corner of the main highway from the Lleyn peninsula with the highway to Sarnbach, an excellent position for trade. The building was in good repair, built of stone with slate roofs, the walls had been cemented over and pebble dashed.
The ground floor had a small portico with a tessellated floor, a hall passage with a tiled floor, a small parlour with a slab floor and a beer cellar. There were two large kitchens – a drinking kitchen with tiled floor and a back kitchen where the floor was partly earth and partly flagged; the pantry had an earth floor. The beer cellar floor was of stone and timber.
The chamber floor – or first floor – had three bedrooms and a box room.
The outbuildings round a large yard consisted of a two stalled stable with a cobble and earth floor, a coal house, a lean-to wash house and a privy. There was a good sized kitchen garden.

The surveyor recommended that £258 17s. 6d. be awarded for the compensation claim calculated from the figures of trade, rentals and depreciation over an average of three years. He specified that a cost of £1 5s. 0d. be charged for the removal of two fixed settles in the kitchen and for painting over the lettered sign, 5s. 0d.

S, CaRO and GRe. 1892-1895 listed as trading.

**Llantysilio** 1886 Richard Parry is listed as licensee and a farmer at the White Horse Inn. PO.

**Nefyn, Moriah Street** 1868 listed as trading in Slater's Directory.
1892-1895 listed as trading. CRe.

# MERTHYR TYDFIL

**Cefn-coed-y-cymmer, 64 High Street** 1849 listed as trading with Thomas Jones as licensee.
1863, 29th April – *The Merthyr Telegraph* reported that:

> 'David Williams, of the White Horse, was charged with permitting drunkenness on the 7th instant . . . the Bench considered it a bad case, and as it was the defendant's third offence, he was fined £2 and 11s. 3d. costs, with a warning that if again convicted, he would be deprived of his license.'

1871-1875 Thomas Thomas is listed as licensee.
The inn is also shown on the 1875 Ordnance Survey map of this date. Wo.

### Merthyr Tydfil
Before 1755 this area was a collection of small villages which grew into one another when the iron and coal mining industries became established:

> 'The town stands upon the river Taff in a bleak and steril country, and though built without any regularity of plan is large and populous . . . About half a century ago it was an insignificant village, and has risen to its present importance solely in consequence of the iron works first established by Mr. Anthony Bacon, who upon a 99 years' lease of a tract of land 8 miles long by 5 broad, commenced to work coal and iron here about 1755.'

In the early 19th century the iron works were the largest in the world. The working conditions of the employees being hot and dusty, their need for ale was considerable – hence there being over 500 public houses in the town. Most of the beer was sold before 7.30 in the morning when workers would buy weak beer to take to work. Many men were employed to dig, by hand, reservoirs in the Brecon Beacons and they would carry the beer up the mountains.

James Bell in his Gazetteer of England and Wales of 1834 described 'Merthyr-Tidvil' as the post town.

The following two villages are now part of Merthyr Tydfil.

**Pen-y-darren** The inn is shown on the 1875 Ordnance Survey Map.

**Twynyrodyn, Windsor Terrace** 1835, 2nd May – *The Cambrian* reported the marriage of Richard Jones of the White Horse to Margaret Richards of the Crown Inn, Merthyr.

1844 listed as trading by Pigot and again in 1849 by Hunt and Company's Directory, but the White Horse is not in the census or map of 1851. This area was developed between the mid 1850s and 1870s; it is thought the present building dates from 1855, planning records only started the following year.

Ponies were all important to the local coal and iron industries. The first verse of Meic Stephens poem 'Ponies, Twynyrodyn' written in 1997 gives a graphic description of these stalward animals:

> *'Winter, the old drover, has brought*
> *these beasts from the high moor's hafod*
> *to bide the bitter spell among us,*
> *here, in the valley streets.*
> *Observe them, this chill morning, as*
> *They stand, backsides against the wind,*
> *In Trevithick Row. Hoofs, shod with ice,*
> *Shift and clatter on the stone kerb.*
> *Steam is slavering from red nostrils,*
> *Manes are stiff with frost and dung.'*

1864, 1st July – *The Cambrian* newspaper reported on the following court case at the Glamorganshire Quarter Sessions:

'John Thomas (23), described as a shoemaker, was charged with maliciously wounding Mary Reardon, on the 9th April . . . Mary Reardon, the wife of an Irishman, said she was at the White Horse on the above-named day, when she saw the prisoner, with whom she had some beer. There were some words between them, when the prisoner made a snatch at her and bit off part of her lower lip, which he spat under the grate, and said, "There's your Irish lip." To the policeman the prisoner said, "If I did do it she must have been kissing me at the time." David Gay was called for the prisoner, and he said the prosecutrix[7] kissed the prisoner, and that annoyed him. Guilty. Nine months' hard labour.'

c.1860–c.1880 The Merthyr Tydfil Honourable Society of Cymmrodonion met here. 1875 Worrall's Directory lists the inn as at 33 Twynyrodyn with William Lewis as licensee. It is also shown on the Ordnance Survey map of this date.

1871 PO; 1895-1926 K. Carolyn Jacob, Merthyr Tydfil Public Library.

# MONMOUTH

**Abergavenny, 14 Frogmore Street** A coaching inn. 1787 Widow Powell is the earliest recorded publican, although Walter Powell's will dated 1785 records him as an 'innholder of Abergavenny'.

1818 Abergavenny Museum holds admission tickets of this date including a theatre ticket, admittance to the White Horse Dance.

1825, 1st July – a poster informed the inhabitants of Abergavenny and Monmouth of a Feat of Pedestrianism to be undertaken by J. Townsend, who proposed to walk sixty six miles a day for six days starting from the White Horse. He planned to walk to the Robin Hood in Monmouth, back to the White Horse and after a few minutes rest he would set out again for the Robin Hood, and then return through Abergavenny to the Lamb and Flag at Llanwenarth and back to the White Horse. Mr. Townsend wished to show the people of Abergavenny and Monmouth what he could do and 'would be thankful for any donations people might give him'.

1834, 8th March – *The Cambrian* reported the death at the age of 85 of John Watkins, the brother of David Watkins of the White Horse. The following year on 20th June the newspaper reported the marriage between Anne Wilkins of the White Horse Inn and Mr. George of Garthen Dolyrus Works.

1880 an octogenarian of the town wrote:

"I remember a rather singular incident in connection with the internment of the body of a man who went by the name of 'Jack o' Breed', in St. Mary's Church yard. The man had been an ostler for many years at the White Horse, and was a bit of a character in his way. The body had not been buried long – not more than a day or two – when it was noticed that the grave had been disturbed, and further investigation showed that the grave only contained the shroud and coffin without the lid. A hue and cry was raised, and it was ascertained that some young men in the town, several of whom belonged to the medical profession and a few choice and daring spirits had abstracted the body and taken it to the residence of one of the party, for, I suppose, the purpose of dissection. It was said that poor old Jack had himself sold his body to the doctors over and over again for a trifling consideration, and that the removal of the body from the grave was merely the fulfilment of which was thought to be a fair and just bargain. However, the resurrectionsts were brought to see the grievous mistake they had made and the body was replaced in the coffin."

c.1901 a photograph held by Abergavenny Museum shows the White Horse with the caption:

'Andrew Buchan & Co. Ltd. Celebrated Ales and Stout. Wines and Spirits is our Speciality. Good Stabling Etc.'

1919, 21st July – a handwritten receipted bill for car hire between the White Horse Hotel and Monmouth from S.E.S. Baker – taxi cab proprietor – is held by the Abergavenny Museum.

1965 the pub was demolished – it had always been prone to flooding from the Cybi Brook. AMu. 1822-1844 P; 1868 S; 1880-1895 PO; 1906-1937 K.

**Chepstow, 17 Moor Street** Pigot's Directory of 1822 describes a walled, irregularly built small sea port at the mouth of the River Wye – a great part of the wall was still standing at this date. At one time there was extensive trade in timber, but this had decreased and ship building had taken over as the main industry:

'The situation of the Wye which empties itself into the bosom of the Severn, about two and a half miles below the town, is highly interesting and picturesque' . . . The tide at this place rises to an almost incredible height; it has been known as high as 60 feet – a phenomenon not witnessed in any other part of the kingdom.'

The inn was situated outside the town wall within sight of the Town Gate, the original building is thought to date from the 17th century when it is believed to have been trading as The Bell – the iron back plate in the taproom fireplace had the initials 'M.R.' and the date1666.
1681, 24th September – mentioned as 'The Bell' in Court Rolls when John West and Alice Arberoth

were charged with 'throwing muck and dust into the mouth of the back lane adjoining to the Bell . . . '

1692 the inn was sold by Nathan Rogers to Giles Pope who changed the name to the 'Greyhound'.

By 1793 the name had been changed to 'White Horse'. On 30th May this year the Chepstow Friendly Society of Tradesmen and Others was founded here after Parliament passed a law requiring clubs meeting in public houses to register – the landlord was often the treasurer.

1814 William Davies was listed as landlord.

1824, 20th March – *The Cambrian* newspaper reported the Meeting of the Magistrates held on the previous Thursday:

> '. . . Wm. Davis, of the White Horse Inn . . . was convicted before Thos. Lewis, Esq. and the Rev. Jas. Ashe Gabb, in the mitigated penalty of 40 shillings and costs, for suffering tippling in his house during the hours of Divine Service . . . At the same meeting, John Thomas, the younger, was convicted of cutting underwood in the Duke of Beaufort's woods, in the parish of Newchurch, Monmouthshire, and fined 40s.; and, on non-payment, committed to the House of Correction at Usk, for one month, and to be once whipped.'

Presumably as a result of this conviction, his wife became the landlady.

1861, 2nd January – the *Chepstow Mercury* reported:

> 'On Saturday evening last, an Inn of this town was entered by one of those ferocious animals, the Gorilla or man eater, which attacked one of the company, who managed to keep the animal at bay for some time but not without injury to himself, for the brute while on the ground seized the inside of the man's thigh with his tusks and most severely lacerated him. He managed to avoid capture and is supposed to have returned to the company of wild animals, which infest Hewelsfield Common.'

It is thought that once the Greyhound in the High Street closed, the White Horse was renamed 'Greyhound'. The inn traded under this sign until 1971 when it closed. WGA and IWa. 1822 P.

**Monmouth, 55 Cinderhill Street** This street had a number of pubs and beerhouses, there was a White Horse trading between 1719 and 1792 which is thought to have been on this site, close to the entrance of Goldwire Lane. The White Horse was flourishing in c.1840.

1842 a policeman, Evans, was described as 'not sober' by witnesses at the White Horse.

1849 Listed as trading with William Morgan as licensee.

1853 the pub was described in *The Merlin* as:

> 'the house of pollution, drunkenness and all but murder . . . a number
> of that hitherto well-conducted regiment, the Monmouth Militia,
> led astray by bad characters of the other sex, after an evening spent in
> drunkenness and debauchery, returned at four in the morning to The
> White Horse (which is famed for its notorious character) demanding
> admission. On meeting refusal, bayonets were drawn, one man was
> stabbed in the arm, and a regular riot ensued until eight drill sergeants
> appeared with drawn swords to quell the disturbance'.

The building was demolished during the 20th century.

H&C, D&K, HH and Monmouth Library. 1868 S; 1880-1895 PO; 1906 K.

# NEATH PORT TALBOT

**Aberavon, Pwllyglaw** 1846, 27th February – *The Cambrian* newspaper reported
that Richard David of the White Horse, Pwllyglaw was charged by PC Wright
with selling beer during the hours of divine service. WGA. 1868 S.

**Cwmavon** 1853 John Elkington took the White Horse with his wife Ann; they
had been married for four years. John only lived here from time to time, his work
on the railways taking him a long way away from home, leaving his wife to run
the business. She continued to live there after Mrs. Thomas became landlady to
whom she paid a rent of £10 per annum.

1857, 22nd May – *The Cambrian* newspaper reported the theft of a cravat from
Mrs. Ann Elkington, at the White Horse:

> 'A labourer employed at the Cwmavon Works, named Stephen Davies,
> was on Friday last charged before the Neath Bench of Magistrates with
> having stolen a cravat, the property of Mrs. Ann Elkington, landlady
> of the 'White Horse'. The prosecutrix deposed to missing her property
> as far back as the 7th of March last; but a few days since she saw it on the
> neck of a nephew of the prisoner. A servant-girl named Sarah Francis
> proved washing the cravat and placing it on a chair to dry on the
> day mentioned. Prisoner was then sitting in the room. The property
> having been thus traced to the possession of the prisoner, the Bench
> considered the charge of felony proved, and he was therefore sent to
> the House of Correction for fourteen days' hard labour.' WGA.

1859 there was no White Horse listed.

1864, 1ˢᵗ July – *The Cambrian* reported an appeal against an order of justices concerning the removal of Ann Elkington and her six children, all under fourteen years of age, from a tenement near Cowbridge. The family no longer lived at the White Horse and were paupers. The eldest son, aged thirteen, worked in an office in Cowbridge. In 1863 John Elkington had sent £1 worth of postage stamps in an envelope addressed to one of his young sons, who gave some of them to his mother.

The Court quashed the order of removal with costs.

**Glyn-Neath** Thought to be the same as the White Horse at Pontneddfechan – the river Nedd – or Neath – runs between the two towns, it is also the county boundary. *See Rhondda.* WGA. 1868 S; 1895 K.

**Neath, Wind Street** 1835 listed by Pigot with Leyson Jones as innkeeper.

*The Cambrian* reported the following events concerning the White Horse:

> 1844, 23ʳᵈ March – The death of the landlord, Mr. Edward Evans, he was aged 52.

> 1850, 5ᵗʰ July – Thomas Edwards, landlord, was charged 'with drawing beer at illegal hours. The information was laid by a navvy who was charged with being drunk and disorderly, who this day did not make his appearance, and the case was dismissed.'

> 1853, 24ᵗʰ June – P.C. Thomas Davies charged the landlord, Thomas Richards, with having kept his house open after 11 o'clock at night. The complainant stated that he visited the defendant's house at 11.40 on Saturday night last, and there found four men and eight women sitting around the table with some beer before them. He was fined 10 shillings, including costs.

> He was also fined for opening on Sunday before 9 am. As he was an old man he was fined the costs only of 6 shillings. At the same time he was charged with serving out of hours, again he had to pay costs only.

> 1854, 26ᵗʰ May – it was reported that Thomas Richards was charged once again with opening after hours.

> 1855, 27ᵗʰ July – White Horse was advertised for sale by auction to be held at the Castle Hotel on Wednesday, 15ᵗʰ August at three o'clock in the afternoon:

> 'Lot 4. All that large Public-house, called the White Horse . . . and now let to Mr. Evan Evans, at the present low rent of £18 per annum.'

1856 according to Pearce's Swansea Directory Thomas Richards was still licensee.
1863, 24ᵗʰ July – the White Horse was advertised for auction – still let to Evan Evans, Esq.
1868 listed as trading in Slater's Directory with Thomas Thomas as innkeeper.

**Pontneddfechan, High Street**

# NEWPORT

**Newport, Cross Street** This was a very small street with six or eight beerhouses situated close to the banks of the River Usk.

During the early 19ᵗʰ century the town grew very rapidly and its infrastructure had not kept up with the sudden increase of people – it was considered safer to drink beer than water, this included children. Apparently local constables did not venture into the Cross Street area which was mainly inhabited by Irish immigrants.

The street no longer exists but Cross Lane is situated on the site.

Alan Ware – local historian.

# PEMBROKESHIRE

**Haverfordwest, Dew Street** 1871 listed as trading with William Jones as innkeeper.

There are leases held by Pembrokeshire Record Office dated 1876 and 1897 and a valuation dated 1921.

1910 by this date the premises were owned by the Swansea Brewery.

The inn, which had become a wine bar by 2005, was situated directly opposite the public library. 1844 P; 1868 S; 1895-1914 K.

**Kilgetty, Station Road** 1866 the railway line between Tenby and Whitland was built with a new station, sidings and cattle pens half a mile east of Begelly – this was named Kilgetty after the local manor house.

1869 trading as the Railway Inn with Thomas Harries as landlord.

c.1890 a blacksmith, John Merriman, was the landlord who was alleged to have specialized in shoeing pit ponies; his smithy was next door to the inn.

These pit ponies were immortalised in 1997 by Meic Stephens in his poem 'Ponies, Twynyrodyn':

*Long before fences and tarmac, they*
*were the first tenants of these valleys,*
*their right to be here is freehold.*

*These beasts are our companions,*
*dark presences from the peasant past,*
*these grim valleys our common hendre,*
*exiles all, until the coming thaw.*

The inn, which has always been a popular social centre, continued to trade as the 'Railway' until c.1965 when Billy Buchan, a Scotsman, was the licensee and renamed it the White Horse – partly because of the old smithy and partly because he enjoyed the whisky of that name. At about this time the inn was refurbished by The Felinfoel Brewery Co. Ltd. who still owned it in 1981. KJ. and Pem.

**Ludchurch** Listed by Keith Johnson in his book 'The Pubs of Narberth, Saundersfoot & South-east Pembrokeshire'.

# POWYS

The old County of Montgomeryshire was incorporated into Powys in 1974. Daniel Defoe (1660-1731) in his guide-book 'A Tour through the Whole Island of Great Britain' wrote in 1724:

'This county is noted for an excellent breed of Welsh horses, which, though not very large, are exceeding valuable, and much esteem'd all over England.'

**Brecon, Dorlangoch** 1871 listed with Mrs. Ann Price as licensee.
Between 1930 and 1940 it became a private residence. 1868 S; 1895-1926 K.

**Builth Wells, High Street** Appearing small from the street, this traditional hostelry stretches back a long way.
1871 listed with Thomas Morgan as innkeeper.
2007 the hotel was flourishing; the inn sign depicting a rearing white charger. 1844 P; 1895-1926 K.

**Derwenlas** Listed by The Brewers' Society. Situated on the River Dovey, Derwenlas was a busy port in the last century. There were three public houses; only the names of two are known – the third could have been 'The White Horse'. MDC.

**Machynlleth, Maengwyn Street** 1632 toll books indicate a thriving market around the site, but there is no proof of an inn at this date; although by c.1780 the inn had been built here.

In the 1830s 'The carriers offering a goods service . . . Two services ran to Tywyn on a Monday. One was operated by . . . and the other by Robert Roberts from the White Horse'.

*The Cambrian News* of 3rd May 1912 reported that on Sunday, 28th April the inn was burnt down, to be rebuilt within the year. WD. 1834-1844 P; 1868 S.

# RHONDDA CYNON TAFF

**Cefnpennar, 2 Canal Row (a hamlet near Aberdare)** 1871 listed as trading with Daniel Griffiths as licensee.

In the 1891 census David Jones is listed as licensee and a coal merchant; he was aged 52, his wife Mary was 49, and he had four daughters – the eldest was a stocking knitter, the second a music teacher and the third a school teacher; he also had four sons – the eldest being a coal miner.

MT. 1895-1914 listed under Cwmbach. K.

**Llantrisant** *c.1900 We are told that there were thirty two pubs, one church and forty houses in Llantrisant, We have therefore had difficulty in locating the site of this White Horse!*

1818, 31ˢᵗ January – *The Cambrian* advertised an auction to be held on behalf of Mr. John Aubrey at the Cross Keys Inn of 'Llantrissent' on Friday, 13ᵗʰ February between the hours of two and four in the afternoon:

> 'Lot 1 – All that Public-house, Court Garden, Stable and other conveniences thereto belonging, called by the name of the White Horse, situate in a central part of the said town; and also a Field thereto adjoining, and now let therewith at the clear yearly rent of £25.'

1824, 17ᵗʰ September – a Quarter Sessions alehouse recognizance lists Thomas Richard as innkeeper.

1836, 17ᵗʰ September – *The Cambrian* reported the death of Thomas Richards, the landlord of the White Horse. WGA, Llantrisant Post Office and Gla. 1844 P; 1849 H&C; 1868 S.

**Newbridge** This is the old name for Pontypridd.

**Pont-nedd-fechan / Pont Neath Vaughan, High Street** Thought to be the same as Glyn-Neath. *See Neath Port Talbot.*

1871 listed as trading with Mark Tucker as licensee.

1874, 3rd July – *The Cambrian* advertised the White Horse public house for sale by auction at the Castle Hotel, Neath on Friday, 24th July.

1875 John Francis is listed as licensee.

2007 listed as trading under 'Old White Horse'. WGA. 1906 K.

## Pontypridd

There was an early small White Horse Inn which was situated close to the ford and to the site of the 20th century health centre. It was demolished in the 18th century and the Old Maltsters Arms was built on the site of the original cellar incorporated into the foundations.

The inn was known to have a ghost in the cellar:

> On one occasion when a maid went to collect a jug of ale, she felt an unusual chill. Setting her candlestick down, she started to fill her jug when she noticed a misty apparition of a man amongst the barrels – she fled up the steps; but returning to the cellar, the man was still there; he had disappeared by the time she returned with her mistress. Having been laughed at by the drinkers in the bar, the following night the intrepid girl went down into the cellar alone, once again the ghost appeared, her candle blew out but she managed to relight it. The ghost beckoned her and told her that he had lost all his money on cock-fighting bouts and had died after he lost everything. He said he had dieted and dressed many a cockerel and, when cock-fighting became illegal – in 1849 – he had hidden a set of two-inch spurs in a secret place near a waterfall in the dark, wooded valley of Cwm Pistyll Goleu at Llanwonno. He could only rest after the spurs were found and thrown into the River Taff to stay there for ever. The resourceful maid found the spurs and threw them into the river; the ghost, now placated, never reappeared.

1809 the Baptists rented a room over the brewhouse while their chapel was being built.

1851-1875 the landlords were William Emanuel and John Rees – who was a hay dealer.

A new Maltsters Arms was built in c.1880 near to the old bridge.

***Bridge Street / Llanover Street*** Originally a small whitewashed building, situated by the humpbacked bridge over the canal.

The Methodists met here in the long room above the brewhouse, it was later used by other denominations. One night there was great confusion when the floor gave way while the Reverend Griffith Hughes of Groeswen was preaching!

1844 listed as trading at Newbridge by Pigot – Newbridge being the old name for Pontypridd.

1849 listed as trading at Newbridge with Edward Jones as licensee.

1852 William Edwards was listed as landlord, by 1859 he had become the licensee.

1868 listed by Slater's Directory as trading.

1871 listed as trading with William Davies as licensee.

c.1890 the building was extended and converted into the Queens Hotel which was demolished in c.1970 when the A470 road was altered.

DPo, H&C, RCW(2) AND Gla.

# SWANSEA

**Morriston, Woodfield Street** 1869 listed as trading with Mary Hopkin as licensee.

**The Mumbles** 1877, 23rd February – *The Cambrian* newspaper reported the wedding at Swansea on 18th February of Richard Hobbs and Mrs. Mary Williams of The White Horse Inn.

**Swansea, Gower Street** 1835 listed by Pigot with William Richards as innkeeper.

# TORFAEN

**Blaenavon, 1-3 White Horse Cottages, King Street** Formerly the White Horse public house, a coaching inn with a large first floor assembly room with stables and coach house below.

1828 shown on the Ordnance Survey map with the west and east ground floor wings.

1843 the building is shown on the Tithe Plan.

1867, 4th February – a notice survives advertising a 'grand evening concert' to be held in the assembly room.

1880 shown on the Ordnance Survey map.

The building is two-storied and stuccoed with a double fronted wing. The windows were late 19th to early 20th century, which have recently been replaced with plastic. Inside there are 19th century moulded cornices, six plaster ceiling roses and boarded dado in the old assembly room; a flat in the former stable wing has a range and bread oven. RCW. 1868 S; 1880-1895 PO; 1906-1937 K.

**Garndiffaith** 1881-1902 listed as a beerhouse, originally owned by Richard Look of Elvaston near Ross on Wye. The Annual Value of the premises in 1881 was £20 when William Saunders was licensee, but in 1891 it had decreased to £17.

By 1897 Ashton Gate Brewery of Bristol had become the owner and the licence had been transferred to Elizabeth Saunders.

1900, 24th February – John Williams became licensee.

By 1937 the White Horse was no longer listed. GWR Ref: C/P.R.3 – C/P.R.6.

**Talywain** Between 1881-1902 it was listed as a beerhouse.

1881-1890 the annual value of the premises was £17.10s.

1889, 28th December – Mary James took over the licence from Richard Enoch James.

By 1891 the annual value of the premises had risen to £19.

1897, 3rd July – Mary James was fined 30 shillings for permitting drinks out of hours, the licencee not endorsed.

By 1937 the White Horse was no longer listed.

GWR Ref: C/P.R.3 – C/P.R.6. 1906 K; 1926 K.

# VALE OF GLAMORGAN

**Cowbridge** 1822 P.

**Coychurch, Church Terrace** The house is situated in the main street of the village.

1868 listed as trading by Slater.

1896 the brewer's, S.A. Brain & Co. Ltd., hold a document confirming that this was a public house named The White Horse.

1895 listed as trading in Kelly's Directory under Bridgend.

1926 again listed as trading by Kelly.

1953 the house was purchased by the brewers, S.A. Brain & Co. Ltd. of Cardiff.

2007 The White Horse with its restaurant was still flourishing.

# WREXHAM

**Acrefair** 1891-1918 RPD.

**Minera** 1895 Simon Lloyd and Job Davies were occupiers for the next five years.

1897, 18th January – Simon Lloyd of the White Horse was convicted of selling liquor on a Sunday, he was fined 20 shillings with 8 shillings and 6 pence costs.

1900 recorded as trading and owned by Edward Tunnah.

By 1918 the inn had closed. CCR. 1891 RPD.

**Overton-on-Dee, High Street** Adjoining an 18th century house.

1874 listed as trading.

1883 the White Horse advertised 'Celebrated Welsh Ales and Stout'.

1892 the owner, Jane Parsonage of Aldford leased the premises to Soames & Co., brewers of Wrexham, the licensee was John Roberts.

By 1903 the brewery had bought the inn and the licensee was Elizabeth Roberts.

There are spirit stock books dated from 1889 to 1953 which are held by the Flintshire Record Office.

The inn was rebuilt in the early 20th century. The alterations included a function room on the first floor and five bedrooms on the second floor. The ground floor contained a public bar, snug, commercial room, coffee room, smoke room and kitchen, scullery, etc. with the cellars below.

The outbuildings in the yard included stabling for six horses, a harness room, a coach house, a shippon[8], and a manure heap.

2006, January – temporarily closed.

Wo, FRO – Ref. D/DM/1219/1-3 and NMW. 1886 PO; 1891 RPD; 1895 S; 1918 RPD. Hotel.

**Rhosllanerchrugog, Market Street / High Street** In 1886 and 1895 the White Horse was also listed as a brewery. 1856-1868 S; 1891 and 1918 RPD.

**Wrexham, 21 Holt Street** 1835-1844 P; 1850 S; 1859 White Horse Inn WD; 1875 Wo; 1886 PO; 1895 S.

## Notes

1. Arthur de Cardonnel Rhys 6th Baron (1836-1911).

2. According to James Bell writing in 1843, the holy well threw up 84 hogsheads every minute, it had never been known to freeze, nor was it increased or diminished by the drought or moisture of the seasons.

3. A match fought between cocks. Up to forty cocks were placed in the pit and the bird which disposed of most of its opponents was the winner. Cockfighting ended in 1822 at the introduction of the Cruelty to Animals Act. KD & DW and SOE.

4. 1812 Henry Paget (1768-1854), became 2nd Earl of Uxbridge and in 1815, 1st Marquis of Anglesey. He lost his leg at the Battle of Waterloo.

5. William Lewis Hughes, MP for Wallingford in Oxfordshire, was created Baron Dinorben of Kinmel in 1831. [Kinmel Hall was designed by the architect Thomas Hopper (1776-1856)]. He was the eldest son of The Rev. Edward Hughes who had built up a prodigious estate of about 85,000 acres in north Wales, due to his large copper mine. He purchased Kinmel Hall in 1786 and later became ADC to Queen Victoria. After his death in 1852, he was succeeded by his son, William, who died unmarried only eight months later, leaving the title extinct.

6. The 'officer', or 'angel', was John Angel, a post-boy who lived his entire life in Holywell; for the rest of his life he kept the hat and light blue jacket he wore the day he was postillion to the Princess and always took a holiday on the anniversary of the event.

7. A female law officer appointed to conduct criminal prosecutions on behalf of the crown or state or in the public interest – now seldom used. SOE.

8. Dialect for a cattle shed, a cow house – from Old English 'scypen'. SOE.

# TRADE
# DIRECTORIES

The Trade Directories throughout the book have been the skeleton upon which we have been able to hang the confirmation of the many dates that the pubs have been trading. The main 19th century publishers of these directories have been Pigot, Slater, White and Kelly however many taverns and inns were trading long before this. Lillywhite was only concerned with inns in London.

### James Pigot (1769-1843)

A Manchester engraver whose directories for the North of England ran from 1811-1843; his information was collected from personal contact by his well-trained agents, some of whom later produced directories of their own. From 1833 he was partnered by Isaac Slater, a fellow Manchester engraver.

### Isaac Slater (1803-1883)

As a Manchester engraver he became apprenticed to James Pigot in 1818, to become his partner in 1833. He continued the business alone until he retired in 1879 when the firm was taken over by Kelly.

### William White Senior (03.01.1799-03.09.1868)

Born in Sheffield; his Directories were centred on Northern England. Aged 18 he joined Mr Edward Baines of Leeds who gave up publishing four years later. William Senior continued running the firm until 1864 when he died of apoplexy at his home in Sheffield.

### William White Junior (1832-1870)

As a young man he joined his father in publishing. Later, in 1860, he became influential in Sheffield freemasonry as Master of Britannia Lodge.

### Frederic Festus Kelly / Keley (1803-1883)

Living in Ealing, London, he became 'Chief Inspector of Inland Letter Carriers'. In 1835 he bought the rights of Critchett's London Directory. He was highly criticised for using employees of the Post Office to gather information, but by 1847 he was forced to abandon this ploy, though he retained the title of 'Post Office directory' for some time. Having taken over the businesses of Pigot and Slater in c.1879 his national dominance was assured.

# REFERENCES

| | | |
|---|---|---|
| **fl.** | = | Flourished. |
| **MSS** | = | Manuscripts. |

## A

| | | |
|---|---|---|
| **AA** | = | Automobile Association, 'Book of British Villages' 1981. |
| **AAD** | = | Dent, A.A. and Machin Goodall, D., 'The Foals of Epona; Galloway, the Dales and the Fells. 1962. |
| **AAT** | = | Automobile Association, 'Treasures of Britain'. 1968. |
| **ABM** | = | Admiral Blake Museum, Bridgwater, Somerset. |
| **AC** | = | Crosby, A., 'A History of Woking'. 1982. |
| **ACM** | = | Armagh County Museum. |
| **ACr** | = | Crawford, A., 'Birmingham Pubs'. 1986. |
| **ACR** | = | Anglesey County Record Office. |
| **AD** | = | Dent, A., 'Horses in Shakespeare's England'. A.J. Allen 1987. |
| **AER** | = | Richardson, A.E., 'The Old Inns of England'. 1935. |
| **AEW** | = | Wilson, A.E. – local historian who bequeathed his research papers to the City of Coventry Archives. |
| **AG** | = | Groom, Arthur, 'Old Coaching Inns and their successors', 'The London. Midland and Scottish Railway and Travel and Transport in four centuries'. |
| **AGe** | = | Gell, A., 'Riseley, Our Village', 'Hostelries and Public Houses'. |
| **AH** | = | Heal, Sir Ambrose (1827-1959), 'London Shop Signs'. English furniture designer; trained as a cabinet maker before joining the family firm, Heals, in 1893. |
| **AHa** | = | Hackney, A., 'Wallington Once in a Millenium'. 1999. |
| **AHi** | = | Hickman, A., 'Thame Inns Discovered'. |
| **AHL** | = | Allen's History of Liskeard. |
| **AHu** | = | Hughes, A., 'Down at the Old Bull and Bush Goods and Chattels of some of Horsham's Innkeepers, 1611-1806'. 1997. |
| **AJM** | = | MacGregor, A.J., 'The Alehouses and Alehouse Keepers of Cheshire 1629 - 1828' (1992), 'The Public Houses of Northwich'. |
| **AKF** | = | Knox Flowerdew, A., 'Rickinghall in the County of Suffolk'. Undated. |
| **AM** | = | 'The History of the Ashford Cattle Market Company'. 1998. |
| **AMi** | = | Mitchell, A., 'Esher – A Pictorial History'. |
| **AMu** | = | Abergavenny Museum – 'Abergavenny Pubs' |
| **AR** | = | Richardson, A., 'English Inns Past & Present'. |
| **AR2** | = | Richardson, A., 'Old Inns of England'. 1934. |
| **AS** | = | Shepley, A., 'Pub Walks in Lancashire'. 1994. |
| **A-S** | = | Alka-Seltzer Guide to the Pubs of Essex and Sussex. 1974. |
| **ASi** | = | Singer, Anita, 'The History of Bempton and Buckton'. c.2002. |

# B

| | | |
|---|---|---|
| **B** | = | Brewer, Dr. E.C., 'Brewer's Dictionary of Phrase & Fable'. Fourteenth Edition, 1989. |
| **B2** | = | Brewer, Dr. E.C., 'Brewer's Dictionary of Phrase & Fable'. Millennium Edition 2000. |
| **B&A** | = | Birch & Armistead, 'Yesterday's Town: Chesham'. 1977. |
| **B&M** | = | Becket, S. and Madgett, J., 'The History of Tasburgh'. December 1993. |
| **B&N** | = | Baines and Newsome Directory. |
| **B&W** | = | Brown, M and Willmott, B 'Brewed in Northamptonshire'. |
| **BaD** | = | Baines's Directory. |
| **BB** | = | Buckley's Brewery. |
| **BBC** | = | British Broadcasting Corporation – 'History on Your Doorstep'. 1982. |
| **BBD** | = | Baker's Biography Dramatica. 1912. |
| **BBe** | = | Bertram, B, 'French Resistance in Sussex'. 1996/97. |
| **BBPR** | = | Bedford Borough Police Register. |
| **BBV** | = | Automobile Association – 'Book of British Villages'. 1981. |
| **BBY** | = | Basford Bystander 1994. Issue 51. Marson, L., 'Mount Street – New Basford'. Issue 52. Potter, D., 'The Golden Hill'. |
| **BC** | = | Cox, B., 'English Inn and Tavern Names'. 1994. |
| **BCA** | = | Birmingham City Archives. |
| **BCL** | = | Bradford Central Library – Local Studies. |
| **BCo** | = | Cousins, B – 'Old Wincanton'. 1992. |
| **BCR** | = | Bedfordshire County Record Office. |
| **BCT** | = | Carpenter Turner, B., 'Winchester'. 1980 |
| **BD** | = | Billing's Directory. |
| **BDE** | = | Bryant's Dictionary of Painters and Engravers, 1915. |
| **BDK** | = | Bagshaw's History, Gazetteer and Directory of Kent. |
| **BDR** | = | Baine's Directories of North and West Ridings. |
| **BER** | = | Bury St. Edmund's Record Office. |
| **BG** | = | Blount's Glossographia. 1656. |
| **BH** | = | Brighton and Hove Local Studies Library. |
| **BHS** | = | Brewery History Society, I.P. Peaty. Woolstone – Journal No. 86 p.41, Winter 1996. |
| **BJ** | = | *Bath Journal*. |
| **BL** | = | Lilleywhite, Bryant, 'London Signs' (Manuscript version). 1966. |
| **BLA** | = | Biographical Dictionary of Living Authors. 1816. |
| **BLAR** | = | Bedfordshire & Luton Archives & Record Service. |
| **BlM** | = | Blomfield's Norfolk. 1805-1810. |
| **BLS** | = | Bradford Local Studies. |
| **BMT** | = | Bushey Museum Trust. |
| **BN** | = | Nixon, B., 'Old Inns and Beer Houses of Bury St. Edmund's', pp.23-24. 1996. |
| **BNP** | = | Room, A, 'Brewer's Names People Places Things'. 1992. |

| | | |
|---|---|---|
| **BP** | = | 'The Bury Post'. |
| **BPe** | = | Pepper, B., 'The Best Pubs in Yorkshire'. 1990. |
| **BPL** | = | City of Birmingham Public Library. |
| **BPo** | = | Poole, B., 'Coventry: Its History and Antiquities' |
| **BR** | = | Boyle, R. (1627-1691). |
| **BreC** | = | Breckland Council – Attleborough Office. |
| **BRO** | = | Bristol Record Office. |
| **BrS** | = | The Brewers' Society - R.J. Webber. |
| **BS** | = | Spiller, B., 'Victorian Public Houses'. |
| **BSE** | = | The Homeland Handbooks Vol. 56 – "Bury St. Edmund's, with its Surroundings". 1907. |
| **BSM** | = | (a) Brewers' Society – 'Inn Signs, their History and Meaning', 1939. (b) Brewers' Society Magazine. |
| **BUD** | = | Bulmer's Directory. |
| **BuL** | = | Buckinghamshire Local Studies. |
| **BW** | = | Wood, B.N.D. – Part author of 'The Wheatsheaf to the Windmill' 1984. |
| **BWr** | = | Wright, B. – 'Fragments of Capel'. 1995. |

# C

| | | |
|---|---|---|
| **C** | = | Chamber's Twentieth Century Dictionary. 1959 Edition. |
| **C&A** | = | Charlton and Archdeacon's Directory. |
| **C&An** | = | Charlton and Anderson's Directory. |
| **C&L** | = | Cockell and Laming, 'The Guide to Collecting Whitbread Inn-Signia'. 1996. |
| **CA** | = | Aldin, C., 'Old Inns'. |
| **CAd** | = | Addison, C., 'King's Meaburn Through the Ages'. 2000. |
| **CaRO** | = | Caernarfon Record Office. |
| **CAS** | = | Cumbria Archive Service. |
| **CB&CA** | = | Birch, C. and Aristead, C., 'Padbury Through The Years'. 1999. |
| **CBC** | = | Charnwood Borough Council. |
| **CBS** | = | Centre for Buckinghamshire Studies. |
| **CC** | = | Cheek, C., 'Architectural Decoration in Aberystwyth During the Victorian and Edwardian Periods'. May 1985. |
| **CCA** | = | Cheshire and Chester Archives and Local Studies. |
| **CCC** | = | Cambridgeshire County Council. |
| **CCL** | = | Calderdale Central Library. |
| **CCM** | = | Carisbrooke Castle Museum. |
| **CCR** | = | Clwyd County Record Office. |
| **CD** | = | Dickens, C. (1812-1870), 'Pickwick Papers'. 1836. |
| **CDA** | = | Calderdale District Archives. |
| **CDH** | = | Morgan, R., 'Chichester - A Documentary History'. 1992. |
| **CFG** | = | Gwilt, C.F., 'Inns and Alehouses of Bridgnorth'. |
| **CFT** | = | Tebbutt, C.F., 'Bluntisham-cum-Earith'. 1941. |

| | | |
|---|---|---|
| **CGH** | = | Harper, C.G., 'The Old Inns of Old England and a Picturesque Account of the Ancient and Storied Hostelries of our own Country'. Volume II. 1906. 'The Dover Road'. 1907. |
| **CH** | = | Trundle, J. – reprinted by Hindley, C., 'The Old Book Collectors Miscellany', Vol. ii. |
| **ChR** | = | Cheshire Record Office. |
| **CHW** | = | Evelyn White, C. H., 'Old Inns and Taverns of Ipswich'. 1885. |
| **CJF** | = | Ferret, 'Fulham Old and New'. 1900. |
| **CL** | = | Courage Limited. |
| **CLH** | = | Central Library, Halifax, West Yorkshire. |
| **CLI** | = | Coventry Libraries and Information Services (Mrs. P. Wills). |
| **CLS** | = | Croydon Local Studies Library. |
| **CLu** | = | Lucas, C., 'Fenman's World'. 1930. |
| **Col** | = | Collins English Dictionary. Third Edition Updated. 1994. |
| **CovA** | = | City of Coventry Archives – Accession 1691. |
| **CoW** | = | City of Westminster Archives. |
| **CPM** | = | Moritz, C.P., 'Journeys of a German in England in 1782'. Translated and edited by R. Nettel. 1965. |
| **CR** | = | Reeve, C., Keeper of Fine and Decorative Arts at Manor House Museum, Bury St. Edmund's. |
| **CRe** | = | Caernarfon Record Office. |
| **CRL** | = | Colne Reference Library. |
| **CRO** | = | Cumbria Record Office. |
| **CSD** | = | Canterbury Street Directory. |
| **CT** | = | Chaucer, G., 'The Canterbury Tales', c.1387. Translated into Modern English by Nevill Coghill. 1974, The Folio Society. |
| **CW** | = | Wade, C., 'More Streets of Hampstead', Parliament Hill to Pond Street. Camden History Society 1973. |
| **CWa** | = | C. Walton, 'The Changing Face of Doncaster'. 1980. |
| **CoW** | = | City of Westminster Archives. |

# D

| | | |
|---|---|---|
| **D&K** | = | Davies, E. and Kissack, K., 'The Innsa and Friendly Societies of Monmouth'. 1981. |
| **D&W** | = | Dunkling and Wright – 'Pub Names of Britain'. 1994. Early references do not give sources. |
| **DA** | = | Doncaster Archives. |
| **DAS** | = | Dorset Archive Service. |
| **DC** | = | Cheason, D., 'The Waterbeach Chronicle'. 1884. |
| **DCA** | = | Derbyshire County Archives. |
| **DCL** | = | Donegal County Library. |
| **DD** | = | Defoe, D., 'A Tour Through the Whole Island of Great Britain'. Three Volumes. 1724-1726. |

| | | |
|---|---|---|
| **DEL** | = | De Laune, 'Present State of London'. |
| **Dev** | = | Devon Record Office. |
| **DF** | = | Foster, D., Collection, City of Westminster Archives. |
| **DFe** | = | Felton, D, 'Speaking of Stanton'. 2003. |
| **DGJ** | = | Jackson, D.G., 'Guide to Free Houses and Inns of Character in the West Riding'. 1973. |
| **DGL** | = | Lamb, D., 'A Pub on Every Corner'. 1996. |
| **DH** | = | Halstead, D., 'Yarns, Recollections and Historical Gleanings: Old Licenced Houses in Haslingden'. c.1930s. |
| **DHa** | = | Hancock, D., 'Pub Walks in Kent – Forty Circular Walks Around Kent Inns'. 1994. |
| **DHC** | = | Dorset History Centre. |
| **DL** | = | Directory of Leeds. |
| **DM** | = | *Derby Mercury.* |
| **DMa** | = | Martin, D. and Mastin, B., 'An Architectural History of Roberts-bridge' 1974. |
| **DNB** | = | The Concise Dictionary of National Biography. 1920 Edition. |
| **DP** | = | Smith, B., 'Pubs of Dover' – revised edition 1982. |
| **DPB** | = | Blaine, D.P., 'Encyclopaedia of Rural Sports'. 1840. |
| **DPo** | = | Powell, D., 'Victorian Pontypridd'. 1996. |
| **DRO** | = | Denbighshire Record Office. |
| **DT** | = | Denis Tye, 'A Village Remembered'. |
| **DTel** | = | *Daily Telegraph.* |
| **DTr** | = | Trumper, D., 'The Twentieth Century Shrewsbury'. 1999. |
| **DUB** | = | Dublin City Archives. |
| **DW** | = | Williams, D., 'Bridgwater's Inns - Past and Present'. 1997. |
| **DWB** | = | Devenish Weymouth Brewery Ltd. |
| **DWH** | = | Wyn Hughes, D., Article 1295H in 'Hertfordshire Countryside'. |

# E

| | | |
|---|---|---|
| **EB** | = | Encyclopaedia Britannica - 14th Edition 1929. |
| **EC** | = | Endowed Charities (County of London), IV. |
| **ECA** | = | Essex County Archives. |
| **ECB** | = | Bryan, E.C. – 'Willaston's Heritage' Second Edition revised by David Morris. 1997. |
| **ECL** | = | Exeter Central Library. |
| **ECM** | = | 'An Ephemeris of Celestial Motions' (1699 – 1705) |
| **ECT** | = | Talbot-Booth, E.C., 'The British Army. Its History, Customs, Traditions & Uniforms'. |
| **ED** | = | Bennett, D., 'Encyclopaedia of Dublin'. 1991. |
| **EE** | = | Ennion, E. – 'Adventurers Fen'. Revised Edition. 1949. |
| **EFP** | = | *Exeter Flying Post.* |
| **EMG** | = | Gardner, E.M., 'Midhurst Court Rolls. West Sussex Record Office |

SNQ13, p267.

| | | |
|---|---|---|
| **EP** | = | Parry, E., 'Cambrian Mirror'. |
| **EPN** | = | English Place Name Society, 'The Place Names of the East Riding'. Vol. XIV. |
| **ERA** | = | East Riding of Yorkshire Archives. |
| **ERD** | = | Delderfield, E.R., 'Introduction to Inn Signs'. |
| **ERG** | = | Green, E.R., 'Pubs in the Gravesend Area'. |
| **ERO** | = | Essex Record Office. |
| **ERQ** | = | East Riding Quarter Sessions. |
| **ESR** | = | East Sussex Record Office. |
| **ET** | = | Tasker, E., 'Tasker Streets'. 1974. |
| **EvE** | = | *Evening Echo.* |

## F

| | | |
|---|---|---|
| **F&D** | = | Fowles, J. and Draper, J. – 'Thomas Hardy's England'. 1984. |
| **FD&EC** | = | Duke, F. and Rev. E. Cox, 'In and Around Steyning'. |
| **FG** | = | Graham, F., 'Hexham a Short History and Guide'. 1973. |
| **FLH** | = | Fleur de Lis Heritage Centre, Faversham. |
| **FP** | = | Fowler, E.G. and Pierce, Iona, 'Llangynhafal – A Parish and Its Past'. 1983. |
| **FR** | = | Frederic Robinson Ltd., Brewers of Stockport. |
| **FRO** | = | Flintshire Record Office. |
| **FVL** | = | *Folkestone Visitors List & Society Journal.* 10th June 1891. |

## G

| | | |
|---|---|---|
| **G** | = | Guildhall Library. |
| **GA** | = | Gwenedd Archives. |
| **GAC** | = | Cooke, G.A., 'Topographical & Statistical Description of the County of York'. |
| **GB** | = | Bryan, G., 'Chelsea in the Olden and Present Times'. 1869. |
| **GBo** | = | Boudier, G, 'A-Z of Enfield Pubs'. Part Two. 2002. |
| **GC** | = | Clinch, G., 'Mayfair and Belgravia: Being an Historical Account of the Parish of St. George, Hanover Square'. 1892. |
| **GCH** | = | Calvert Holland, G., 'The Vital Statistics of Sheffield'. 1843. |
| **GCW** | = | Williamson, G.C. – Editor of 'Trade Tokens Issued in the 17th Century' by William Boyne (d.1893). 1967. |
| **GEM** | = | Minter, G and E., 'Discovering Old Lepton and Kirkheaten' Second edition. |
| **GEP** | = | General Evening Post (London). |
| **GHM** | = | Morgan, G.H., 'The Romance of Essex Inns'. 1967. |
| **GK** | = | Kelly, G., 'The White Horse Public House, Old Buckenham – A History'. 1985. |
| **GL** | = | Gosport Library. |

| | | |
|---|---|---|
| **Gla** | = | Glamorgan Record Office. |
| **GLH** | = | 'A List of All Licensed Houses in the County of Gloucester'. 1891. |
| **GM** | = | *Gentleman's Magazine.* |
| **GPG** | = | Good Pub Guide. |
| **GPO** | = | General Post Office Record Department. |
| **GRe** | = | Gwynedd Record Office. |
| **GrK** | = | Greene King – brewers. |
| **GRM** | = | 'A Guide to the Royal Mile in Edinburgh'. |
| **GRO** | = | Gloucestershire County Record Office. |
| **GWR** | = | Gwent Record Office. |

# H

| | | |
|---|---|---|
| **H&C** | = | Hunt and Company's Directory. |
| **H&H** | = | *Hampstead and Highgate Express* 5.12.1952. |
| **H&W** | = | Hall and Woodhouse Ltd., Brewers. |
| **HA** | = | Halifax Archives. |
| **HaT** | = | Hampshire Treasures – New Forest Volume. |
| **HB** | = | The Heavitree Brewery Ltd. |
| **HAH** | = | Harben, Henry A., 'A Dictionary of London'. 1918. |
| **HCA** | = | Humberside County Archives. |
| **HCR** | = | Hertfordshire County Record Office. |
| **HEB** | = | Bocking, H.E.–'The Inns and Public Houses of King's Lynn'. 1950. |
| **HEP** | = | Popham, H.E., 'Taverns in the Town'. 1937. |
| **HFA** | = | Hammersmith and Fulham Archives. |
| **HG** | = | Grieve, H., 'The Sleepers and The Shadows: Chelmsford; a town, its people and its past'. |
| **HGH** | = | Halifax Guardian Historical Almanack. |
| **HH** | = | Hall, H. – 'Lindfield Past and Present'. 1959. |
| **HHD** | = | Hampstead and Highgate Directory. |
| **HHu** | = | Hurley, H., 'The Pubs of Monmouth, Chepstow and the Wye Valley'. 2007. |
| **HIT** | = | Heritage of the Ile Trust, 'Images of the Past'. 1983. |
| **HL** | = | Lea, Hermann, 'Thomas Hardy's Wessex'. 1887. |
| **HLD** | = | Douch, H.L.; Royal Cornwall Museum, 'Old Cornish Inns and Their Place in the Social History of Cornwall'. 1966. |
| **HLS** | = | Huddersfield Local Studies Library. |
| **HM** | = | Murray, H., 'A Directory of York Pubs, 1455-2003'. 2003 & 2004. |
| **HMR** | = | Richardson, H.M., 'Burwell A Stroll Through History'. 1990. |
| **HoW** | = | House of Whitbread. |
| **HP** | = | Humphreys, R., 'A Pictorial Study of Hawkinge Parish'. |
| **HPl** | = | 'Haselbury, Plucknett and There-abouts in Earlier Days. Glimpses of Local People and Everyday Life from Old Documents'. |
| **HPM** | = | H.P. Maskell, 'Old Country Inns'. 1910. |

| | | |
|---|---|---|
| **HRO** | = | Hampshire Record Office. |
| **HT** | = | Horsfall Turner, 'Ancient Bingley'. |
| **HTD** | = | Holden's Triennial Directory. |
| **HW** | = | Walker, H., 'This Little Town of Otley'. 1974. |

# I

| | | |
|---|---|---|
| **IAA** | = | The Irish Architectural Archive. |
| **IaW** | = | Wakeford, I., 'Bygone Woking'. 1983. 'Woking as it was'. 1985. |
| **ICL** | = | Islington Central Library. |
| **IOE** | = | Return of Licensed Houses in the Isle of Ely. 1906. |
| **IH** | = | Hales, I., 'Old Maidstone's Public Houses'. 1982. |
| **IoW** | = | Isle of Wight County Record Office. |
| **IPP** | = | Peaty I.P., 'Essex Brewers'. 1992. |
| **IR** | = | Rabey, I., 'The Book of St. Columb and St. Mawgan. 1979. |
| **IRO** | = | Ipswich Record Office. |
| **ISt** | = | Strange, I., 'Time! Gentlemen, Please! Dorchester Pubs Remembered'. 1995. |
| **IS** | = | Stevens, I., 'The Story of Esher'. 1966. |
| **IW** | = | Watkin, I., 'Oswestry'. 1920, T. Owen. |
| **IWa** | = | Waters, I., 'Inns and Taverns of the Chepstow District'. |
| **IY** | = | Yearsley, I., 'A History of Southend'. |

# J

| | | |
|---|---|---|
| **J&J** | = | Jolliffe, G., and Jones, A., 'Hertfordshire Inns and Public Houses, an Historical Gazetteer'. 1955. |
| **JA** | = | Ashby, J., 'Character of Coventry'. 2001. |
| **JAL** | = | Langford, J.A., 'A Century of Birmingham Life or, A Chronicle of Local Events, From 1741 to 1841'. Vol. 1. |
| **JB** | = | Bell, J., 'A New and Comprehensive Gazetteer of England and Wales'. 1834. |
| **JBa** | = | Bailey, J. and Nigel Harris (composer), 'Ego Aethelredus – The People of Ile Minster from 995 to 1995 A.D'. 1995. |
| **JC** | = | Cockayne, J., 'Walsall Pubs'. 1984. |
| **JD** | = | Dodd, J. – a landlord of 'The Baker's Arms' in the village of Droxford, Hampshire. |
| **JDs** | = | Davies, J., 'Bacup's Hotels, Inns, Taverns and Beerhouses'. c.1995. Also Rawtenstall – Vol. 2. |
| **JES** | = | Smith, J.E., 'St. John the Evangelist, Westminster'. 1892. |
| **JFC** | = | Curwen, J.F., 'Kirkbie Kendall'. |
| **JGB** | = | Bishop, J. G., 'A Peep into the Past: Brighton in Olden Times'. 1892. |
| **JGH** | = | Horton, J.G., 'A Survey of the Inns of Boston and Notes on the Old Licensing Laws.' 1988. |
| **JH** | = | Haden, Jack – one time local newspaper reporter. |

| | | |
|---|---|---|
| **JH&GS** | = | Harris, J. and Stantan, G.M., 'A History of Winkfield'. 2001. |
| **JHF** | = | Futcher, J.H., 'Public Houses in Victorian Farnham and Liquor Licensing Laws', Volume 6. 2001. |
| **JHM** | = | Macmichael, J.H., 'The Story of Charing Cross'. 1906. |
| **JJB** | = | Baddeley, Sir J.J., 'Cripplegate'. 1921. |
| **JJT** | = | Tohill, J.J., 'Pubs of the North'. 1990. |
| **JM** | = | Morse, Sir Jeremy – the great-great grandson of the Norfolk brewer George Morse. |
| **JOH** | = | Halliwell, J.O., 'The Nursery Rhymes of England'. 1842. |
| **JP** | = | Preshous, J., 'Bishop's Castle Well-Remembered'. 1990. |
| **JPa** | = | Parratt, J., 'Licensed Farnham, a tour of pubs past and present'. 1994. |
| **JPe** | = | Janet Pennington. Local Historian. |
| **JPJ** | = | Pryce-Jones, J., 'Historic Oswestry'. 1982. |
| **JPr** | = | The Reverend J. Pridden – a list of Suffolk Inns. 1786. |
| **JR** | = | Rocque, John, 'Plan of the Cities of London and Westminster and Borough of Southwark'. 1746. |
| **JRB** | = | Burden, J.R., 'Freemasonry in Bury St. Edmund's', p.5. 1954. |
| **JRi** | = | Richards, J., 'The History of Batham's Black Country Brewers'. 1993. |
| **JS** | = | Shepperson, J. – local historian. |
| **JSk** | = | Skeggs, J., Somerset local historian. |
| **JSL** | = | Journal of the Sun Life Society. Spring 1999. An article by Captain Bright. |
| **JSR** | = | Rudd, J. and S., 'Old Inns and Alehouses of Twyford'. |
| **JSS** | = | Jacksons Stopps & Staff – Estate Agents. |
| **JSt** | = | Street, J., 'The Delights of Droxford' – Hampshire May 1966. |
| **JSY** | = | Strype, John (1643-1737) – Editor of Stow, J., 'Survey of London and Westminster, 1598'; 1603. |
| **JT** | = | Timbs, John, 'Curiosities of London – Clubs and Club Life in London', 1855 and 1872. |
| **JTe** | = | Tetley, J., 'Heritage Inns and Pubs of Character'. 1987. |
| **JTK** | = | Thirkettle, J., 'The Best of British Pubs'. |
| **JW** | = | James Woodforde, 'The Diary of a Country Parson 1758-1802'. 1924 Edition. |
| **JWi** | = | Williams, J., 'Alehouses'. Published by Warbleton Historical Society 2002. |
| **JWC** | = | J.W. Cameron & Co. Ltd., Brewers. |
| **JWL** | = | J.W. Lees & Co., Brewers. |

# K

| | | |
|---|---|---|
| **K** | = | Kelly's Directory. |
| **KBD** | = | Kelly's Bristol Directory. |
| **KCC** | = | Kent County Council T205 and T233. |
| **KCL** | = | Borough of Kensington and Chelsea Libraries. |

| | | |
|---|---|---|
| **KD&DW** | = | Davies, K. and Wilkes, D., 'Holywell High Street Past and Present'. Published by Friends of the Greenfield Valley 1989. |
| **KE** | = | Keltic Enterprises Ltd. 1976. |
| **KG** | = | Gilliver, K. 'Last Orders Gentlement Place: A History of Pubs and Beerhouses in the Swadlincote Area of Derbyshire'. 2004. |
| **KH** | = | Heslton, K., 'A History of Sunbury's Pubs'. 1988. |
| **KJ** | = | Johnson, K., 'The Pubs of Norberth, Saundersfoot and South East Pembrokeshire'. 2004. |
| **KL** | = | Kettering Library. |
| **KL&W** | = | King's Lynn and Wells Libraries. |
| **KP** | = | Page, Ken, 'Thirsty Old Town'. The Story of Biggleswade Pubs. 1997. |
| **KPC** | = | 'Kidd's Picturesque Pocket Companion to Brighton' mid 1830s. |
| **KSGH** | = | K.S.G. Hinde, OBE – Local Historian. |
| **KT** | = | Turner, K., 'Images of England – Birmingham Pubs'. 1999. |
| **KU** | = | Keele University. |

# L

| | | |
|---|---|---|
| **L** | = | Leigh's Road Book of England and Wales. 1826. |
| **L&M** | = | *Licensee and Morning Advertiser.* |
| **LA** | = | Lambeth Archives. |
| **LAO** | = | Lincolnshire Archives Office. |
| **LCC** | = | Leicestershire County Council. |
| **LCG** | = | Dodsley, R., 'London – A Complete Guide to All Persons Who Have Any Trade or Concern with the City of London'. 1760. |
| **LCL** | = | Lloyd, L.C., 'Inns of Shrewsbury'. 1942. |
| **LE** | = | The London Encyclopaedia 1983. |
| **LFP** | = | Pearson, L.F., 'The Northumberland Pubs'. 1989. |
| **LG** | = | Greenall, L., 'A History of Kettering'. 2003. |
| **LGL** | = | London Guildhall Library. |
| **LH** | = | Larwood and Hotten, 'The History of Signboards'. Revised edition 1951. First published 1898. *Jacob Larwood was the pseudonym for Herman Diederik Johan van Schevichaven.* |
| **LHP** | = | Living History Publications – 'Epsom Town Downs and Common'. 1979. |
| **LHS** | = | Local History Society/Group/Club/Association. |
| **LHu** | = | Hutchins, L., 'Esher and Claygate Past'. |
| **LiL** | = | Lincolnshire County Library, Boston. |
| **LL** | = | 'The History of Hunting', The Lonsdale Library Vol. XXIII. |
| **LM** | = | Marson, L, 'Mount Street – New Basford'. The Basford Bystander Issue 51 1994. |
| **LLH** | = | Lewisham Local History Centre. |
| **LLS** | = | Leeds Lical Studies Library. |
| **LMa** | = | Lindal and Marton Heritage Part 1. www.lindal-in-furness.co.uk |

| | | |
|---|---|---|
| **LMA** | = | London and Middlesex Archaeological Society. |
| **LMu** | = | Murgatroyd, L., 'A Pub-Crawl through Time!'. 2000. |
| **LP** | = | Lloyd Price, 'Holywell in the Early part of the Nineteenth Century'. A lecture given on 4th December 1912. |
| **LPT** | = | Thompson, L.P., 'Old Inns of Suffolk'. 1946. 'Ipswich Inns, Taverns and Pubs'. 1991. |
| **LR** | = | Licence Registers. |
| **LRO** | = | Leicestershire Record Office. |
| **LT** | = | 'Little Totham: The Story of a Small Village'. 2005. |
| **LTM** | = | London Transport Museum. |
| **LVR** | = | Licensed Victuallers' Records. |

# M

| | | |
|---|---|---|
| **M** | = | *Manchester Evening Chronicle* 24th April 1963. |
| **MA** | = | *Morning Advertiser.* |
| **MAC** | = | Craven, M., 'Inns and Taverns of Derby'. 1992. |
| **MALS** | = | Medway Archives and Local Studies Centre. |
| **MB** | = | Brown, M., 'Old Pubs of Leighton Buzzard and Linslade'. 1994. |
| **MBD** | = | Matthew's Bristol Directory. |
| **MBe** | = | Bennett, M., 'The White Horse of Oswestry'. Shropshire Magazine December 1979. |
| **McM** | = | McMullen and Sons Limited. Brewers. |
| **MC** | = | Mercer and Crocker's Directory. |
| **MCh** | = | Miller Christy, 'The Trade Signs of Essex'. 1887. |
| **MCL** | = | Mayo County Library. |
| **MD** | = | Dickinson, M., '17th Century Tokens of the British Isles and Their Value'. 1986. |
| **MDA** | = | Mannex Directory – Adlington and District. |
| **MDC** | = | Montgomeryshire District Council. |
| **MDG** | = | Minnitt, S.C.; Durnel, J; Gunstone, A.J.H., 'Somerset Public House Tokens'. 1985. |
| **MDM** | = | Mirams, M.D., 'Kent Inns and Inn Signs'. 1987. |
| **MF** | = | Monson-Fitzjohn, G.J., 'Quaint Signs of Olde Inns'. 1926. |
| **MG** | = | Gardener, Meryl - Irish Historian. |
| **MGH** | = | Heenan, M. G., 'Canterbury Pubs Past and Present'. *Kentish Gazette*, 23.3.1978. |
| **MJ** | = | Mincher, M. and Jarvis, M., 'The Public Houses of Launceston'. Monograph No. 12. Lawrence House Museum 2003. |
| **MJB** | = | Baddeley, M.J.B., 'The English Lake District', Fifteenth Edition. |
| **MJC** | = | Cutten, M.J., 'Some Inns and Alehouses of Chichester'. 1964. |
| **MJE** | = | Leveritt, N. and Elsden, J.J., 'Aspects of Spalding 1790 to 1930'. |
| **MkD** | = | Dean, M – former owner of The Gin Palace, 36 Market Road, Islington. |

| | | |
|---|---|---|
| **ML1** | = | Malton Library. |
| **ML2** | = | Meath County Library. |
| **MLK** | = | Kilvert, M.L, 'A History of Framlingham'. 1995. |
| **MLS** | = | Margate Library - Local Studies. |
| **MM** | = | Michael Marshman – Wiltshire County Local Studies and Trowbridge Reference Librarian. |
| **MoD** | = | Morris, D., 'Two early Breweries in Mile End Old Town, Stepney 1700-1780'. Published in 'Brewery History No. 102, Winter 2000'. |
| **Mor** | = | Morland & Co. Ltd., Brewers. |
| **MR** | = | Rendell, M., 'Haselbury Plucknett. Portrait of a Village'. Undated. |
| **MS** | = | Stuart, Marie W., 'Old Edinburgh Taverns'. 1952. |
| **MSo** | = | Soanes, M., 'Hostelries', The Corton History, Bulletin 12. September 1998. |
| **MSt** | = | Sturley, M., 'The Breweries and Public Houses of Guildford'. Part 2. 1990. |
| **MT** | = | Merthyr Tydfil Central Library. |
| **MTE** | = | Marston Thompson & Evershed Ltd. Brewers. Mr. I. D. Tantrum. |
| **MW** | = | McGrath and Williams, 'Bristol Inns and Alehouses in the mid-Eighteenth Century'. |
| **MWF** | = | Washington Fletcher, M.H., 'Inns and Inn Signs of Dudley'. 1953. |
| **MX** | = | Mannix and Whellan Directory. |

# N

| | | |
|---|---|---|
| **N** | = | Nicholson, 'Cambrian Travellers' Guide'. |
| **NaL** | = | Nantwich Library. |
| **NC** | = | *The Nuneaton Chronicle*. September 1899. |
| **NCB** | = | Benson, N.C., 'Dunstable in Detail'. 1986. |
| **NCD** | = | National Commercial Directory. |
| **ND** | = | 'Nuneaton Diary' Memorandum Book of Occurrences at Nuneaton Vol. 2., May 1825 – November 1845, p.97. |
| **NH** | = | Harris, N. (composer) and J. Bailey, 'Ego Aethelredus – The People of Ile Minster from 995 to1995 A.D'. 1995. |
| **NI** | = | Nightingale. |
| **NL** | = | Nuneaton Library and Information Centre. |
| **NMW** | = | National Monuments Record for Wales – Cadw List of Buildings of Special Architectural Interest, No. 27 Aberystwyth, 24.11.87. |
| **NoR** | = | Norfolk Record Office. |
| **NPP** | = | 'Northamptonshire Past & Present', Vol. VII, No. 4. 1987. |
| **NQ** | = | Notes and Queries - 3rd S. Vol. VI Oct. 29 1864. |
| **NRO** | = | Northamptonshire Record Office. |
| **NT** | = | Tiptaft, N., 'Inns of The Midlands'. 1951. |
| **NTD** | = | Norfolk Trade Directory. |
| **NYR** | = | North Yorkshire County Record Office. |

# O

| | | |
|---|---|---|
| **O&M** | = | Ogilby and Morgan, 'A Large and Accurate Map of the City of London'. 1677. |
| **OA** | = | Oxfordshire Archives. |
| **OCE** | = | Drabble, M., 'Oxford Companion to English Literature'. 1985. |
| **OCM** | = | Mayo, O.C., 'The Pampisford Archive'. |
| **ODNB** | = | Oxford Dictionary of National Biography Archive. 2004. |
| **OPI** | = | Opie, I. and P., 'The Oxford Dictionary of Nursery Rhymes'. 1951. |
| **OS** | = | Ordnance Survey. |
| **OSJ** | = | Museum and Library of The Order of St. John of Jerusalem. |
| **OUP** | = | Oxford University Press. |
| **OWL** | = | Old Whittington Library – Local History. |
| **OxD** | = | Withycombe, E.G., 'Oxford Dictionary of English Christian Names' – 1948 Edition. |
| **OxN** | = | Ekwall, E., 'The Concise Oxford Dictionary of English Place-Names'. 1936. |

# P

| | | |
|---|---|---|
| **P** | = | Pigot's Directory. |
| **Par** | = | Parkes, J., 'History of Tipton'. 1915. |
| **PC** | = | Cullen, Dr. P., 'Shoreham-by-Sea & Sussex: Colloquial Words'. |
| **PD** | = | Porter's Directory of Whitehaven, etc. |
| **PDD** | = | Postal Directory of Denbighshire. |
| **PE** | = | Earle, Peter, 'The Best of British Pubs'. 1982. |
| **PEB** | = | Baines, P. and E. - 'History & Gazetteer of the County of York'. |
| **Pem** | = | Pembrokeshire Record Office. |
| **PEv** | = | Everill, Pat, 'Pubs and Publicans of Cheslyn Hay'. |
| **PG** | = | Gillanders, P., 'South Shields'. 1974. |
| **PH** | = | Hudson, P. – of Hudson History, publishers. |
| **PJ** | = | Jennings, P., 'The Public House in Bradford 1770-1970'. 1995. |
| **PJM** | = | Munday, P.J. – 'Esher – 'A Pictorial History'. 1983. From the Journal of the Framlingham and District Local History and preservation Society. |
| **PJS** | = | Stannare, P.J., 'Inns of Framlingham, Past and Present'. 1959. |
| **PME** | = | Enderby, P. M. – City Archivist card index held by Canterbury Local Studies Library. |
| **PNC** | = | Preservation Society of Newport County. |
| **PNW** | = | Place Names of Wiltshire. |
| **PO** | = | Post Office Directory. |
| **POT** | = | Post Office Trades Directory. |
| **PR** | = | 'Paterson's Roads'. 1829. Edited by Edward Mogg. |
| **PRO** | = | Public Record Office. |
| **PSD** | = | Pearce's Swansea Directory. |

| PT | = | P. Tuffrey, 'Doncaster District Old Inns and Taverns'. 1986. 'Doncaster Old Inns and Taverns. 1990. |
| PWD | = | Parson & White Directory of Cumberland, Westmorland, Furness and Cartmel. |

# R

| R | = | Robson's Directory. |
| R&W | = | Riden, P. and Webb M., 'Towcester Tenements'. 2000. |
| RAB | = | Baldwin, R.A., 'The Gillingham Chronicles'. 1998. |
| RBB | = | Trust Deed of The Rock Brewery Brighton Limited – 1st April 1912. |
| RBD | = | A Calendar to the Records of the Borough of Doncaster, Volume 4 'Courters of the Corporation'. 1902. |
| RCG | = | *Royal Cornwall Gazette.* |
| RCH | = | Royal Commission on Historical Monuments. |
| RCT | = | Rotary Club of Tarporley – 'Tarporley Then and Now'. 2000. |
| RCW | = | Royal Commission on Ancient and Historical Monuments in Wales – Aberystwyth Files. |
| RCW2 | = | Publications and Outreach Branch. |
| RD | = | Reader's Digest Association Limited – 'The most amazing places to visit in Britain'. 2006. |
| RH | = | Richardson, N. and Hall, R., 'The Pubs of Swinton and Pendlebury'. c.1981. |
| RHo | = | Holmes, R., 'Ely Inns'. 1984. |
| RHL | = | 'Robbin's History of Launceston'. |
| RK | = | Keverne, Richard, 'Tales of Old Inns'. 1939. |
| RL | = | Long, R., 'Haunted Inns of Sussex'. 2001. |
| RLT | = | Federation of the Retail Licensed Trade, Northern Ireland. |
| RM | = | Maybey, R., 'Home Country'. |
| RMo | = | Morgan, R., 'Chichester, A Documentary History'. 1992. |
| RN | = | Neep, R. – Archive CD Books Project. |
| RO | = | Romsey Online. |
| RP | = | Purdy, R.J.W., 'Hautbois Magna', Norfolk Archaeology, vol. 16, 1907. |
| RPa | = | Parker, R., 'A Guide to Foxton'. c.1987. |
| RPD | = | Returns of Public and Beer Houses in Denbighshire (Ref: QSD/CB/3/1,5). |
| RPH | = | Hudson, R. and P., 'Take a Closer Look at Settle' 2000; and 'Take a Closer Look at Richard Preston and The Folly in Settle'. 2001. |
| RS | = | Shoesmith, R., 'The Pubs of Hereford City'. 1994. |
| R.S.O. | = | See *S.O.* |
| RSA | = | 'The Journal of the Royal Society of Antiquaries of Ireland', p.97. 1910. |
| RSS | = | Summerhays, R.S., 'Encyclopædia for Horsemen'. Fourth Edition 1966. |
| RSu | = | Sulima, R., 'Around Tamworth in Old Photographs'. |
| RW | = | 'Public Houses, Inns and Hotels of the former brewers, Richard |

Whitaker, Ltd. of Halifax. 1967.

| | | |
|---|---|---|
| **RWC** | = | Carter, R. W. – Hon. Curator of the Chard Museum. |
| **RWIB** | = | 'Recommended Wayside Inns of Britain'. |

# S

| | | |
|---|---|---|
| **S** | = | Slater's Directory. |
| **SA** | = | *The Staffordshire Advertiser.* |
| **SAC** | = | Sussex Archaeological Collections. |
| **SAL** | = | *Salisbury Journal.* |
| **SAN** | = | Transactions of the Shropshire Archaeological and Natural History Society. Vol VIII. 1885. |
| **SaS** | = | O.G.W. Smith - Samuel Smith Old Brewery, Tadcaster. |
| **SBC** | = | Spelthorne Borough Council – 'A walk through the interesting village of Sunbury-on-Thames'. |
| **SBG** | = | Baring-Gould, S., 'Cornish Characters and Strange Events.' 1909. |
| **Sc** | = | Scrutons, 'Old Bradford Pubs and Inns'. 1907. |
| **SCA** | = | Suffolk County Arts and Libraries Department. |
| **SCL** | = | Southend Central Library. |
| **SCM** | = | Brown I.J., Shropshire Caving and Mining Club Account No. 13. 'An Interim Report on the Lincoln Hill Limestone Mines near Ironbridge, Shropshire'. 1981. |
| **SD** | = | Somerset Directory. |
| **SDM** | = | Stevens Directory of Maidstone. |
| **SG** | = | *Shields Gazette.* |
| **SGB** | = | Burnay, S.G., 'Time, Gentlemen please – 300 + years of pubs and breweries in East and West Ilsley'. 2003. |
| **ShA** | = | Shropshire Archives. |
| **SHC** | = | Surrey History Centre, Woking. |
| **ShM** | = | *Sherborne Mercury.* |
| **SHM** | = | Miller, S.H. – 'Handbook to the Fenland'. 1889. |
| **SJ** | = | Stevens, J., F.S.A. |
| **SJe** | = | Jenkins, S., 'England's Thousand Best Churches'. 2000. |
| **SL** | = | Survey of London. |
| **SLS** | = | Sheffield Local Studies Library. |
| **SLSL** | = | Slough Local Studies Library. |
| **SM** | = | *Suffolk Mercury.* |
| **SoL** | = | Sevenoaks Local Studies Library. |
| **S.O.** & *R.S.O.* | = | These were abbreviations adopted by H.M. Post Office to represent Sub-Office and Railway Sub-Office; if these initial letters, with the name of the County, were added to the addresses of letters in lieu of the usual name of the Post Town, the delivery of such letters was accelerated. By February 1906 the letters R.S.O had been abolished. |
| **SOE** | = | The Shorter Oxford English Dictionary on Historical Principles. 1933. |

| | | |
|---|---|---|
| **SPB** | = | Barker, J., 'Shrewsbury Public Houses'. A list of Licence Returns for Shrewsbury. c1902. |
| **SRL** | = | Southampton Reference Library. |
| **SRO** | = | Somerset Record Office. |
| **SS** | = | Stanley Smith – Local Historian. |
| **SSL** | = | South Shields Central Library. |
| **Sta** | = | Staffordshire Record Office. |
| **STR** | = | Strype, J. (1643-1737). Ecclesiastical historian who corrected and enlarged Stow's 'Survey of London'. 'A Map of St. James's Parish, Westminster, Book VI p.83. |
| **STT** | = | Street, 'Ghosts of Piccadilly'. |
| **SUF** | = | Suffolk Record Office. |
| **SUR** | = | Surrey Record Office. |
| **SVH** | = | Holroyd, E.V., 'Stage Coach and Highwayman'. 1963 thesis. |
| **SWA** | = | *Sussex Weekly Advertiser*. |

# T

| | | |
|---|---|---|
| **T** | = | Thom's Directory. |
| **T&H** | = | Tunstead and Happing Licence Register. |
| **T&I** | = | Telegraph and Independent of Sheffield. |
| **T&W** | = | Tyne and Wear Archives Service. |
| **TCA** | = | Trinity College Archives. |
| **TDL** | = | Towcester and District Local History Society. |
| **TF** | = | Flynn, T., 'A History of the Pubs of Eccles'. |
| **TFB** | = | Bulmer, T.F., 'Directory of East Cumberland'. 1884. |
| **THB** | = | Todmorden and Hebden Bridge Historical Almanack. |
| **THL** | = | Tower Hamlets Library. |
| **TN** | = | *The Times* newspaper. |
| **TNu** | = | Nurse, T., 'The Public Houses of Whittington'. Part One, Old and New Whittington. 2006. |
| **TS** | = | Smith, T., 'The Kettering Album'. |
| **TSJ** | = | Jennings, T.S., 'The Story of Great Paul'. |
| **TWW** | = | Whitley, T.W., 'Humorous Reminiscences of Coventry Life, Coventry Coaching and Coach Roads, and Other Local Works'. 1888. |

# U

| | | |
|---|---|---|
| **UBD** | = | Universal British Directory. |

# V

| | | |
|---|---|---|
| **VCH** | = | Victoria County History. |
| **VFI** | = | Vintners' Federation of Ireland, 1982. |
| **VS** | = | Surtees, V., 'A Second Self'. 1990. |

# W

| | | |
|---|---|---|
| **W** | = | White's Directories. |
| **WA** | = | Whitbread Archive. |
| **WAd** | = | Addison, Sir William, 'Understanding English place-names', 1978. |
| **WB** | = | West Briton. |
| **WBJ** | = | Branch-Johnson, W., 'Hertfordshire Inns'. 1962/63. |
| **WBL** | = | Whaley Bridge Library. |
| **WBR** | = | Redfarn, W.B., 'Old Cambridge'. 1876. |
| **WC** | = | Wright and Curtis, 'The Inns and Pubs of Nottinghamshire'. 1995. |
| **WCM** | = | Winchester College Muniments. |
| **WCO** | = | Oulton, W.C., 'The Traveller's Guide'. 1805. |
| **WCR** | = | Warwickshire County Record Office. |
| **WCSL** | = | Westcountry Studies Library, Exeter. |
| **WD** | = | Wyn Davies, D., 'The Town of a Prince – A History of Machynlleth'. 1991. |
| **WDG** | = | White, W., 'History, Gazetteer and Directory'. |
| **WDH** | = | Wetherby and District Historical Society, 'The Story of the Hostelries'. 1995. |
| **WGA** | = | West Glamorgan Archive Service. |
| **WGT** | = | Whitbread Grand Trade Ledger, W/21/1, 1790-1794, 71/4. |
| **WH** | = | Hargrove, William, 'History and Description of the Ancient City of York'. |
| **WHa** | = | Hazlitt, W, 'Table Talk', Vol. 2, 'The Ignorance of the Learned'. 1821. |
| **WhB** | = | *What's Brewing.* |
| **WHD** | = | White Horse Distillers Ltd. |
| **WLH** | = | Walsall Local History Centre. |
| **WLS** | = | Wiltshire County Local Studies Library. |
| **WmB** | = | Booth, W., 'Here's to Boroughbridge'. |
| **WmC** | = | Cobbett, W., 'Rural Rides', Vols. I and II. Published 1908. |
| **WN** | = | *Whitbread News.* |
| **Wo** | = | Worrall's Directory. |
| **WP** | = | Whitbread Property. |
| **WPCC** | = | The Waverton Parochial Church Council – 'Waverton, A History of its People and Places. December 2002. |
| **WPD** | = | W. Parsons Directory. |
| **WPO** | = | White's Post Office Directory. |
| **WrD** | = | Wrexham Directory. |
| **WRG** | = | Wreglesworth, Richardson and Gall, 'The Pubs and Breweries of Macclesfield'. 1981. |
| **WRO** | = | Wiltshire and Swindon County Record Office. |
| **WSL** | = | William Salt Library, Stafford. |
| **WSP** | = | Walkerley's Sussex Pubs. |
| **WSR** | = | West Sussex Record Office. |

**WWH** = Hutching, W.W., 'London Town Past and Present', Vol. II. 1909.

**WYA** = West Yorkshire Archive Service, DD/RE, pp10 and 12.

# Y

**YHD** = History Directory & Gazetteer of the County of York. Vol. 1.

# Notable Visitors Index

Addison, Joseph  217-18, 226
Andrew, Prince, Duke of York  134
Austen, Jane  91

Barnes, Professor Robert  49
Beare, Ellen  93
Boswell, James  503, 507
Browne, Hablot Knight *(Phiz)*  52
Bradman, Sir Donald  355
Bulkeley, Sir Richard *(19th C.)*  521
Button, Billy *(the clown)*  226

Charles I, King  28, 100, 506
Charles II, King  21, 227, 385
Chaucer, Geoffrey  155, 157
Churchill, Sir Winston  134
Cook, Captain James  451
Courtenay, Sir William *see Thom / Tom, John Nichols*
Coverdale, Miles  50
Cranmer, Thomas  49
Cromwell, Oliver  206, 415
Cruickshank, George  231
Cumberland, William Augustus, Duke of *(18th C.)*  370

Dick, William  507
Dickens, Charles  52, 231-2, 344-5, 357, 452
Dodsley, Robert  205, 209, 228, 241
Dugdale, Sir William  429

Edward VII, King *(when Prince of Wales)*  139
Edward VIII, King *(after abdication)*  345
Edward the Black Prince  163
Eisenhower, Gen. Dwight  98, 134
Eldon, Lord *see Scott, Sir John*
Empey, Sir Reg  494

Fisher, Douglas  114
Foote, Samuel  507

George II, King  25, 343
Gwynne, Nell  206

Hardy, Thomas  104, 105
Hawkwood, Sir John  117
Hazlitt, William  231
Henry VIII, King  29, 218, 254
Highmore, Thomas  227-8

Jenner, Dr. Edward  122
Johnson, Dr. Samuel  228, 507

Kent, Duchess of *(1832)*  520
King, McKenzie  134

Langtry, Lillie *(The Jersey Lily)*  139
Louis XVIII, King of France  343-4

MacCulloch, Horatio  507
Mackie, Sir Peter  507
Marlowe, Christopher  156
Marlowe, Dorothy  157
Mathews, Charles  231
Mercer, John  174
Meynell, Hugo  184
Middleton, Frank  185
Moritz, Karl Philipp  40, 182-3, 288
Mostyn, Sir Edward *(19th C.)*  521
Munnings, Sir Alfred  253-4, 347

Nelson, Lord Horatio  114, 134, 343
Norfolk, Duke of  343, 386

Oates, Titus  207

Peel, John  88
Pepys, Samuel  201, 206, 210, 218, 220, 234, 358
Phiz *see Browne, Hablot Knight*
Pitt Lennox, Lord William  230
Pollard, James  231

St. Quintin *(family)*  435, 437
Scott, Sir John *(later Lord Eldon)*  207
Scott, Sir Walter  229, 506
Scott, William *(later Lord Stowell)*  207
Scott, William Bell  285-6
Shakespeare, William  237
Simpson, Wallace  345
Smuts, General  134
Stephens, Meic  525, 531
Stowell, Lord *see Scott, William*
Strype, John  235
Stubbs, George  159, 327

Thom / Tom, John Nichols *(Courtenay, Sir William)*  73-4, 155-6
Torrington, Viscount *(18th C.)*  91
Turner, J.M.W.  222
Turpin, Dick  123, 239-40, 429-30, 476, 479

Villiers, Sir George *(father of 1st Duke of Buckingham)* 301
Victoria, Queen 25, 520
Vyner, Sir Robert 221

Walpole, Horace 229, 230, 377
Watts, Henry 122
Webster, John 453
Wedgwood, Josiah 327, 413
Wesley, John 174, 190, 464
Westminster, Duke of 521
Westminster, Marquis of 521
Willesby, Thomas 192
Williams Wynn, Sir Watkin 521
Wright, Steve 241

# Places Index

**A**

Aberavon 529
Aberdeen 505
Abergavenny 526-7
Abergwili 513
Abertridwr 512
Aberystwyth 515-16
Abingdon 291
Acrefair 536
Adwalton 463
Ainsworth 127
Alconbury 47
Aldergrove 491
Alford 187-8
Almondsbury 119
Alresford 133
Alton 133
Amblecote 397-8
Amotherby 441
Ampfield 133
Ampleforth 441
Angmering 377
Appleby 188
Appleby-in-Westmoreland 81
Ardsley 463
Arlesey 25
Armagh 491
Armley 463-4
Arundel 377
Ashbourne 91
Ashby-de-la-Zouch 181
Ashdon 107
Ashill 245
Ashton 134
Ashton-under-Lyne 127
Ashwellthorpe 245-6
Athboy 501
Atherstone 391
Attleborough 246
Aylesbury 43
Aylsham 247

**B**

Bacup 169-70
Baddesley Ensor 391
Badingham 333
Badwell Ash 333-4
Bagillt 518-19

Bakewell 91-2
Bala 523
Baldock 145
Ballincollig 497
Ballyhaunis 501
Ballynahinch 492
Ballyshannon 498
Balmedie 505
Balsall Common 398-9
Bamber Bridge 170
Bampton 97-8
Banbury 291-2
Bangor 523
Barking 224
Barnby Moor 287
Barnsley 455
Barnstaple 98
Barton, Cambridgeshire 47
Barton, Lancashire 170
Basingstoke 134
Baston 188
Bath 313-14
Bayford 314
Beaconsfield 43
Beadlam Nawton 441-2
Beaminster 103
Bearsted 155
Beaufort 511
Beaumaris 511
Beccles 334
Beckernet 81
Bedford 25
Bedminster 119
Bedworth 391
Belfast 491
Bempton 429
Berwick-upon-Tweed 285
Bethnal Green 215
Bettws Gwerfil Goch 517
Beverley 429-31
Beyton 334
Bicester 292
Bicker 188
Biddestone 415
Bideford 98
Biggleswade 26-7
Billingshurst 378
Bilsington 155

Bilston 399-400
Binfield 37
Bingley 464
Birch Vale 92
Birchills 400-1
Birmingham 401-3
Birstall, Leicestershire 181
Birstall, West Yorkshire 465
Bishop's Castle 303
Bishop's Stortford 145
Blackburn 170
Blackfriars 216
Blaenavon 535
Blakeney 247-8
Blandford Forum 103
Bletchley 43
Blunham 27-8
Bluntisham 47
Blyth 287
Bodle Street Green 367-9
Bognor Regis 378
Borough 225
Boroughbridge 442
Borrowby 442
Borstal 155
Boston 188
Boughton 248
Bourne End 145-6
Bow 201
Bradeston 248
Bradfield St. George 334
Bradford 465-6
Bradford-on-Tone 314-15
Bramham 466
Brancaster Staithe 248
Brandon 334
Bratton 415-16
Braughing 146
Breage 71
Brearton 442
Brecon 532
Brentford 222
Brentwood 107
Bressingham 248
Brewham 315
Briantspuddle 103
Bridge 156
Bridgemary 134
Bridgnorth 303
Bridgwater 315
Bridport 103

Brierley Hill 403
Brigg 189
Briggate 249
Brighton 369-71
Briningham 249
Brisley 250
Bristol 119-21
Bromley 225
Bromsgrove 143
Bromyard 143
Broom 28
Broughton 277
Broughton Astley 181
Broughton-under-Blean 155-6
Brownhills 404
Brundall 250
Bruton 315
Bryn Mawr 511
Bryndu 511
Buckden 48
Buckland, Hertfordshire 146
Buckland, Kent 160
Buckover 122
Builth Wells 532
Bungay 334-5
Bures St. Mary 335
Burford 292
Burley-in-Wharfedale 466
Burley on the Hill 301
Burnham 43
Burnham Green 146-7
Burnley 170
Burringham 189
Burslem 327
Burton upon Trent 327
Burwell 48
Bury 127
Bury St. Edmunds 335
Bushey 147
Buxton 250

C
Caernarfon 523
Cainscross 122
Caistor 189
Calne 416
Cambridge 49-50
Cannington 315
Canterbury 156-8
Capel Garmon 516
Capel St. Mary 336

Carbrooke 250
Cardiff 512
Cardigan 516
Carlisle 81
Carmarthen 513
Carrowdore 492
Castle Pulverbach 303-4
Castleford 466
Catfield 251
Cavendish 336-7
Cawston 251
Cefnpennar 533
Chadwell Heath 215
Chantry 316
Chapel End 391-2
Chapeltown 455
Chard 316-17
Charfield 122
Charing Cross 240
Charlton 203
Chatham 158
Cheadle 327-8
Chearsley 43
Checkley 328-9
Chedgrave 251
Chelmsford 107-9
Chelsea 203-4
Cheltenham 122
Chepstow 527-8
Chesham 43-4
Cheslyn Hay 329
Chester 63
Chesterfield 92
Chesterton 50
Chichester 378-9
Chilgrove 379
Chilham 158-9
Chipping Ongar 109
Chipping Sodbury 122
Chislehurst 240-1
Chorleywood 147
Church Fenton 442
Churston Ferrers 98
Churton-by-Aldford 63
Cilcaen / Cilcain 519
Cippenham 37-8
Cleator Moor 81-2
Clerkenwell 234-5
Cley-next-to-the-Sea 251
Cliff 173-4
Clitheroe 170-1

Clun 304
Coggeshall 109
Colchester 109-10
Coleford 123
Coleshill 392
Colne 171-2
Coltishall 252
Comberton 50
Compton Bassett 416
Congleton 63-4
Conway 516
Cootehill 497
Corby 277
Corey's Mill 147
Cork 497
Cornhill 203
Corton 337-8
Coseley 404
Cottenham 50-1
Coventry 404-6
Cowbridge 536
Coychurch 536
Cradley 406-7
Craigellachie 505
Cranbrook 159
Cranworth 252
Crawshawbooth 172-3
Cricklade 416
Cromer 252-3
Crostwick 253-4
Croughton 277
Crowland 189
Cullingworth 466
Curdworth 392
Cwmavon 529-30

**D**
Dalrymple 508
Dalton-in-Furness 82
Darlington 77-8
Dartford 159
Daventry 277
Dawley 304
Deal 159
Dean Priors 136
Deighton 466-7
Denbigh 517
Dengie 110
Denholme 467
Deptford 204
Derby 92-4

Dersingham 254
Derwenlas 532
Desford 181-2
Devizes 417
Devonport 99
Dickleburgh 254
Dinton 44
Disley 64
Diss 254
Ditchling 371-2
Doncaster 456
Dorchester 103
Dorking 357-8
Douglas 487
Dover 160
Dovercourt 110-11
Downham in the Isle 54-5
Downham Market 254
Downton 417-18
Driffield 431
Drogheda 500-1
Droxford 134
Dublin 498-9
Dudley 408
Dulcote 317
Dulverton 317
Dundalk 501
Dundee 508
Dundry 317
Dungannon 494
Dungiven 493
Dunkineely 501
Duns Tew 292
Dunstable 28-9
Dunston Fen 189

**E**
Eamont Bridge 82
Earl Stonham 338
Easebourne 379
Easington 431
Easingwold 442
East Barsham 254-5
East Bergholt 338
East Dean 123
East Dereham 255
East Ham 216
East Harling 255
East Ilsley 38-9
East Rudham 255
East Runton 255

East Ruston 255
East Stour / Stower 103
Easton 338
Eaton Bray 29-30
Eaton Socon 51-2
Eccles 127
Edenbridge 160-1
Edenfield 173
Edgefield 256
Edgworth 173
Edinburgh 506-8
Edington 418-20
Edmonton 210-11
Edwardstone 338
Egham 358
Eglinton 493-4
Egwlysilan 512
Elgin 505
Ellingham 256
Ely 52
Emley 467
Emmet Green 39-40
Empingham 301
Epping 111
Epsom 358
Esher 358-9
Ettington 392
Evershot 104
Exeter 99
Exford 317-18
Exning 338-9
Eythorne 161

**F**
Falkingham 189
Falsgrave 442
Fareham 135
Farnham 359-60
Farnham Royal 44
Farnworth 128
Farringdon Hill 99
Faversham 161
Fazeley 329-30
Felixstowe 339
Fenstanton 52
Ferryhill 79
Finchampstead 40
Finglesham 161
Finningham 339-40
Fisherton-Anger 420-1
Fittleworth 379

Flamstead End 148
Flitton 30
Flitwick 30
Flordon 256
Folkingham 189
Forest Hill 292
Foulsham 256-7
Foxton 52
Framlingham 340
Frampton Mansell 123
Freethorpe 257
Friendly 467
Frogmore 148
Frome 318
Froxfield Green 135

**G**
Gainsborough 189-90
Gamlingay 52
Garndiffaith 535-6
Garvestone 257
Gayton 257
Gaywood 257
Gilberdyke 431-2
Girton 52-3
Glasgow 508
Glastonbury 318
Glyn-Neath 530
Godshill 135
Gomersal 468
Goodshaw 173
Gorleston-on-Sea 257-8
Gort 499
Graffham 380
Gravesend 161-2
Great Aycliffe 79
Great Baddow 111
Great Barrow 64
Great Broughton 64
Great Chesterford 111
Great Cransley 277
Great Dockray 87
Great Driffield 431
Great Dunmow 111
Great Eastcheap 214
Great Edstone 443
Great Finborough 340-1
Great Fransham 258
Great Harwood 173-4
Great Hautbois 258
Great Horton 468-9

Great Kingshill 44
Great Malvern 143
Great Maplestead 111
Great Massingham 258
Great Neston 65
Great Sampford 111
Great Waldingfield 341
Great Yarmouth 258-9
Green Tye 148
Greenwich 214
Grimsby 190
Guestwick 259
Guildford 360

**H**
Haddenham 53
Hadleigh 341
Hadley 215-16
Halesworth 341-2
Halifax 469
Halstead 111-12
Ham 318
Hambrook 123
Hampstead 209
Hanley 330
Hapton 259
Harbeldown 162
Harborne 408-9
Harefield 203
Harlech 523
Harleston 260
Harlow 112
Harpenden 148
Harrold 30
Harrow on the Hill 223
Hartfield 372
Harvel 162
Harwood 128
Hascombe 360-1
Haselbury Plucknett 318-19
Haslemere 361
Haslingden 174-5
Hatching Green 148
Hatfield Broad Oak 112
Hatfield Heath 112
Haughley 342
Haverfordwest 531
Haverhill 342
Hawkesbury 123
Hawkhurst 162
Haynes 30

Hayton 432
Headcorn 162
Headford 499
Headington 293
Headley 361
Heath Charnock 175
Heath Hill 305
Heavitree 99
Hebden Bridge 469-73
Heckington 190
Hedgerley 44
Helmshore 175
Hempstead 260
Henbury 123
Hendrerwydd 517-18
Henley-in-Arden 392
Henley on Thames 293-6
Hereford 143-4
Hermitage 40
Herne Bay 162
Hertford 148
Hertingfordbury 148-9
Hetton-le-Hole 389
Hexham 285-6
Hickling 260
High Crosby 82
High Cross 149
High Ongar 112
High Wycombe 44
Higham Ferrers 277-8
Highworth 421
The Hill 89
Hindon 421
Hinton St. Mary 104
Hitcham 342
Hitchin 149
Hockliffe 30
Holbrook 342
Hollinwood 128
Holme 473
Holme-next-the-Sea 260
Holtye 372
Holywell 519-22
Honiton 99
Horsham 380
Horsham Saint Faith 260-1
Horwich End 94
Houghton Regis 31
Howden 432-3
Howden Clough 473-4
Huddersfield 474

Hull 433-4
Hulme 128
Hundon 343
Hungerford 319
Hunslet 474
Huntingdon 53
Huntington 443
Hurst Green 372
Hurstpierpoint 380
Husborne Crawley 31
Hutton Cranswick 434

**I**

Ickleton 53
Ilchester 319
Ilford 202
Ilminster 319-20
Ingoldisthorpe 261
Ingrow 475
Ipswich 343-5
Irlam 128
Iron Acton 123
Ironbridge 305
Irthlingborough 278
Isleham 53-4
Islington 222-3
Ivy Bridge 99

**J**

Jackson Bridge 475-6

**K**

Kearsley 129
Kedington 345-6
Kelstedge 94
Kelvedon 112
Kemerton 144
Kendal 82-4
Kenninghall 261
Kersey 346
Kettering 278
Kettlestone 261
Keynsham 320
Keysoe Row 31
Keyston 54
Kidderminster 144
Kiftsgate 123
Kilcullen 500
Kilgetty 531-2
Kilkhampton 71
Kimbolton 54

Kimpton 149
Kings Lynn 262
Kings Meaburn 85-6
Kings Sutton 296
Kingsbridge 84-5
Kingsthorpe 278
Kingston upon Hull 433-4
Kington 144
Kington St. Michael 421
Kirkbymoorside 443
Kirkheaton 476
Kirriemuir 508
Kirton 346-7
Knaresborough 443
Knockcloghrim 494
Knutsford 64

**L**
Lakenham 262
Lambeth 205
Lambourne 40
Lancaster 175-6
Langport 320-1
Lanreath 71
Launceston 71-3
Lavenham 347
Laxfield 347
Laxton Grange 435
Layer Marney 112
Leamington Spa 392-3
Ledston 476
Leeds 476-7
Leicester 182-3
Leigh 129
Leire 182
Leiston 347
Leominster 144
Lepton 477-8
Leverstock Green 149-50
Lewes 372
Lewisham 218-19
Lifton 100-1
Lightcliffe 478-9
Limehouse 241
Lincoln 191
Lindfield 380-1
Linslade 32
Linton 54
Liphook 135
Liskeard 73
Little Chishill 54

Little Cressingham 262
Little Downham 54-5
Little Lever 129
Little Totham 113
Littleport 55
Litton Cheney 104
Liverpool 243
Llandeilo 514-15
Llandyrnog 518
Llanelidan 518
Llanengan 523-4
Llanerchymedd 511
Llanfair-Dyffryn-Clwyd 518
Llangadog 515
Llangefni 511
Llangynhafal 518
Llantrisant 533
Llantysilio 524
Lofthouse 479
Loftus 69
London 199-242
London Colney 150
Long Melford 347
Longford 200
Longham 263
Longton 330
Longwick 44
Loughborough 183-4
Louth 191
Lowick 278-9
Lowsonford 393
Ludchurch 532
Luddenden Foot 479
Luddesdown 162
Ludlow 306
Lurgan 491
Lutterworth 184
Lye 409
Lyme Regis 104
Lyndney 123

**M**
Macclesfield 65
Machynlleth 533
Maiden Newton 104-5
Maidenhead 40
Maidstone 162-3
Mainsgate 86-7
Maldon 113
Malton 444-5
Manchester 129

Manea 55
Manningsheath 381-2
Maplehurst 382
March 55
Marchwood 135
Mare Hill 382
Mareham-le-Fen 191
Mark 321
Market Deeping 191
Market Harborough 184
Marrick 445
Marshchapel 191
Martock 321
Marton 87
Melbourne 94
Melcombe Regis 105
Menai Bridge 511
Mendham 347
Merthyr Tydfil 524-6
Mickley 445
Middlemarsh 105
Middleton 129
Middlewich 65
Midhurst 382-3
Milford 422
Milford-on-Sea 136
Milton 56
Milverton 321
Minehead 322
Minera 536
Minety 422
Minster-in-Thanet 163
Minterne Magna 105
Misson 287
Mistley 114
Mitcheldean 123-5
Mitchelstown 497
Mochdre 516
Mold 522
Monmouth 528-9
Morcott 301
Moretonhampstead 101
Moreton-in-Marsh 125
Morley 479
Morriston 535
Morton-on-the-Hill 263
Moulton Seas End 192
Mountrath 500
The Mumbles 535
Mundesley 263
Mundon 114

**N**
Nafferton 435-8
Nantwich 65
Neath 530-1
Neatishead 263-4
Nefyn 524
Neston 65
Netherton 409
Netley Marsh 136
Netteswell 114
New Basford 288
New Brighton 522
New Buckenham 264
Newark 287
Newbridge 533
Newbury 40
Newcastle 389
Newcastle under Lyme 330
Newmarket 523
Newport, Essex 115
Newport, Isle of Wight 136
Newport, Shropshire 306
Newport, Wales 531
Newtownmountkennedy 501
Nine Ashes 115
Normanton 479
North Cave 438
North Ockendon 225
North Stoke 296
North Thoresby 192
Northallerton 446
Northampton 279
Northop 522
Northowram 479
Northwich 66
Norton, Northamptonshire 279
Norton, West Midlands 410
Norton Heath 115
Norwich 265-7
Nottingham 288
Nuneaton 393-4
Nutbourne 383

**O**
Oadby 184
Oakington 56
Okehampton 101
Old 280
Old Buckenham 267-8
Old Chesterton 50
Old Ford 234

Old Leake 192
Old Sharlston 479-80
Old Whittington 94-5
Old Woking 361-2
Oldbury 410
Oldham 129
Openshaw 130
Ormskirk 176
Osmond-Thorpe 480
Oswestry 306-8
Otham 163
Otley, Suffolk 347
Otley, West Yorkshire 480-1
Otterbourne 136
Ottringham 438-9
Over Stowey 322
Overstrand 268-9
Overton-on-Dee 536-7
Oving 383
Owston Ferry 192
Oxford 296-7

**P**
Padbury 45
Padiham 176-7
Painswick 125
Pampisford 56
Paslow Common 115
Paulerspury 280
Pelton 79
Pendlebury 130
Penmachno 517
Penrith 87
Pentraeth 511
Pentwyn 512
Pershore 144
Perth 508
Peterborough 280
Petworth 383
Pickering 446
Pickwell 184
Pilgrims Hatch 115
Pilning 125
Pimlico 202, 238
Pirton 150
Pleshey 115
Pokesdown 105
Pontesbury 308
Pontneddfechan / Pont Neath
   Vaughan 531, 533-4
Pontypridd 534-5

Poole 105
Portsea 136
Portsmouth 136
Potter Street 115
Potterhanworth 192
Potters Bar 150
Poulton-le-Fylde 177
Prescot 243
Preston 177
Prestwich 130
Priors Dean 136
Pudsey 481

**Q**
Quarry Bank 410
Queen Camel 322
Queensbury 481
Quidhampton 422
Quorndon 184

**R**
Radford 288
Rainham 163
Ramsden Heath 115-16
Ramsey 57
Rastrick 481
Rawtenstall 177-8
Rayleigh 116
Reach 57
Reading 40
Redbourn 150-1
Redcar 69
Reigate 362
Rendham 347
Rhes-y-cae 522
Rhosllanerchrugog 537
Rhuddlan 518
Rhyl 518
Richmond, London 242
Richmond, North Yorkshire 446
Rickinghall Superior 348
Ridgewell 116
Ripley, North Yorkshire 446
Ripley, Surrey 362
Ripon 446
Risby 348-9
Riseley 33
Rishworth 481
Robertsbridge 372-3
Rochdale 130
Rochford 116

Rockland 269
Roffey 384
Rogate 384
Romsey 137-8
Rosedale Abbey 446
Rostherne 66
Rotherhithe 234
Rottingdean 373-5
Rowington 394
Roydon, Essex 116
Roydon, Norfolk 269
Royston 151
Ruddington 288
Rudgwick 384
Rugby 394
Rugeley 330
Rumwell 322
Ruthin 518

S
Saffron Walden 116
Saham Toney 269
St. Albans 151
St. Asaph 518
St. Columb Major 73-4
St. David's 99
St. Helen's 243
St. Helier 21
St. Ives, Cambridgeshire 57
St. Ives, Cornwall 74
St. James 349
St. John's 81
St. Lawrence 164
St. Martin's Lane 99
St. Saviour (St. Sauvier) 21
St. Sidwell's Street 99
Saintfield 492-3
Salehurst 375
Salisbury 423-4
Salle 269
Sandhurst 125
Sandway 164
Sandwith 88
Sandy 515
Scales 88
Scarborough 447
Scole 269
Scopwick 192
Seagrave 185
Seal 164
Seamer 447

Sedgefield 79
Sedgeford 269
Sedgley 411
Selly Oak 411
Settle 447
Sharlston 479-80, 482
Sheerness 164
Sheffield 457
Shenley 151
Shepshed 185
Shepton Mallet 322
Sherbourne 394
Shere 362-3
Shifnal 308
Shillington 33
Shipdham 270
Shipston On Stour 394
Shrewsbury 309-10
Shropham 270
Sible Hedingham 116-17
Sibton 349
Silsoe 33
Silverdale 330
Silverstone 280
Sittingbourne 164-5
Skirpenbeck 439
Slaugham 384-5
Sleaford 192
Smarden 165
Smethwick 411
Snainton 447
Snettisham 270
Soham 57
Soho 235
Soudley 125
Soughton 522-3
South Bersted 385
South Cheriton 322
South Harting 385
South Norwood 236
South Shields 389
Southampton 138-9
Southend-on-Sea 117
Southill 33-4
Southminster 117
Southsea 139
Southwark 213, 218, 235, 239
Southwold 349
Spalding 192-3
Spilsby 193
Spitalfields 235, 240

Spofforth 447-8
Spratton 280-1
Stafford 330
Stagsden 34
Staithes 448
Stalybridge 130-1
Stamford 193
Stamford Bridge 439
Stanton, Derbyshire 95
Stanton, Staffordshire 330-1
Starling Green 117
Staunton 126
Stetchworth 57
Stevenage 151
Stewkley 45
Steyning 385-6
Stockport 131
Stockton-on-Tees 69
Stogumber 322
Stogursey 322
Stoke Albany 281
Stoke Ash 350
Stoke-by-Nayland 350
Stoke in Guildford / Stoke-
   next-Guildford 363
Stokesley 448
Stone Street 351
Stonesfield 297
Stony Stratford 45
Storrington 386-7
Stotfold 34
Stourpaine 105-6
Stow Bedon 270
Stowmarket 351
Stratford 223
Stratford-le-Bow 201
Stratford-upon-Avon 394-5
Stretham 57
Stretton 301-2
Strichen 506
Sudbury 351-2
Sun Street 117
Sunbury-on-Thames 363-4
Sundridge 165
Susworth 193-4
Suton 271
Sutton, Cheshire 66
Sutton, Norfolk 271
Sutton, West Sussex 387
Sutton Coldfield 411
Swaffham 271

Swanage 106
Swanland 439
Swansea 535
Swavesey 57-8
Swefling 352
Swinton 131
Syleham 352-3

T
Tadcaster 448
Tafarnaubach 512
Talywain 536
Tamworth 331
Tansor 281
Tarporley 66
Tasburgh 271
Tattingstone 353
Taunton 322-3
Tea Green 152
Tetbury 126
Thame 298
Thanington 165
Thelnetham 353
Thetford 271-2
Thirsk 448-9
Thornbury 126
Thornton, East Yorkshire 439
Thornton, West Yorkshire 482
Throckenhall 194
Thruxton 139
Thurlton 272
Tibthorpe 439
Tickhill 458
Tilbrook 58
Tipton 411
Tiverton 101-2
Tonbridge 165
Tovil 165
Towcester 281-2
Tower Hamlets 160
Towyn 517
Trelawnydd 523
Triangle 482
Tring 152
Trowbridge 424
Trowse Newton 272
Truro 74
Tunbridge Wells 165
Tunstall 331
Twerton 323
Twyford 40-1

Tyldsley 131

**U**
Uffington 298
Uphill 166
Upper Poppleton 449-50
Upper Stoke 166
Upton 272-3
Uttoxeter 332
Uxbridge 200

**W**
Wadsley Bridge 458
Wakefield 482-3
Wall Heath 411
Wallington 139
Walpole St. Peter 273
Walsall 411-12
Walshaw 131
Waltham Abbey 117
Wantage 298
Wareside 152
Wargrave 41-2
Warley 483
Warwick 395
Washford 323
Waterbeach 58-9
Waterfoot 178-9
Watford 152
Waverton, Cheshire 66-7
Waverton, Cumbria 89-90
Weasenham St. Peter 273
Wednesbury 412
Welbourn 194
Welling 241
Wellingborough 282
Wellington 323
Wellow 289
Wells 323-4
Welton 282
Welwyn 152
Wem 310
West Bilney 273
West Boldon 389-90
West Bromwich 412
West Chiltington 387
West Dean 387
West Dereham 273
West Ham 204
West Meon 139
West Rounton 450

West Row 354
West Tarring 387
West Teignmouth 102
West Wickham 59-60
Westbourne 387
Westbury-on-Trym 126
Westerham 167
Westhoughton 131
Westleton 354
Westminster 213, 237
Weston 152-3
Wetherby 483-4
Wetheringsett Green 354
Whalley 179
Whaplode Drove 194
Whepstead 354
Whitby 450-2
Whitchurch 45-6
White Mill 513
Whitkirk 484
Whittlesey 60
Whitwell 140
Whitwick 185
Whitworth 179
Wickhambrook 354-5
Wickwar 126
Widford 118
Wierton 166-7
Wigan 179
Willaston 67
Willenhall 412
Willesborough Lees 167
Willesden 204
Willingham 60
Wilstead 34
Wincanton 324
Winchester 140-1
Winkfield Row 42
Winterbourne 126
Winterbourne Bassett 424
Winwick 67
Wirral 67
Wisbech 60
Witcham 60-1
Witham 118
Withersfield 355
Wiveliscombe 324
Woburn 34-5
Wokingham 42
Wolverhampton 412-13
Woodbridge 355

Woodbury Salterton 102
Woodford 282-3
Woodhouse Carr 484
Wooler 286
Woolley Moor 95-6
Woolstone 298
Woolton 243
Worcester 144
Worksop 289
Worstead 249
Worthen 310-11
Worting 141
Wortley 484
Wrangle 194-5
Wrawby 195
Wrexham 537
Wrockwardine Wood 311
Wroxton St. Mary 298
Wymeswold 185-6
Wymington 35
Wymondham 274

**Y**
Yeovil 324
Yieldon 35
York 452-3

# THERE IS A TAVERN IN THE TOWN

Adapted from a Cornish folksong by the author